Jersey Cattle in America
Part One

by John S. Linsley

with an introduction by Jackson Chambers

Self Reliance Books

Get more historic titles on animal and stock breeding, gardening and old
fashioned skills by visiting us at:

Introduction

I am pleased to present another title in the "Cattle" series.

The work is in the Public Domain and is re-printed here in accordance with Federal Laws.

As with all reprinted books of this age that are intended to perfectly reproduce the original edition, considerable pains and effort had to be undertaken to correct fading and sometimes outright damage to existing proofs of this title. At times, this task is quite monumental, requiring an almost total "rebuilding" of some pages from digital proofs of multiple copies. Despite this, imperfections still sometimes exist in the final proof and may detract from the visual appearance of the text.

I hope you enjoy reading this book as much as I enjoyed making it available to readers again.

Jackson Chambers

PYRRHA. 3d, 11850.
FARMERS MAID. 18219.
JEWELL BEAUTY. 2d. 1701.

ROSEBELL 2d. 11722   LADY MADELINE. 10526.
SURPRISE OF M S. 10928.
DANDELION 2321.   DANDY BOY. 7834.
DANDELION, 3d. 21889.   DANDELION. 4th. 27000.

"HOLLY GROVE" HERD.

JOHN I HOLLY. Plainfield. N J

TO

THE MEMBERS

OF THE

AMERICAN JERSEY CATTLE CLUB,

THROUGH WHOSE DISCERNMENT AND ENTERPRISE WISE PROVISION HAS BEEN
MADE TO SECURE TO THE AGRICULTURISTS OF AMERICA THE PERPETUITY
AND PURITY OF THE UNRIVALLED BREED OF JERSEY CATTLE,
THIS VOLUME IS RESPECTFULLY INSCRIBED.

JOHN S. LINSLEY.

ERRATA.

Page 10,   SALT UNDISSOLVED IN BUTTER..................408

Page 270,   MAY.

1st line, 15 lbs. Best Mixed Hay.

Page 271,   JUNE.

2d line, 20 lbs. Green Rye or Rye Grass.

Page 271,   SEPTEMBER.

3d line, 30 lbs. Green Barley.

4th line, 20 lbs. Millet.

5th line, 20 lbs. Wheat.

Page 272.   OCTOBER.

3d line, 50 lbs. Green Barley.

4th line, 10 lbs. Green Wheat.

DECEMBER.

1st line, 20 lbs. Best Early Hay.

JANUARY.

1st line, 10 lbs. Green Oat Hay.

Page 273,   FEBRUARY.

2d line, 15 lbs. Green Millet Hay.

MARCH.

1st line, 10 lbs. Green Clover Hay.

APRIL.

1st line, 15 lbs. Green Millet Hay or 50 lbs. Green Rye.

A CHEAP WINTER RATION.

1st line, 20 lbs. Green Corn Stover.

Page 274,   STANDARD WINTER RATION.

2d line, 10 lbs. Rowen Hay.

Page 275,   RATION ONE MONTH BEFORE CALVING.

1st line, 15 lbs. Best Timothy Hay.

OR THIS.

1st line, 15 lbs. Rowen Hay.

BUTTER TESTS.

Page 653, Fillpail 2d 24,388.................................26 lbs. 2 oz.

JERSEY FOUNTAINS.

HOMER H. 3683 omitted from page 565, Page 742.

PREFACE.

THE object of this work is to set forth fully and clearly the special merits and rare qualities of the beautiful breed of Jersey cattle; to show how these qualities have been developed, their mode of perpetuity, and their still further possible improvement.

It is intended to be thoroughly practical and progressive, as well as suggestive of a higher standard in all that pertains to agriculture, cattle-breeding, and the arts of dairying.

In a work treating of such a wide variety of topics, it has been necessary to consult many authors and make numerous studies and compilations.

The author has drawn from the writings of many eminent authorities, including the *Encyclopædia Britannica;* Morton's *Encyclopædia of Agriculture; Chambers' Encyclopædia; Reports of Connecticut Experiment Station; Reports of New York Experiment Station; Reports of Agricultural Bureau, Washington, D.C.; The Marriage of Near Kin,* by Alfred Henry Huth; *The Butter Tests of Jersey Cows,* by Campbell Brown; *Feeding Animals,* by E. W. Stewart; *Guenon on Milch Cows,* by Thomas J. Hand; *The Atmospheric System,* by Thomas B. Butler; *The Country Gentleman; The Jersey Bulletin; The New York Tribune;* the writings of J. Le Couteur, John Thornton, and George E. Waring, Jr.; also the sale catalogues and herd catalogues of breeders.

Acknowledgment is made of the kindness of Major Henry E. Alvord, manager of Mr. Valentine's Houghton Farm at Mountainville, N. Y., for reports and chemical tests. Special thanks are due to the hearty and substantial support of all those who have contributed portraits of cattle to illustrate the text, and butter records, and render the work attractive to lovers of the Jersey.

The medical and sanitary treatment herein suggested, the author hopes, may be the means of saving the lives of many valuable animals.

JOHN S. LINSLEY, M.D.

NEW YORK, April, 1886.

CONTENTS.

PART FIFTH.

ILLUSTRATIONS.

SILHOUETTES.

PORTRAITS OF JERSEY BULLS.

PORTRAITS OF JERSEY COWS.

CHARTS, DRAWINGS, DIAGRAMS, PLANS AND IMPLEMENTS.

INTRODUCTORY.

OUR DOMAIN.

THE American people are now preparing a continent to be the dwelling-place, before another century shall have passed, of more than five hundred millions of people.

To us the nations of the earth are looking for the solution of many of the problems of political and social economy and questions that relate to the welfare of the human race.

One of the most important elements determining our material prosperity and our permanent progress is an enlightened system of agriculture.

By the condition of a nation's agriculture we may judge of its advancement in the path of civilization.

Not yet is the Golden Age of American Agriculture.

Looking backward to the austere and gloomy barbarism of our Anglo-Saxon ancestry, beyond a thousand years ago, we exclaim, How great the transition! Looking at the progress of a century, or a generation, we are filled with self-gratulation.

But when we consider how much we lack, in knowledge, in method, in purpose—when we try to picture the possibilities of the future of American agriculture, we are impressed with the idea that we are only at the threshold of the way of enlightenment and progress.

In the contest of wresting from the soil an abundant supply of food, clothing, and all the necessaries of physical existence, and at the same time means of leisure, cultivation, refinement, and mental growth for the multitude, we are required to deal with a problem which has not yet been solved. At the very beginning of study we are forcibly convinced of the wastes that are continually draining the resources of a nation—waste of vital force in a thousand ways, waste of material from negligence or from ignorance, waste through unprofitable labor and lack of system.

In our agriculture we need new ideas and new methods. We must apply the lessons we have learned from history and from experience. We must also learn to anticipate the wants of the near future.

There must be an economy of vital force, a profitable system of fertilizing, more thorough tillage, improved sanitary buildings for the farmer and his cattle, and a practical system of education in all schools, from the primary to the university. But, at last, the basis of our agriculture consists in the races of cattle we cultivate. Without cattle it would be impossible to have any civilization. The cattle must fatten the ground and feed the race of men that live upon it. Agriculture is the mother of all arts, and the cow is the mother of agriculture. Not only are cattle the essential element upon which agriculture depends for existence, but a progressive agriculture requires that the races of cattle adopted by a people must be of the highest excellence to insure prosperity.

In the promotion of this most important but most neglected of all human industries, it is the patriotic duty of every successful business man to devote a portion of his wealth.

The inventor, poet, physician, artist, merchant, miner, lawyer, statesman, soldier, editor, banker, manufacturer—each and all that have accumulated a competency, should hold a portion of the soil for a cultivated farm, a home which should be made as productive and as beautiful as possible, a veritable Paradise, with fruitful fields, orchards, and groves, and herds of the choicest cattle.

As Americans we rejoice in the memory of our famous men, and that many of the best of them were farmers. Washington, Jefferson, the inventor of the first mould-board plow upon mathematical principles; Clay, Webster, Greeley, the great editor; Bryant, the poet; Garfield, the beloved President, and many more whom we love, revere, and honor, have left us a wholesome and worthy example of doing what they could for agriculture. They loved the country home and its pure attractions. They loved the art which, above all other arts, is designed to make home happy.

The American farmer is desirous to excel. He wants to have the best of everything that pertains to his calling. When he shall ascertain what is best for his present or prospective need, he will bend his energies to secure it, if practicable.

It is of the first necessity that he supply himself with the breed of cattle best suited to his needs—cattle that shall help to make farming a source of material prosperity, joyous health, and perpetual pleasure. Let the cattle, then, be worthy of our choice and have a large place in our esteem. What we think of our cattle, how we shall use them and make them serve us and our national prosperity, how we shall improve, transform, and perfect them for our purpose, how kindly we shall treat them and care for them, how they shall influence our life, our comfort, our health, our happiness, our usefulness, our sentiments, our philanthropy, will be told to the ages that come after us. Let the historian, the painter, and the poet have a share in this record, for they are to set forth in a new era of enlightenment a consummation of excellence that shall far exceed in beneficence all the earlier ages of the world's effete civilizations. The coming ages will not foster so much a pride of war and barbaric splendor. The patriot's boast in the new era shall not be like that of Henry V. of his soldiers, in battle, "whose blood is fet from fathers of war proof," but a prouder exclamation will be that of all Americans—" We are the sons of fathers who made the name of their country glorious by the culture of the arts of peace!'"

INFLUENCE OF CATTLE UPON LITERATURE.

All the lovers of choice cattle are glad to read about them. If the songs and sayings of those who have best expressed the sentiment of mankind in all that relates to cattle, the dairy, and the charms of country life should be gathered, they would make many delightful volumes.

Beginning with the oldest literature, we have in the writings of Moses the brief but sublime account of the creation of the world, with its plants, its cattle, and man, who is given the dominion over all cattle and all the earth's productions, and a lordly self-control.

There we have sketches of the patriarchs, of the religious sentiment of the world's best men ; the history of sacrifice, confession of a moral stain that needed forgiveness and a divine cleansing, by and through a Substitute who was typified in the victim.

There we read of Abraham, who was " very rich in cattle, in silver and gold ;" of the strife between his herdsmen and his nephew Lot, so amicably settled as ever to show himself the typical peace-maker among neighbors ; his entertainment of the three angels with a calf tender and good, dressed with milk and butter.

We read of Isaac, his son, " who became very great, and had possession of flocks and herds ;" of Jacob, the most famous cattle-breeder of the ancient world, who made his father-in-law rich, and then enriched himself out of his wages of spotted cattle.

What a perennial charm has the story of Joseph—his wonderful interpretation of the strange dream of Pharaoh, his purchase of all the cattle of the Egyptians for his brethren while he ruled in Egypt and furnished all the world with wheat ! Then the deliverance, the laws of sacrifice, the promise to be led to " a goodly land that floweth with milk and honey." The songs of Moses, and his great poem, the Book of Job, contain many allusions to cattle.

The record of the capture of the Ark of Jehovah in the Book of Samuel, the miraculous disasters that befell its captors, and their device for returning the Ark to the Israelites, by a new cart drawn by two milch cows, that left their calves and went lowing all the way straight to the land of the Jews, is one of the most wonderful of the events in the history of that most wonderful of nations.

Asaph the Seer,* in his sacred psalms, sings of the majesty of Jehovah and his dominions :

---

* Bible Union Version, by T. J. Conant, D.D.

PSALM L.—THE CATTLE BELONG TO GOD.

"I am God, thy God
  Not for thy sacrifices will I reprove thee ;
  And thy burnt-offerings are continually before me.
  I will not take a bullock from thy house,
  Nor he-goats from thy folds.
  For mine is every beast of the forest,
  The cattle on a thousand hills.
  I know every bird of the mountains,
  And the beasts of the field are before me.
  If I were hungry I would not say it to thee ;
  For the world is mine, and the fulness thereof.
  Will I eat the flesh of bulls,
  And drink the blood of goats ?
  Sacrifice to God thanksgiving,
  And pay to the Most High thy vows,
  And call upon me in the day of trouble ;
  I will deliver thee, and thou shalt honor me."

PSALM LXXIII.—OUT OF EGYPT.

"And he removed as a flock his own people,
  And guided them as a herd in the wilderness ;
  And he led them on safely, and they feared not,
  But their enemies the sea overwhelmed."

The sacred psalms of David the King are full of poetic beauty and the melody of praise to Jehovah.

PSALM LXV.—GOD THE GIVER OF PROSPERITY.

"Thou hast visited the earth, and made it overflow [with plenty] ;
  Thou greatly enrichest it.
  The river of God is full of water.
  Thou preparest their grain, for so dost thou prepare the earth ;
  Drenching its furrows, settling its ridges ;
  Thou makest it soft with showers,
  Its springing up thou dost bless.
  Thou hast crowned the year with thy goodness ;
  And thy footsteps drip with fatness ;
  The pastures of the wilderness they drip,
  And the hills gird themselves with gladness.
  The pastures are clothed with flocks,
  And the valleys are robed with grain ;
  They shout together, yea, they sing."

PSALM CIV.—GOD'S BENEFICENCE IN CREATION.

" He sends out springs among the valleys ;
   They run among the mountains.
   They give drink to every beast of the field ;
   The wild asses quench their thirst.
   Above them dwell the fowls of heaven ;
   From among the branches they utter a voice.
   He waters the mountains from his chambers ;
   The earth is sated with the fruit of thy working.
   He causes grass to grow for the cattle,
   And herbs for the service of man,
   Bringing forth food out of the earth.

       *   *   *   *   *   *

   They all wait for thee,
   To give their food in its season.
   Thou givest to them ; they gather ;
   Thou openest thy hand, they are sated with good ;
   Thou hidest thy face, they are troubled ;
   Thou withdrawest their breath, they expire.
   And return to their dust ;
   Thou sendest forth thy breath, they are created ;
   And thou renewest the face of the ground."

PSALM CVII.—GOD'S CARE FOR HIS PEOPLE.

" He turns the wilderness into a pool of water,
   And a dry land into water-springs,
   And there he makes the hungry dwell,
   And they found a city for a habitation.
   And they sow fields, and plant vineyards,
   And produce fruits of the yearly increase.
   And he blesses them, and they multiply greatly,
   And their cattle he makes not few."

PSALM CXLIV.——A PRAYER FOR DIVINE BLESSING OF PEACE AND SECURITY.

" So that our sons may be as plants,
   Full grown in their youth ;
   Our daughters as corner pillars,
   Sculptured after the structure of a palace ;
   Our garners full, supplying of every kind ;
   Our flocks multiplying by thousands,
   By tens of thousands, in our fields ;
   Our oxen laden ;
   No breaking in, nor going forth,
   And no outcry in our streets.
   Happy the people to whom it is thus ;
   Happy the people whose God is Jehovah !"

PSALM CXLVIII.—SONG OF PRAISE.

" Praise Jehovah from the earth ;
  Ye sea monsters and all deeps ;
  Fire and hail, snow and vapor,
  Stormy wind fulfilling his word ;
  Ye mountains and all hills,
  Fruit-trees and all cedars ;
  Beasts, and all cattle,
  Creeping things and winged birds ;
  Kings of the earth, and all peoples,
  Princes and all judges of the earth ;
  Young men, and also maidens,
  Old men, with children ;
  Let them praise the name of Jehovah."

Solomon says :

" I had great possessions of great and small cattle above all that were in Jerusalem before me."

Among his three thousand proverbs we note :

" Be thou diligent to know the state of thy flocks, and look well to thy herds."
" A righteous man regardeth the life of his beast : but the tender mercies of the wicked are cruel."
" Where no oxen are, the crib is clean ; but much increase is by the strength of the ox."

In the historic temple built by Solomon to the worship of Jehovah was a great brazen laver, or sea, resting on twelve gigantic brazen statues of oxen, in groups of three, looking north, south, east, and west. In the dedication of the temple twenty-two thousand oxen were offered among the sacrifices.

Isaiah, the most fervid and exalted in spirit of all the Hebrew poets, shows us the coming of the Giver of grace and truth, and a restoration of spiritual blessings, graphically typified by milk and honey, pleasant fields, and the feet of cattle. Habakkuk, too, in a sublime poem upon the majesty of God and his providence, intersperses like figures to portray the blessings of the day of prosperity ; while the prophet Joel, by the desolate garners, by broken-down barns and withered corn, by groaning beasts and perplexed herds, by dried-up rivers and fire-devoured pastures, describes drouth and famine.

A PASTORAL ANTHOLOGY.

The Egyptians deified and worshipped the bull, and the cow was their symbol of the goddess of Love. Homer,* the greatest of Greek poets, makes frequent allusions to cattle, and many of the finest portions of the Iliad are thus illustrated.

Agamemnon, at the head of his armies on the plains before the city of Troy, is described as

---

* Translation of Alexander Pope.

> " majestically tall,
> Towers o'er his armies and outshines them all ;
> Like some proud bull that round the pasture leads
> His subject herds, the monarch of the meads."

They sacrifice a steer to Jove in honor of the prowess of Ajax, and at the feast which follows, in which they eat the roasted flesh,

> " Before great Ajax placed the mighty chine."

Agamemnon, in his desire to appease the wrath of Achilles, makes a list of the rich presents and honors he will bestow, among them seven cities, and all the rich lands appertaining :

> " Along the verdant margin of the main,
> There heifers graze and laboring oxen toil."

When Patroclus is killed by Hector, the Spartan king Menelaus guards his body from capture :

> " Thus round her new-fallen young the heifer moves,
> Fruit of her throes and first-born of her loves,
> And anxious (helpless as he lies and bare)
> Turns and re-turns her with a mother's care."

The terrible fighting of Ajax Telemon, the Great, and Ajax Oileus, the Swift, side by side, in the fourth battle, is likened as follows :

> " So when two lordly bulls, with equal toil,
> Force the bright plowshare through the fallow soil,
> Joined to one yoke, the stubborn earth they tear,
> And trace large furrows with the shining share.
> O'er their huge limbs the foam descends in snow,
> And streams of sweat down their sour foreheads flow."

The shield of Achilles, as wrought by the god Vulcan, is of silver, brass, tin, and solid gold—

> " There shone the image of the master mind,
> There earth, there heaven, there ocean he designed."

The sun, the moon, the stars, two cities, two armies, golden gods, two spies, flocks, herds, battles, a field with plowmen, grain fields, vineyards with maids and youths.

> " Here herds of oxen march erect and bold,
> Rear high their heads, and seem to low in gold,
> And speed to meadows, on whose sounding shores
> A rapid torrent through the rushes roars ;
> Four golden herdsmen as the guardians stand,
> And nine sour dogs complete the rustic band.
> Two lions rushing from the wood appeared
> And seized a bull, the master of the herd ;

He roared ; in vain the dogs, the men, withstood ;
They tore his flesh, and drank the sable blood.
The dogs, oft cheered in vain, desert the prey,
Dread the grim terrors, and at distance bay.
Next this, the eye the art of Vulcan leads
Deep through fair forests and a length of meads,
And stalls, and folds, and scattered cots between,
And fleecy flocks that whiten all the scene.
A figured dance succeeds. . . .
The gazing multitudes admire around.
Thus the broad shield complete, the artist crowned
With his last hand, and poured the ocean round ;
In living silver seemed the waves to roll,
And beat the buckler's verge, and bound the whole."

An expression favorite with the great Homer, and showing his appreciation of the beauties of the bovine race of Greece, was,

" Goddess of the cow's fair eyes."

Hesiod, another Greek poet, is described by Elizabeth Barrett Browning as

" Hesiod old,
Who, somewhat blind and deaf and cold,
Cared most for gods and bulls."

In the Norse mythology, as recorded in the songs and legends which form the Icelandic Edda, "the giant Ymir and his shapeless progeny, Whirlwinds of the North and Terrors of the Deep, the enemies of the Sun and of Life, are succeeded by Aedhumla the Cow, who is formed of melting snow, and she, licking the white frost from the rocks, brings to light *Buri*, a Man ! The sons of Man kill the giant Ymir, and from his flesh is formed the earth, from his bones the hills, from his skull the sky, from his blood the sea, and from his brains the clouds."

In the twelfth century Bernard of Clugny wrote a Latin hymn suggested by the verse of Moses, " a land flowing with milk and honey." In 1851 the hymn was translated into English by J. M. Neale. It is one of the most joyous and inspiring lyrics ever written—

" Jerusalem the golden,
With milk and honey blest,"

a view of that goodly land of everlasting peace and pleasure.

Among all the nations of antiquity, the Jews were the greatest lovers of cattle ; but since their dispersion they seem to have lost that instinct, and now the Anglo-Saxon has become the leading race of cattle fanciers, and English literature is rich with its allusions to rural felicity. Shakespeare, in the third part of King Henry VI., Scene V., makes the king desire a farmer's life :

"O God! methinks it were a happy life
  To be no better than a homely swain;
  To sit upon a hill, as I do now,
  To carve out dials quaintly, point by point,
  Thereby to see the minutes how they run,
  How many make the hour full complete:
  How many hours bring about the day;
  How many days will finish up the year;
  How many years a mortal man may live.
  When this is known, then to divide the times:
  So many hours must I tend my flock;
  So many hours must I take my rest;
  So many hours must I contemplate;
  So many hours must I sport myself;
  So many days my ewes have been with young;
  So many weeks ere the poor fools will yean;
  So many years ere I shall shear the fleece:
  So minutes, hours, days, weeks, months, and years,
  Passed over to the end they were created,
  Would bring white hairs into a quiet grave.
  Ah! what a life were this! how sweet! how lovely!"

The prince of Christian poets, John Milton, invoking Mirth, invites her to show him all pleasant sights and give him all joyous sounds of rural life:

"While the plowman near at hand
  Whistles o'er the furrowed land,
  And the milkmaid singeth blithe,
  And the mower whets his scythe,
  And every shepherd tells his tale
  Under the hawthorn in the dale."

Again Milton pictures rural delights to the unaccustomed senses—"Paradise Lost," Book IX.:

"As one who, long in populous city pent,
  Where houses thick and sewers annoy the air,
  Forth issuing on a summer's morn to breathe
  Among the pleasant villages and farms
  Adjoined, from each thing met conceives delight,
  The smell of grain, or tedded grass, or kine,
  Or dairy, each rural sight, each rural sound."

A plaintive allusion in "Paradise Lost" touches us with a strange pathos:

"Thus with the year
Seasons return, but not to me returns
Day, or the sweet approach of even or morn,

> Or sight of vernal bloom or summer's rose,
> Or flocks or herds, or human face divine;
> But cloud instead, and ever-during dark
> Surrounds me."

How beautiful is his description of the Angel Raphael and his visit to Paradise! Radiant with the splendor of

> " Downy gold and colors dipped in heaven,"

he comes to the garden, and is entertained as a guest by Adam and Eve in their bower. Eve prepares a feast—

> " fruit of all kinds,
> Nect'rous draughts between from milky stream.
> . . . . . For drink the grape
> She crushes, inoffensive must, and meaths
> From many a berry; and from sweet kernels pressed
> She tempers dulcet creams."

Robert Herrick in quaint verse, thanking God for his little house and the blessings of garden and field, says:

> "The while the conduits of **my kine**
> Run cream for wine."

Thomas Tickell pictures in fancy a country home, in which these lines occur:

> " A rill shall warble 'cross the gloomy grove—
> A little rill o'er pebbly beds conveyed
> Gush down the steep, and glitter through the glade.
> What cheering scents these bordering banks exhale!
> How loud that heifer lows from yonder vale!
> That thrush, how shrill !"

Alexander Pope, at twelve years, thus describes " The Quiet Life ":

> " Happy the man whose wish and care
>   A few paternal acres found,
> Content to breathe his native air
>   In his own ground.
> Whose herds with milk, whose fields with bread,
>   Whose flocks supply him with attire,
> Whose trees in summer yield him shade,
>   In winter, fire."

James Thomson, in his " Castle of Indolence," gives many pleasing pictures. I select one:

> " In health the wiser brutes true gladness find.
>   See how the younglings frisk along the meads,
> As *May* comes on, and wakes the balmy wind;
>   Rampant with life, their joy all joy exceeds;
> Yet what but highstrung health this dancing pleasaunce breeds ?"

In his " Spring " he loves to

> " wander o'er the dewy fields "

and

> " Through the verdant maze of sweet-brier hedges
> Taste the smell of dairy."

He was tender-hearted to all animals—

> " To merit death ?   You who have given us milk in luscious streams !"

He describes well the restlessness of a pastured bull, and the contest when two of them meet :

> " And groaning deep the impetuous battle mix ;
> While the fair heifer, balmy breathing near,
> Stands kindling up their rage."

In his " Summer " (after a thunder storm) :

> " 'Tis beauty all, and grateful song around,
> Joined to the low of kine and numerous bleat
> Of flocks thick nibbling through the clovered vale."

Alexander Hume thus pictures the " Summer Day " :

> " The burning beams down from his face
>   So fervently can beat,
> That man and beast now seek a place
>   To save them from the heat.
>
> " The herds beneath some leafy tree,
>   Amid the flowers they lie ;
> The stable ships upon the sea
>   Send up their sails to dry."

Thomas Gray, in his " Elegy," gives many a perfect verse :

> " The lowing herd winds slowly o'er the lea,
>   *   *   *   *   *
> And drowsy tinklings lull the distant folds.
>   *   *   *   *   *
> How jocund did they drive their team afield !"

Oliver Goldsmith, in the " Deserted Village," thus describes sights and sounds at Auburn :

> " Sweet was the sound when oft at evening's close
> Up yonder hill the village murmur rose ;
> There as I passed, with careless steps and slow,
> The mingling notes came softened from below :
> The swain responsive as the milkmaid sung ;
> The sober herd that lowed to meet their young ;
> The noisy geese that gabbled o'er the pool :
> The playful children just let loose from school ;
> The watch-dog's voice that bayed the whispering wind,
> And the loud laugh that spoke the vacant mind—
> These all in sweet confusion sought the shade,
> And filled each pause the nightingale had made."

William Cowper, the pensive Puritan poet, expressed a strong sympathy for man or beast suffering from cruelty. He was a lover of animals. How fair is the view of Ouse—the river he so loved—and the fields along its banks:

> " Slow winding through a level plain
> Of spacious meads, with cattle sprinkled o'er.
>
> *   *   *   *   *
>
> A breath of unadulterate air,
> The glimpse of a green pasture, how they cheer
> The citizen and brace his languid frame !
>
> *   *   *   *   *
>
> The heart is hard in nature, and unfit
> For human fellowship, as being void
> Of sympathy, and therefore dead alike
> To love and friendship both, that is not pleased
> With sight of animals enjoying life,
> Nor feels their happiness augment his own.
>
> *   *   *   *   *
>
> The very kine that gambol at high noon,
> The total herd receiving first from one,
> That leads the dance, a summons to be gay,
> Though wild their strange vagaries and uncouth
> Their efforts, yet resolved with one consent
> To give such act and utterance as they may
> To ecstasy too big to be suppressed—
> These and a thousand images of bliss
> With which kind nature graces every scene,
> Where cruel man defeats not her design,
> Impart to the benevolent, who wish
> All that are capable of pleasure, pleased,
> A far superior happiness to theirs—
> The comfort of a reasonable joy."

James Beattie, in " The Minstrel," gives us this pleasing line :

> " Crowned with her pail, the tripping milkmaid sings."

Robert Burns, in his matchless picture of " The Cotter's Saturday Night," thus describes the simple meal, when Jennie's lover comes in to spend the evening :

> " But now the supper crowns their simple board,
>   The halesome parritch, chief o' Scotia's food ;
> The soupe* their only hawkie† does afford,
>   That 'yont the hallan‡ snugly chows her cood :
> The dame brings forth, in complimental mood,
>   To grace the lad, her weel-hained kebbuck,§ fell,‖
>   An' aft he's prest an' aft he ca's it guid,
> The frugal wife, garrulous, will tell
> How 'twas a towmond auld, sin' lint was i' the bell."

---

* Milk.   † Cow.   ‡ Partition wall.   § Cheese.   ‖ Sharp or biting.

The following gem is from "The Farmer's Boy," by Robert Bloomfield:

" A little farm his generous master tilled,
 Who with peculiar grace his station filled,
 By deeds of hospitality endeared.
 Served from affection—for his worth revered ;
 A happy offspring blest his plenteous board,
 His fields were fruitful and his barns well-stored,
 And four-score ewes he fed, a sturdy team,
 And lowing kine, that grazed beside the stream.
 Unceasing industry he kept in view,
 And never lacked a job for Giles to do.
         *   *   *   *   *
 The clattering dairymaid immersed in steam,
 Singing and scrubbing midst her milk and cream,
 Bawls out, ' Go fetch the cows ! '
         *   *   *   *   *
 Straight to the meadows then he whistling goes ;
 With well-known halloo calls his lazy cows ;
 Down the rich pastures heedlessly they graze,
 Or hear the summons with an idle gaze ;
 For well they know the cow-yard yields no more
 Its tempting fragrance nor its wintry store.
 Reluctance marks their steps, sedate and slow,
 The right of conquest the only law they know ;
 The strong press on, the weak by turns succeed,
 And one superior always takes the lead,
 Is foremost wheresoe'er they stray,
 Allowed precedence, undisputed sway ;
 With jealous pride her station is maintained,
 For many a broil that post of honor gained.
         *   *   *   *   *
 Forth comes the maid, and like the morning smiles ;
 The mistress, too, and followed close by Giles.
 A friendly tripod forms their humble seat,
 With pails bright scoured and delicately sweet.
 Where shadowing elms obstruct the morning ray
 Begins the work, begins the simple lay ;
 The full-charged udder yields its willing stream,
 While Mary sings some lover's amorous dream.
 And crouching Giles, beneath a neighboring tree,
 Tugs o'er his pail, and chants with equal glee ;
 Whose hat, with battered brim, of nap so bare,
 From the cow's side purloins a coat of hair—
 A mottled ensign of his harmless trade,
 An unambitious, peaceable cockade.
 As unambitious, too, that cheerful maid ;
 With joy she views her plenteous reeking store,
 And bears a brimmer to the dairy-door,

Her cows dismissed, the luscious meads to roam
Till eve again recall them loaded home."

Here is " Country Life," from the pen of William Wordsworth—" March " :

" The cock is crowing,
    The stream is flowing,
    The small birds twitter,
    The lake doth glitter,
The green field sleeps in the sun
    The oldest and youngest
    Are at work with the strongest ;
    The cattle are grazing,
    Their heads never raising,
There are forty feeding like one.

" Like an army defeated,
    The snow hath retreated,
    And now doth fare ill
    On the top of the bare hill ;
The plowboy is whooping—anon—anon !
    There's joy on the mountain,
    There's life in the fountain,
    Small clouds are sailing,
    Blue sky prevailing,
The rain is over and gone !"

A contrast to the sad experience of the homesick farmer in the city, by the same author, " The Farmer of Tilsbury Vale " :

" To London—a sad emigration, I ween—
With his gray hairs, he went from the brook and the green,
And there with small wealth but his legs and his hands,
As lonely he stood as a crow on the sands.
    *   *   *   *   *   *
In the throng of the town like a stranger is he,
Like one whose own country's far over the sea
And nature, while through the city he hies,
Full ten times a day takes his heart by surprise.
    *   *   *   *   *   *
'Mid coaches and chariots, a wagon of straw,
Like a magnet, the heart of old Adam will draw.
With a thousand soft pictures his memory will teem,
And his hearing is touched with the sound of a dream.
    *   *   *   *   *   *
Up the Hay-market hill he oft whistles his way,
Thrusts his hands in a wagon and smells at the hay ;
He thinks of the fields he so often hath mown,
And is as happy as if the rich freight were his own.
    *   *   *   *   *   *

But chiefly to Smithfield he loves to repair ;
If you pass by at morning you'll meet with him there.
The breath of the cows you may see him inhale,
And his heart all the while is in Tilsbury Vale."

James Hogg gives us this pretty song :

" Come, all ye jolly shepherds
   That whistle through the glen ;
I'll tell ye o' a secret
   That courtiers dinna ken :
What is the greatest bliss
   That tongue of man can name ?
'Tis to woo a bonnie lassie
  When the kye come hame,
    When the kye come hame,
    When the kye come hame,
    'Tween the gloamin' an' the mirk,
  When the kye come hame.
     *   *   *   *   *
" When the blackbird bigs his nest
   For the mate he lo'es to see,
And on the tapmost bough,
   Oh, a happy bird is he !
There he pours his melting ditty,
   And love is a' the theme ;
And he'll woo his bonnie lassie
  When the kye come hame.

" When the blewart bears a pearl,
   And the daisy turns a pea,
And the bonnie lucken gowan
   Has fauldit up his ee,
Then the lavrock, frae the blue lift,
   Draps down and thinks nae shame
To woo his bonnie lassie
  When the kye come hame.
     *   *   *   *   *
" When the little wee bit heart
   Rises high in the breast,
And the little wee bit starn
   Rises red in the East,
Oh, there's a joy sae dear,
   That the heart can hardly frame,
Wi' a bonnie, bonnie lassie,
  When the kye come hame."

Felicia Hemans, in " The Switzer's Wife," has this melodious couplet :

" And when the herd's returning bells are sweet
In the Swiss valleys, and the lakes grow still."

The following beautiful stanza is taken from John Keats's "Ode on a Grecian Urn":

> "Who are these coming to the sacrifice?
> To what green altar, O mysterious priest,
> Lead'st thou that heifer lowing at the skies,
> And all her silken flanks in garlands drest?
> What little town by river or seashore,
> Or mountain-built with peaceful citadel,
> Is emptied of its folk this pious morn?
> And, little town, thy streets forevermore
> Will silent be, and not a soul to tell
> Why thou art desolate, can e'er return."

Samuel Ferguson, in "The Pretty Girl of Loch Dan," has this stanza:

> "She brought us in a beechen bowl
>   Sweet milk that smacked of mountain thyme,
> Oat cake, and such a yellow roll
>   Of butter—it gilds all my rhyme!"

From Alfred Tennyson's "In Memoriam" the following is culled:

> "And brushing ankle-deep in flowers,
>   We heard behind the woodbine veil
>   The milk that bubbled in the pail,
> And buzzings of the honeyed hours."

From "The Gardener's Daughter":

> "The fields between
> Are dewy fresh, browsed by deep uddered kine,
> And all about the large lime feathers, low,
> The lime a summer home of murmurous wings.
>   *   *   *   *   *   *
>   All the land in flowery squares,
> Beneath a broad and equal blowing wind,
> Smelt of the coming summer.   .   .   .
>   The steer forgot to graze,
> And, where the hedgerow cuts the pathway, stood
> Leaning his horns into the neighbor field,
> And lowing to his fellows.   From the woods
> Came voices of the well-contented doves.
> The lark could scarce get out his notes for joy,
> But shook his song together as he neared
> His happy home, the ground.   To left and right
> The cuckoo told his name to all the hills;
> The mellow ouzel whistled in the elm;
> The redcap whistled; and the nightingale
> Sang loud, as though he were the bird of day."

From "The Palace of Art":

> " Or sweet Europa's mantle blew unclasped
>   From off her shoulders backward borne :
> From one hand drooped a crocus ; one hand grasped
>   The mild bull's golden horn."

The pathetic ballad, by Charles Kingsley, "O Mary, go and call the Cattle Home!" is very popular :

> " O Mary, go and call the cattle home,
>   And call the cattle home,
>   And call the cattle home,
>   Across the banks o' Dee !
> The western wind was wild and dank wi' foam,
>   And all alone went she.
>
> " The creeping tide came up along the sand,
>   And o'er and o'er the sand,
>   And round and round the sand,
>   As far as eye could see ;
> The blinding mist came down and hid the land,
>   And never home came she.
>
> " Oh, is it weed, or fish, or floating hair—
>   A tress o' golden hair,
>   O' drowned maiden's hair—
>   Above the nets at sea ?
> Was never salmon yet that shone so fair,
>   Among the stakes on Dee.
>
> " They towed her in across the rolling foam—
>   The cruel, crawling foam,
>   The cruel, hungry foam—
>   To her grave beside the sea :
> But still the boatmen hear her call the cattle home
>   Across the sands o' Dee."

The following is a part of " The Milkmaid's Song," by Sidney Dobell:

> " Wheugh ! wheugh ! He's whistling through—
> He's whistling ' The Farmer's Daughter.'
>   Give down, give down,
>   My crumpled brown !
>   He shall not take the road to town,
> For I'll meet him beyond the water.
>   Give down, give down,
>   My crumpled brown !
> And send me to my Harry !
>   The folks o' towns
>   May have silken gowns,
> But I can milk and marry.
>   *   *   *   *   *

I'm too late for my Harry !
And oh, if he goes a-soldiering,
The cows they may low, the bells they may ring,
    But I'll neither milk nor marry.
        Fill pail,
Neither milk nor marry.
    *   *   *   *   *

They may talk of glory over the sea,
But Harry's alive, and Harry's for me,
    My love, my lad, my Harry !
Come spring, come winter, come sun, come snow,
What cares Dolly whether or no,
    While I can milk and marry ?
Right or wrong, and wrong or right,
Quarrel who quarrel, and fight who fight,
But I'll bring my pail home every night
    To love, and home, and Harry !"

No English poet has better depicted the emotions of the human heart or the vicissitudes of rural life than Jean Ingelow. How sweetly is the " old story" told in these few verses selected from " The Maiden with the Milking-Pail ":

" What change has made the pastures sweet,
And reached the daisies at my feet,
    And cloud that wears a golden hem ?
This lovely world, the hills, the sward—
They all look fresh, as if our Lord
    But yesterday had finished them.
    *   *   *   *   *

" I see the pool more clear by half
Than pools where other waters laugh
    Up at the breasts of coot and rail.
There, as she passed it on her way,
I saw reflected, yesterday,
    A maiden with a milking-pail.

" There, neither slowly nor in haste—
One hand upon her slender waist,
    The other lifted to her pail—
She, rosy in the morning light,
Among the water-daisies white,
    Like some fair sloop appeared to sail.

" Against her ankles as she trod
The lucky buttercups did nod ;
    I leaned upon the gate to see.
The sweet thing looked, but did not speak ;
A dimple came in either cheek,
    And all my heart was gone from me.
    *   *   *   *   *   *

" With happy youth and work content,
So sweet and stately on she went,
  Right careless of the untold tale :
Each step she took I loved her more,
And followed to her dairy door
  The maiden with the milking-pail.

   *   *   *   *   *

" And when the west began to glow
I went—I could not choose but go—
  To that same dairy on the hill ;
And while sweet Mary moved about
Within, I came to her without,
  And leaned upon the window-sill.

" The garden border where I stood
Was sweet with pinks and southern wood.
  I spoke—her answer seemed to fail.
I smelt the pinks—I could not see !
The dusk came down and sheltered me,
  And in the dusk she heard my tale.

" O life, how dear thou hast become !
She laughed at dawn, and I was dumb.
  But evening counsels best prevail.
Fair shine the blue that o'er her spreads,
Green be the pastures where she treads,
  The maiden with the milking-pail !"

In " The High Tide on the Coast of Lincolnshire" the poet makes an English
matron who lived five miles from old Boston tell the tragic story, in the quaint
speech of the time (1571), a little more than a half century before people of the
same neighborhood came to settle Massachusetts and Connecticut. The poet
Spenser was then a youth of eighteen and Shakespeare was a boy of seven years.

" The old mayor climbed the belfry tower,
  The ringers ran by two, by three.
' Pull if ye never pulled before—
  Good ringers, pull your best !' quoth he.
' Play uppe, play uppe, O Boston bells !
Ply all your changes, all your swells—
  Play uppe the *Brides of Enderby.*'

" Men say it was a stolen tyde—
  The Lord that sent it, He knows all ;
But in myne ears doth still abide
  The message that the bells let fall.
And there was nought of strange beside
The flight of mews and peewits pied
  By millions crouched on the old sea wall.

" I sat and spun within the doore ;
    My thread brake off, I raised myne eyes ;
 . The level sun, like ruddy ore,
    Lay sinking in the barren skies ;
And dark against day's golden death
She moved where Lindis wandereth,
My sonne's fair wife, Elizabeth.

" ' Cusha ! Cusha ! Cusha !' calling
Ere the early dews were falling.
    Farre away I heard her song,
    ' Cusha ! Cusha !' all along ;
Where the reedy Lindis floweth,
        Floweth, floweth,
From the meads where melick groweth
    Faintly came her milking song.

" ' Cusha ! Cusha ! Cusha !' calling,
' For the dews will soon be falling ;
Leave your meadow grasses mellow,
        Mellow, mellow,
Quit your cowslips, cowslips yellow.
Come uppe, Whitefoot, come uppe, Lightfoot,
Quit the stalks of parsley hollow,
        Hollow, hollow ;
Come uppe, Jetty, rise and follow,
    From the clovers lift your head ;
Come uppe, Whitefoot, come uppe, Lightfoot,
Come uppe, Jetty, rise and follow,
    Jetty, to the milking shed.'

" If it be long—aye, long ago,
    When I begin to think howe long,
Again I hear the Lindis flow,
    Swift as an arrowe, sharpe and strong ;
And all the aire it seemeth me
Bin full of floating bells (sayth shee),
That ring the tune of Enderby.

" All fresh the level pasture lay,
    And not a shadow mote be seene,
Save where full fyve good miles away
    The steeple towered from out the greene ;
And lo ! the great bell, farre and wide,
Was heard in all the country side,
That Saturday at eventide.

" The swannerds where their sedges are
    Moved on in sunset's golden breath ;
The shepherde lads I heard afarre,
    And my sonne's wife, Elizabeth ;

Till floating o'er the grassy sea
Came down that kyndly message free,
The 'Brides of Mavis Enderby.'

" Then some looked uppe into the sky,
 And all along where Lindis flows,
To where the goodly vessels lie,
 And where the lordly steeple shows.
They sayde, 'And why should this thing be,
What danger lowers by land or sea ?
They ring the tune of Enderby !

" ' For evil news from Mablethorpe,
 Of pyrate galleys warping down ;
For shippes ashore beyond the scorpe,
 They have not spared to wake the towne ;
But while the west bin red to see,
And storms be none, and pyrates flee,
Why ring 'The Brides of Enderby' ?

" I looked without, and lo ! my sonne
 Came riding downe with might and main ;
He raised a shout as he drew on,
 Till all the welkin rang again—
'Elizabeth ! Elizabeth !'
(A sweeter woman ne'er drew breath
Than my sonne's wife, Elizabeth.)

" ' The olde sea wall (he cried) is downe,
 The rising tide comes on apace,
And boats adrift in yonder towne
 Go sailing up the market-place !'
He shook as one that looks on death :
'God save you, mother !' straight he saith ;
'Where is my wife, Elizabeth ?'

" ' Good sonne, where Lindis winds away
 With her two bairns I marked her long ;
And ere yon bells beganne to play
 Afar I heard her milking song.'
He looked across the grassy sea,
To right, to left, 'Ho, Enderby !'
They rang 'The Brides of Enderby !'

" With that he cried and beat his breast,
 For lo ! along the river's bed
A mighty eygre reared his crest,
 And uppe the Lindis raging sped.
It swept with thunderous noises loud,
Shaped like a curling, snow-white cloud,
Or like a demon in a shroud.

" And rearing Lindis backward pressed,
    Shook all her trembling banks amaine ;
Then madly at the eygre's breast
    Flung uppe her weltering walls again.
Then banks came down with ruin and rout :
Then beaten foam flew round about ;
Then all the mighty floods were out.

" So farre, so fast the eygre drave.
    The heart had hardly time to beat,
Before a shallow, seething wave
    Sobbed in the grasses at oure feet ;
The feet had hardly time to flee
Before it brake against the knee,
And all the world was in the sea.

" Upon the roofe we sate that night,
    The noise of bells went sweeping by ;
I marked the lofty beacon light
    Stream from the church tower, red and high—
A lurid mark and dread to see ;
And awsome bells they were to me,
That in the dark rang ' Enderby.'

" They rang the sailor lads to guide
    From roofe to roofe who fearless rowed ;
And I—my sonne was at my side,
    And yet the ruddy beacon glowed ;
And yet he moaned beneath his breath,
' O come in life, or come in death,
O lost ! my love, Elizabeth.'

" And didst thou visit him no more ?
    Thou didst, thou didst, my daughter deare ;
The waters laid thee at his doore,
    Ere yet the early dawn was clear.
Thy pretty bairns in fast embrace,
The lifted sun shone on thy face,
Downe drifted to thy dwelling-place.

" That flow strewed wrecks about the grass,
    That ebbe swept out the flocks to sea ;
A fatal ebbe and flow, alas !
    To manye more than myne and me.
But each will mourn his own (she sayth),
And sweeter woman ne'er drew breath
Than my sonne's wife, Elizabeth.

" I shall never hear her more
    By the reedy Lindis shore,

' Cusha ! Cusha ! Cusha !' calling,
Ere the early dews be falling :
I shall never hear her song,
' Cusha ! Cusha !' all along
Where the sunny Lindis floweth,
    Goeth, floweth ;
From the meads where melick groweth,
When the water windeth down,
Onward floweth to the town.

" I shall never see her more
  Where the reeds and rushes quiver,
    Shiver, quiver,
  Stand beside the sobbing river,
  Sobbing, throbbing, in its falling,
  To the sandy, lonesome shore ;
  I shall never hear her calling,
  ' Leave your meadow grasses mellow,
    Mellow, mellow,
  Quit your cowslips, cowslips yellow ;
  Come uppe, Whitefoot, come uppe, Lightfoot ;
  Quit your pipes of parsley hollow,
    Hollow, hollow,
  Come uppe, Lightfoot, rise and follow ;
    Lightfoot, Whitefoot,
   From your clovers lift the head.
  Come uppe, Jetty, follow, follow,
   Jetty, to the milking shed.' "

American poets, too, are appreciative of the beauties of rural life. William Cullen Bryant, in his " Summer Ramble," lulls the soul with that

      " deep quiet that awhile
Lingers the lovely landscape o'er."

" The quiet August noon has come,
  A slumberous silence fills the sky ;
The fields are still, the woods are dumb,
  In glassy sleep the waters lie.
    *   *   *   *   *
' And mark yon soft white clouds that rest
  Above our vale, a moveless throng ;
The cattle on the mountain's breast
  Enjoy the grateful shadow long.
    *   *   *   *   *
" The village trees their summits rear
  Still as its spire, and yonder flock,
At rest in those calm fields, appear
  As chiselled from the lifeless rock."

Henry Wadsworth Longfellow, in his " Rain in Summer," thus expresses " Rest in the Furrow " :

> " In the furrowed land
>   The toilsome and patient oxen stand ;
>   Lifting the yoke-encumbered head,
>   With their dilated nostrils spread,
>   They silently inhale
>   The clover-scented gale
>   And the vapors that arise
>   From the well-watered and smoking soil.
>   For this rest in the furrow after toil
>   Their large and lustrous eyes
>   Seem to thank the Lord,
>   More than man's spoken word."

In his " Evangeline " he thus describes an " Evening in Acadia " :

> " Now recommenced the reign of rest and affection and stillness ;
>   Day, with its burden and heat, had departed, and twilight descending,
>   Brought back the evening star to the sky, and the herds to the homestead.
>   Pawing the ground they came, and resting their necks on each other,
>   And with their nostrils distended inhaling the freshness of evening.
>   Foremost, bearing the bell, Evangeline's beautiful heifer,
>   Proud of her snow-white hide and the ribbon that waved from her collar,
>   Quietly paced and slow, as if conscious of human affection.

> " Patiently stood the cows meanwhile, and yielded their udders
>   Unto the milkmaid's hand ; while loud and in regular cadence
>   Into the sounding pails the foaming streamlets descended.
>   Lowing of cattle and peals of laughter were heard in the farmyard,
>   Echoed back by the barns."

The following is a description of John Alden's bull, from the " Courtship of Miles Standish " :

> " Close to the house was the stall, where, safe and secure from annoyance,
>   Raghorn, the snow-white bull, that had fallen to Alden's allotment
>   In the division of cattle, might ruminate in the night-time
>   Over the pastures he cropped, made fragrant with sweet pennyroyal."

After the wedding :

> " Then from a stall near at hand, amid exclamations of wonder,
>   Alden, the thoughtful, the careful, so happy, so proud of Priscilla,
>   Brought out his snow-white bull, obeying the hand of its master,
>   Led by a cord that was tied to an iron ring in its nostrils,
>   Covered with crimson cloth, and a cushion placed for a saddle.
>   *　*　*　*　*　*　*　*

> " Onward the bridal procession now moved to their new habitation,
>   Happy husband and wife, and friends conversing together.

Pleasantly murmured the brook, as they crossed the ford in the forest,
Pleased with the image that passed like a dream of love through its bosom.
Tremulous, floating in air, o'er the depth of the azure abyss.
Down through the golden leaves, the sun was pouring his splendors,
Gleaming on purple grapes, that from branches above them suspended,
Mingled their odorous breath with the balm of the pine and the fir tree,
Wild and sweet as the clusters that grew in the valley of Eschol.
Like a picture it seemed of the primitive, pastoral ages,
Fresh with the youth of the world, and recalling Rebecca and Isaac.
So through the Plymouth woods passed onward the bridal procession."

John Greenleaf Whittier embodies a sentiment of " Peace " as follows :

"The grain grew green on battle plains,
   O'er swarded war-mounds grazed the cow ;
The slave stood forging from his chains
   The spade and plow ;"

and " Prosperity " in these lines from " The Preacher " :

" The land lies open and warm in the sun,
   Anvils clamor and millwheels run ;
Flocks on the hillsides, herds on the plain,
   The wilderness gladdened with fruit and grain !"

From " Mountain Pictures " :

" So twilight deepened round us.   Still and black
The great woods climbed the mountain at our back ;
And on the skirts where yet the lingering day
On the shorn greenness of the clearing lay,
The brown old farmhouse like a bird's nest hung.
With home-life sounds the desert air was stirred :
The bleat of sheep along the hill we heard,
The bucket plashing in the cool, sweet well,
The pasture bars that clattered as they fell ;
Dogs barked, fowls fluttered, cattle lowed ; the gate
Of the barnyard creaked beneath the merry weight
Of sunbrown children, listening while they swung,
The welcome sound of supper call to hear ;
And down the shadowy lane, in tinklings clear,
The pastoral curfew of the cowbell rung."

" The Barefoot Boy " sighs as fond memory calls up the past :

" O for festal dainties spread
   Like my bowl of milk and bread—
Pewter spoon and bowl of wood
   On the doorstone gray and rude !"

How sweetly he sings of "The Merrimac River"!

> " Sing soft, sing low, our lowland river,
>   Under thy banks of laurel bloom ;
> Softly and sweet, as the hour beseemeth,
>   Sing us the songs of peace and home.
>
>   *   *   *   *   *   *
>
> " Bring us the airs of hills and forests,
>   The sweet aroma of birch and pine ;
> Give us a waft of the north wind laden
>   With sweetbrier odors and breath of kine !
>
>   *   *   *   *   *   *
>
> " And well may we own thy hint and token
>   Of fairer valleys and streams than these,
> Where the rivers of God are full of water,
>   And full of sap are his healing trees."

From " The Voice of the Grass," by Sarah Roberts, these happy lines are taken :

> " Here I come creeping, creeping everywhere ;
>   In the noisy city street
>   My pleasant face you'll meet,
>   Cheering the sick at heart,
>   Toiling his busy part—
> Silently creeping, creeping everywhere.
>
> " Here I come creeping, creeping everywhere,
>   More welcome than the flowers
>   In summer's pleasant hours.
>   The gentle cow is glad,
>   And the merry bird not sad,
> To see me creeping, creeping everywhere.
>
> " Here I come creeping, creeping everywhere ;
>   My humble song of praise
>   Most joyfully I raise
>   To Him at whose command
>   I beautify the land—
> Creeping, silently creeping everywhere."

What child in the land does not love this pretty " Milking Song," by Celia Thaxter ?

> " Little dun cow to the apple tree tied,
>   Chewing the cud of reflection,
> I that am milking you sit by your side,
>   Lost in a sad retrospection.
>
> " Far o'er the fields the tall daisies blush warm,
>   For rosy the sunset is dying ;
> Across the still valley, o'er meadow and farm,
>   The flush of its beauty is lying.

" White foams the milk in the pail at my feet,
   Clearly the robins are calling ;
Soft blows the evening wind after the heat ;
   Cool the long shadows are falling.

" Little dun cow, 'tis so tranquil and sweet !
   Are you light-hearted, I wonder ?
What do you think about—something to eat ?
   On clover and grass do you ponder ?"

And the " Farm-Yard Song," by J. T. Trowbridge :

" Over the hill the farm-boy goes,
   His shadow lengthens along the land,
   A giant staff in a giant hand ;
   In the poplar tree, above the spring,
   The katydid begins to sing.
      The early dews are falling.
   Into the stone-heap darts the mink,
   The swallows skim the river's brink,
   And home to the woodland fly the crows,
   When over the hill the farm-boy goes,
      Cheerily calling,
     ' Co', boss ! co', boss ! co' ! co' ! co' !'
   Farther, farther over the hill,
   Faintly calling, calling still,
     ' Co', boss ! co', boss ! co' ! co' ! co' !'

" Into the yard the farmer goes,
   With grateful heart, at the close of day :
   Harness and chain are hung away ;
   In the wagon-shed stand yoke and plow,
   The straw's in the stack, the hay in the mow,
      The cooling dews are falling.
   The friendly sheep his welcome bleat,
   The pigs come grunting to his feet,
   The whinnying mare her master knows,
   When into the yard the farmer goes,
      His cattle calling—
     ' Co', boss ! co', boss ! co' ! co' ! co' !'
   While still the cow-boy, far away,
   Goes seeking those who have gone astray—
     ' Co', boss ! co', boss ! co' ! co' ! co' !'

" Now to her task the milkmaid goes,
   The cattle come crowding through the gate,
   Lowing, pushing, little and great ;
   About the trough, by the farmyard pump,
   The frolicsome yearlings frisk and jump,
      While the pleasant dews are falling.

The new milch heifer is quick and shy,
But the old cow waits with tranquil eye ;
And the white stream into the bright pail flows,
When to her task the milkmaid goes,
  Soothingly calling,
 ' So, boss ! so, boss ! so ! so ! so !'
The cheerful milkmaid takes her stool,
And sits and milks in the twilight cool,
  Saying, ' So ! so, boss ! so ! so !'

" To supper at last the farmer goes,
The apples are pared, the paper read,
The stories are told, then all to bed.
Without, the cricket's ceaseless song
Makes shrill the silence all night long ;
  The heavy dews are falling.
The housewife's hand has turned the lock ;
Drowsily ticks the kitchen clock ;
The household sinks to deep repose,
But still in sleep the farm-boy goes,
  Singing, calling,
 ' Co', boss ! co', boss ' ! co' ! co' ! co !'
And oft the milkmaid in her dreams,
Drums in the pail with the flashing streams,
  Murmuring, ' So, boss ! so !' "

This collection would hardly be complete without introducing the following
" Reminiscence" from the *Harvard Advocate :*

" We stood at the bars as the sun went down
 Behind the hills, on a summer day ;
Her eyes were tender and big and brown,
 Her breath as sweet as the new-mown hay.

" Far from the west the faint sunshine
 Glanced sparkling off her golden hair ;
Those calm deep eyes were turned toward mine,
 And a look of contentment rested there.

" I see her bathed in the sunlight flood,
 I see her standing peacefully now ;
Peacefully standing and chewing her cud,
 As I rubbed her ears—that Jersey cow !"

PART FIRST.

HISTORY OF JERSEY CATTLE — PRINCIPLES OF BREEDING.

JERSEY.

THE island of Jersey, the native home of the breed of Jersey cattle, is the chief in size of the group called Channel Islands, lying near the coast of France in the English Channel.

Jersey lies west of the province of Normandy about sixteen miles and about the same distance south-west from the island of Guernsey, and is eleven miles in length from east to west and seven and a half miles in breadth.

The surface of the land has a general slope south-eastwardly, being high and precipitous on the north, with table-lands in the central portion intersected by brooks and runnels which flow to the south and east.

The coast is picturesque in savage ruggedness, being high and precipitous on the north, and indented by numerous bays on the east, south, and west.

The climate is mild and equable, and the air moist, and rains frequent. The mean temperature is 50.8°, August being the warmest and February the coolest month, while from mid October to mid December the weather resembles our Indian summer, and is called St. Martin's Summer.

The soil is very rich, deep, and porous from centuries of tillage. Means of fertility are afforded by the large number of cattle, green herbage, and large quantities of sea-weeds collected under strict regulations of the local government.

The island contains 39,680 acres, 25,000 of which are cultivated. The population is nearly 57,000, about 15,000 being English denizens and 2000 Parisians and others, who resort thither for health or the pleasant enjoyment of a very delightful climate and picturesque scenery. Jersey is divided into twelve parishes, and the lands are held in small farms of five to twenty acres.

The productions are the famous Jersey cattle, enormous crops of potatoes, wheat, parsnips, mangolds, carrots, turnips, and a variety of cabbage which has a long, woody stem surmounted by a tuft of broad leaves ; these last grow from six to twelve feet high, and are used for cow fodder. There are numerous orchards and graperies, which produce choice fruit.

The flora and fruits of semi-tropical regions flourish equally well as those of the temperate zone, and include oranges, lemons, and such trees as azalea, oleander, and fuchsia, the last being used for hedges and decoration of buildings.

The grasses are short and luscious, and green all winter.

The Romans occupied the island in the third and fourth centuries, and were so charmed with its natural beauties and climate that they called it Cæsarea, or Cæsar's Isle. Subsequently a mixed population of Gauls, Goths, Danes, and Saxons occupied Jersey until the Norman conquest of England. Jersey was English under William the Conqueror; English under Henry I.; Norman again under Stephen; English again under Henry II., since which time it has been steadfastly loyal to the English crown.

During the last century the people of Jersey have become very prosperous, and now derive a good income by the exportation of their favorite cattle, which are sent to all parts of the world, but chiefly to America, where they are best appreciated and most successfully bred.

ORIGIN OF THE RACE OF JERSEY CATTLE.

" The cattle of this island are superior to the French cattle" (Philip Falle, A.D. 1734). This is the first historical statement I have found regarding the quality of the Jersey race of cattle. The history of their origin is more mythical and legendary than that of the people of Jersey. The cattle are commonly supposed to be a composite race derived from the cattle of Brittany and Normandy, but neither the Brittany nor Cotentin breed equals the Jersey of to-day in productive capacity or beauty of form or color. The Montafu breed of cattle in the mountainous district near Lake Constance is said to resemble much the modern Jersey, as also the cattle of Lombardy; and in the Saguenay region of Canada there are specimens closely resembling the Jersey, the descendants of cattle brought by French emigrants from Brittany.

We know little of the races of cattle of Southern Europe at this day, and much less of their history of one or two thousand years ago. The remote origin of the Jersey is still more problematical. It is well to note the very striking resemblance between the modern Jersey and the Zebu or sacred cattle of India. The beautifully blended silver gray and slate shadings, the delicacy of frame, the fine bone, the yellow skin, the black muzzle, black tongue, and black switch, the almost identical facial expression, the shape and setting of the eye, the small ear, the slender horn, are wonderfully alike in Jersey and Zebu.

Is it too much to conjecture that the patriarch Jacob, in his experiments with the herds and flocks of Laban, whereby he produced and fixed fantastic and grotesque markings of white, also combined the blood of the Zebu bull with that of his historical race of spotted cattle ?

Most writers on the origin of the Jersey attribute the yellow coats, buff points, and white patches to the Normandy or Cotentin race, which is supposed to be the source of the present breed of Guernsey Island, while the solid colors and black points are attributed to the Brittany race, although some assume that there has been an admixture of Norway cattle with the Jersey.

Mr. James P. Swain says: "I consider the cows on the island of Jersey Norman, mixed with another distinct breed, the main characteristics of each being still plainly visible, though growing less so yearly. The original, or highest type, I call the wild Jersey; the other type I consider Norman or Guernsey.

"The wild Jersey has a black nose, black tongue, and mealy muzzle; the other, a buff nose. The wild Jersey's horns are black, pointed, firm, with single curve, forming nearly a semicircle, deeply fluted inside when taken off. The other has weak horns, shelly, yellow, waxy near the head, inclined downward, with double curve, compacted, smooth inside when taken off. The color of the female wild Jersey is chocolate, or mink color, no white spots, and the males nearly black. The others are yellowish, brown and white, star in forehead. The wild Jersey's skin is olive brown; the other, skin very yellow, even to the end of the tail. In the wild Jersey the tail terminates in a small tuft of long hairs, the skin near the end scaly with the accumulation of coloring matter. The other, skin on tail very yellow, even to the end, where there is an accumulation of coloring matter, which the Guernsey men call 'a lump of butter;' the long hair on the tail starts higher up."

Professor Low and Charles W. Elliott support the statement that these "darker colored or wild Jerseys clearly resemble the Norwegian cattle of to-day," and "that these old sea-rovers have taken their cattle to these islands."

But it is the island of Jersey, with its bland climate and centuries of gentle care and management by the women of Jersey, that has produced what is now known as the best butter cow in the world.

One hundred and fifty years ago the Rev. Philip Falle wrote of the Jerseys as above quoted, and it may have required centuries of selection to enable a faithful historian to make this statement.

The Jersey cow is tethered to the ground, being changed five or six times a day to a new station. When she calves she is regaled with toast and with cider, the nectar of the island, to which powdered ginger is added.

Thomas Quayle, who in 1812 wrote a work on the "Agriculture of the Channel Islands," is quoted as saying that "on hearing praises bestowed on any particular cows, they generally, but not always, were found to have a black tinge."

He also states that "the general purity of the breed is guarded by the rooted opinions of the inhabitants rather than by the sanction of law; but hitherto no persevering, systematical experimenter has attempted, by a careful selection of individuals and attention to their crosses, to improve this breed. When a cow is famed as a good milker, her male progeny is preserved ; but this is for a short period, and it is not known whether any other measure whatever has been persevered in to keep up the breed at its present standard."

IMPROVEMENT OF THE JERSEY.

The Royal Jersey Agricultural Society originated in the year 1833 from a desire on the part of some intelligent and progressive men to improve the island cattle and advance their system of agriculture.

Previous to that time laws had been passed by the local legislature prohibiting importation of any cattle from France, the first bearing date of July 16th, 1763. This continued in force until 1789, when the celebrated " Act of the States of Jersey" was passed on the 8th of August of that year. The first article of the Act of 1789 provided that any person introducing any cattle from France should be subject to a fine of £200 sterling, besides the confiscation of the cattle and the boat, and obliged every sailor to be an informer against his master within twenty-four hours, under a penalty of £50 sterling, such fines to go one third to the crown and two thirds to the poor of the parish ; and if the master was insolvent, he was to be imprisoned six months. Article II. required all beef cattle imported to be landed at St. Helier or St. Aubin, under the same penalties for violation.

Article III. required cattle from the adjacent islands to be landed at the same ports, under the same penalties for violation.

Article IV. confiscated every French animal landed contrary to law, and required its immediate slaughter and distribution to the poor of the parish where seized.

Articles V., VI., VII., and VIII. regulated the exportation of Jersey cattle.

The law of March 18th, 1826, increased the fine to £1000 for importing French animals, the fine imposed being repeated for each and every animal. All accomplices were subjected to the same fine. All cattle found on ship or boat within two leagues of the island were confiscated, as well as the boat, and the same fines imposed as for landing cattle.

Three ports were set apart for the introduction of beef cattle.

Still another act was passed in 1864, in harmony with the treaties between France and England. Article III. permitted the importation of French cattle for consumption or in transit. Article IV. prohibited the breeding of foreign cattle on the island. Article VIII. required all French cattle to be branded with the letter F, and to be slaughtered at the port of St. Helier, or re-embarked at the same port. The fine was reduced to £10 sterling for each head of cattle, one third to be paid to the informer, or six months' imprisonment of the principal, if unable to pay the fine. Several attempts have been made to cross the Jersey with the Shorthorn and

Ayrshire breeds, but they were abandoned, and the progeny slaughtered because it was inferior to the Jersey.

Guernsey cattle are not prohibited, and a very few may be found upon the island. Crosses between the breeds produce buff nose and eyes, and the offspring retains a coarseness, at once detected and rejected by the judges at examination for Herd Book or for prizes at fairs. The natural pride that a Jerseyman has in his cow, and his desire to mate her with a prize bull, is an incentive to keep the breed pure.

On the 18th of January, 1834, over fifty years ago, the society drew up their first scale of points. The Jersey cow as she then existed and was described by Colonel Le Couteur and by the judges that officiated at the show was quite a different animal from the Jersey cow of to-day. It was impossible then to find a cow on the island that came near to the ideal by the standard of that time. Two of the best cows were selected from which to make up a scale of points, one of them being considered perfect in forequarters and barrel, the other in her hindquarters. The scale consisted of seven articles and twenty-five counts for a bull and the same number of articles and twenty-seven counts for a cow.

The Jersey cow was described by the judges in the year 1834 as follows:

"1. That the cattle were very much out of condition.

"2. Too slightly formed behind and cat-hammed.

"3. Gait unsightly.

"4. The udder ill-formed.

"5. The tail coarse and thick.

"6. The hoofs large.

"7. The head coarse and ill-shaped.

"8. Many were without that golden or yellow tinge within the ears which denotes a property to produce yellow and rich butter.

"9. Some cows and heifers had short bull necks.

"10. Some had too much flesh or dewlap under the throat.

"11. Some were too heavy in the shoulders.

"The first show was held March 31st, 1834. The prizes amounted to £24. Colonel Le Couteur won the general prize of £3 with a red and white yearling bull. . . . The cultivation of parsnips was advocated. It was resolved to encourage fine bulls with points up to perfection by giving a premium of £10 for perfect bulls, and allowing the owners 2s. a head for each cow that shall have been with calf by such bulls."

"In 1835 the show furnished not only a larger supply, but the animals were of a much finer order as to breed and condition."

"Her Majesty became a patroness in 1837.

"Two shows were held—one in March for bulls and the other in May for cows and heifers. This division of the shows has continued up to the present day. The system of giving points for pedigree commenced in 1838." The scale of points was modified, increasing the number of counts to twenty-eight for bulls and heifers, and thirty for cows. Two new rules required that the owner of a prize bull, by withholding his services from the public, should forfeit his prize-money, and the second that prize heifers must remain upon the island until they had dropped their first calf. The annual reports indicate that improvement in the cattle exhibited was very rapid. After seven years, attention to breeding had almost caused the ancient characteristic defect, the drooping hindquarter, to disappear; also several minor defects; and it only remained to give squareness to the hindquarter and roundness to the barrel to render the Jersey a most beautiful animal."

At the annual dinner Colonel Le Couteur said in a speech: " Let me say to those who are lukewarm to this society to look back ten years. The land foul with weeds, crops inferior, liquid manure wasted, the market ill supplied. What had been effected?

" In cattle, beauty of form and flesh had been added to milking and creaming qualities. More cattle had been decorated this year than on any previous occasion, and the breed had so greatly improved that many of the animals rejected for having less than nineteen points would have been prize cattle when the society was formed, so well were their merits understood. The price of cattle had fully doubled."

The scale of points was revised again in the years 1849, 1851, and 1858.

During these years the reputation of the Jersey had greatly increased in England and America, and a fraudulent trade had sprung up by the French dealers exporting the cattle of Brittany to England as Jerseys, or Alderneys, as they were then misnamed. In 1850 and subsequent years several American gentlemen of wealth and influence began to make importations of Jerseys to the United States; among these were Daniel Buck, Jr., John A. Taintor, and John T. Norton, of Connecticut, and Thomas Motley, of Massachusetts; importations have been almost constant, except during the civil war, since that time. It is believed that the average quality of those early importations has not been excelled in later days, as the Americans tempted the Jerseymen to forfeit their prizes by offering them very liberal sums for decorated bulls and cows. But the American importations gave a new stimulus to Jersey breeders on the island, and the several parishes began to form farmers' clubs, which resulted in a great increase of cattle shows and a larger exhibition for the parent society. The report for 1858 was retrospective: " Thirty years ago the cattle were ill-fed, ill-shaped beasts that knew not the taste of mangolds, carrots, or swedes, scarcely that of hay; whose stabling was wretched, and whose winter food consisted chiefly of straw and a few watery turnips.

" Now they were well fed, improved in quality and symmetry, and well housed.

"The watery turnip, by careful husbandry, has become as rich as cheese. New buildings dotted the island, and general prosperity dawned on the farmer."

The Island Herd Book was started in the year 1866. "The Herd Book is entirely due to the forethought and untiring efforts of Mr. Charles P. Le Cornu. . . . He foresaw, many years before the Herd Book was started, the necessity of some further classification of the animals in a show where upward of two hundred were exhibited, so he determined to work out a unique system of his own. His principle was to sift, as it were, these large gatherings into three classes, by highly commending the best for their quality, symmetry, and constitution, and their butyraceous or milk-flowing properties; commending the second best and rejecting the remainder, or third class; and by examining the approved offspring he hoped in time to root out the bad animals, so that with six or seven registered crosses animals might be bred more to a certainty."

At the May show, 1874, Mr. Charles Nicolle offered a cup for the cow with the best escutcheon according to the Guenon system. The prize is still continued by voluntary contributions. Guenon prizes are also given for bulls.

The keeping of the modern Jersey upon the island from calfhood is as follows:* "When the cow has dropped her calf, there is sprinkled upon it a handful of powdered salt, and the cow licks it off. This bit of salt causes the cow to drink. While she is licking her calf she is milked, and drinks the first milking. The calf being dry, it is placed upon a bed of straw in a small stall. After some hours the cow is again milked, and her milk, mixed with tepid water, is given to the calf. The little animal is fed in this manner three times a day for the first three days. Afterward, for the next three days, the evening milk is kept till morning, the cream taken off, and the remainder, mixed with water quite warm, is served to the calf. The sixth day the keeping of cream for butter-making begins; the milk is skimmed every twenty-four hours, which permits it to become thick and acid. This milk is given to the calf twice every day, not forgetting to mix warm water therewith, and not hesitating to add at the end cooked flour, or even a slice of broken bread taken upon a plate held in the hand, in order to assist the calf in swallowing it. From time to time salt is added to the beverage, and a little hay. At the end of three months, if the weather is fine, the calf is able to go out; it then becomes stronger, and when the milk of the dam begins to diminish, the calf is given twice a day a warm beverage composed of cornmeal and bran.

"When the animal, always submitted to this regimen, attains the age of ten to thirteen months, and it is exceptionally fine, there is no hesitation in continuing this alimentation until it is sold; if the animal is of second quality, the beverages are stopped, in order to habituate it, little by little, to the food of the fields. A heifer

---

* Jersey Cattle, by Henri Johanet, translated by W. E. Simonds.

can be taken to the bull at nine months, but good heifers ought to wait to fifteen months. The cattle are as much as possible left in the open air from the month of May till the first of September. They are tethered in the open field, where the animal takes delight, making a void around itself. When in the stable it receives every day four or five pounds of dry food, and from twenty-five to thirty pounds of roots, the feedings taking place seven or eight times a day. The principal forage plants are, before all, Swedish turnips and parsnips; then come carrots, radishes, field turnips, beets, etc. During a milking period of about three hundred days a good Jersey cow gives daily, at the maximum, twenty-seven litres; at the minimum, eighteen litres of milk. The result in butter is from eight to ten pounds a week; it may be three to five kilogrammes of butter. The Jersey pound is four hundred and eighty-eight grains."

Insufficient attention is paid to the butter quality of island animals. A bull may take all the first prizes—that for best bull on the island, the prize for best Herd Book bull, and the silver cup for best escutcheon, and his dam and grand dam might be very poor butter-makers; so a cow may take the silver cup for best cow on the island, the first Herd Book prize, and the Guenon prize for best escutcheon without having a record for butter-making herself, and not belonging to a line of noted butter-makers; she may have all the fine and fancy points, and produce a large quantity of milk, and still be a very poor butter cow.

It is in America that the breed has begun to be rightly appreciated, and that only recently, because of the practice of testing cows to ascertain their butter-making capacity.

THE AMERICAN JERSEY CATTLE CLUB.

In the year 1868 Colonel George E. Waring, Jr., Samuel J. Sharpless, Charles M. Beach, Thomas J. Hand, and a number of Jersey cattle-breeders held a meeting in Philadelphia, which resulted in the organization of the American Jersey Cattle Club, with about forty members. The number has since increased to more than three hundred, and it is believed to represent more wealth and intelligence than any similar body of men in the world.

The object of the club from its incipiency was to foster absolute purity of breeding, and all the interests accruing from such breeding.

The adoption of a constitution and stringent by-laws and the formation of a "Herd Register" prepared the way for the success which has followed. Up to that time, by a sort of "Irish bull," the Jersey was called an "Alderney," and the Guernsey cattle also went by the same appellation, although the two breeds are very unlike, and neither of them was associated in any way whatever with the island of Alderney, except that Alderney has a mongrel mixture of the two breeds which are not imported to this country. No animal can be registered as imported from Jersey which is not identified by certificates from the agent of the club resident in the island, and no American animal can be registered which is not proven to be the offspring of animals already registered.

The Herd Register is now the standard of pedigree in the United States and Canada, and contains a record of all transfers of cattle, with the names of owners, thus giving a complete history of every animal recorded.

CHARACTERISTICS OF THE MODERN JERSEY.

The Jersey bull or cow of the year 1885 differs widely in form and color from the Jersey bull or Jersey cow of fifty years ago. By the skill of numerous breeders on the Island, in England, and America, as well as by the influences of climate and feed, and also by various hidden causes, very marked changes have been effected in perpetuating features and peculiarities that were once very rare, or by fixing the characteristics of sports and phenomenal animals so as to form distinct families and diverse types.

The Jersey of to-day is the most beautiful of all the bovine races, matchless for symmetry, variety of beautiful colors and shadings, and for that delicacy of frame and fineness of quality which makes the race attractive to the eye and taste of all lovers of bovine beauty. At the same time, the Jersey cow excels all other races in the amount and quality of butter. Since the practice of testing cows for butter has become popular, which is only in recent years, upward of one thousand Jersey cows have produced fourteen pounds of butter in a test of seven days, while the reports show that ninety of these have tested twenty pounds, or upward, in a week; and twelve cows are classed in the list that have produced twenty-five pounds, or upward, in seven days; one cow has made, by official test, forty-six pounds, twelve and a half ounces; another, thirty-nine pounds, twelve ounces; and another, thirty-six pounds, twelve and a quarter ounces of butter in seven days.

The Jerseys have been bred for centuries for their choice quality of milk and butter, and during the last half century, in their native island, in England, and America, much attention has been given to perfection of form and beautiful tints of color and fancy markings.

The breed is classed as medium to small in size; but in America the tendency is to select those of larger development, and to cultivate an increase of the size. The Jersey is of that spare habit of flesh consistent with the best dairy qualities, and the food she eats so assimilated and the secretory powers so highly developed as to fill the udder with all the fats and oils, instead of excreting them or accumulating them about the vital organs or upon the body, as in beeves.

The Jersey is fine in bone, of rare symmetry, and has just enough muscular development for healthful activity and full digestive force. Some individuals indicate a marvellous capacity for changing a large quantity of grain and forage into the best of dairy productions.

JERSEY COLORS.

The young Jersey is colored like the fawn, or young deer. This ground color is, later on, so modified by the second growth of hair as to produce, in different animals, an endless variety of soft, pleasing tints. In describing Jersey colors they are classed as fawns, grays, or browns. The fawns are described with the tint and shading as, for the

METALLIC TINTS:

golden fawn, silver, copper, bronze, steel, slate, brick-dust, granite, and pearl fawns;

COLOR TINTS:

yellow, fawn, red, blue, gray, brown, bay, buff, cinerous, drab, dun, smoky, tan, dusky and blackish fawns, and ivory black;

ANIMAL TINTS:

buckskin, beaver, bison, dove, otter, oriole, fox, mink, moose, mouse, seal, salmon, seashell, sable, and squirrel fawns;

VEGETABLE TINTS:

orange, lemon, banana, apple, strawberry, russet, maize, butternut, mulberry, cane, mahogany, coffee-seed, cinnamon, and chocolate fawns;

DAIRY TINTS:

cream fawn, milky fawn, cheese and butter fawns;

GRAYS:

squirrel, silver, slate, orange, court, French, blue, steel, iron, cinerous or ash, russet, and lavender grays.

One animal may have several of these tints beautifully blended and shaded, as the bull St. Helier 45, bright salmon fawn and silver gray, or the cow Mary Anne of St. Lambert 9770, a light smoky bay fawn. Some bulls have dark markings resembling the spots of a leopard. Many Jerseys have irregular patches of white, the white being soft and sometimes margined with a half-inch border of deep indigo. Jersey Belle of Scituate 7828 was a dark strawberry fawn, with white saddle on withers, and white on hips, sides, belly, and legs. A rich golden waxy dandruff shone under the white and a nankeen color on the udder and escutcheon. Some Jerseys of dark color have a rich cadmium orange tint within the ears, and very conspicuous also on the dewlap and the escutcheon. Many of the best cows have the broad white saddle upon the withers. A Jersey does not depend upon the color of the coat for any degree of the richness of milk or creamy qualities. The great amount of butter fat secreted in the milk is a special trait, highly developed in the whole Jersey race and phenomenally shown in certain individuals and families. Some animals have the special ability to give a rich golden tint to their butter. This desirable trait is generally thought to be positively indicated by a rich golden-orange lining of the ears.

There need be no fashion in color, but in the essential dairy qualities the highest perfection should be sought. One can breed for color and fancy points if he so desires. Wonderful results can be achieved by selection and inbreeding. An ancient breeder was very successful in fixing spots, ringstreak, and specks not only upon bulls and cows, but upon sheep, goats, camels, and asses; and some of the modern breeders of Jerseys have proved that they can breed out the spots without any detriment to the race.

THE SCALE OF POINTS.

From the first organized effort to improve the Jersey a "scale of points" has been deemed a necessity. The scale is supposed to embody in a schedule the descriptions of the ideal Jersey bull and ideal cow.

The first scale of points adopted on the Island of Jersey, January 18th, 1834, is as follows :

SCALE OF POINTS FOR BULLS.

ARTICLES.	POINTS.
1. Purity of breed on male and female sides reputed for having produced rich and yellow butter...	4
2. Head fine and tapering ; cheek small ; muzzle fine, and encircled with white ; nostrils high and open ; horns polished, crumpled, not too thick at the base, tapering, and tipped with black ; ears small, of an orange color within ; eye full and lively...............................	8
3. Neck fine and highly placed on shoulders ; chest broad ; barrel hooped and deep, well ribbed home to hips...............................	3
4. Back straight from the withers to the setting on of tail, at right angles to the tail ; tail fine, hanging two inches below the hock...............	3
5. Hide thin and movable, mellow, well covered with soft and fine hair of a good color...	3
6. Forearm large and powerful ; legs short and straight, swelling and full above the knee and fine below it...............................	2
7. Hind-quarters from the huckle to the point of the rump long and well filled up ; the legs not to cross in walking.......................	2
Perfection...	25

No prize to be awarded to a bull having less than twenty points. .

SCALE OF POINTS FOR COWS.

ARTICLES.	POINTS.
1. Breed on male and female sides reputed for producing rich and yellow butter	4
2. Head small, fine, and tapering; eye full and lively; muzzle fine and encircled with white; horns polished and a little crumpled, tipped with black; ears small, of an orange color within	8
3. Back straight from the withers to the setting on of the tail; chest deep and nearly on a line with the belly	4
4. Hide thin, movable, but not too loose, well covered with fine soft hair of good color	2
5. Barrel hooped and deep, well ribbed home, having but little space between the ribs and hips; tail fine, hanging two inches below the hock	3
6. Fore-legs straight and fine; thighs full and long, not too close together when viewed from behind; hind-legs short, and bones rather fine; hoofs small; hind-legs not to cross in walking	2
7. Udder full, well up behind; teats large and squarely placed, being wide apart; milk veins large and swelling	4
Perfection for cows	27

Two points shall be deducted for heifers.

A heifer will be considered perfect at twenty-five points.

No prize shall be awarded to cows and heifers having less than twenty-four points.

The scale of points had several changes at various times. In 1858 bulls stood at thirty-one and cows at thirty-three articles and thirty-three points, each article counting but one in the scale.

In April, 1875, a new scale was adopted.

RATIO SCALE OF POINTS FOR BULLS.

ARTICLES.	POINTS.
1. Registered pedigree	5
2. Head fine and tapering, forehead broad	5
3. Cheek small	2
4. Throat clean	4
5. Muzzle dark, encircled by light color, with nostrils high and open	4
6. Horns small, not thick at base, crumpled, yellow, tipped with black	5
7. Ears small and thin, and of a deep orange color within	5
8. Eyes full and lively	4
9. Neck arched, powerful, but not coarse and heavy	5
10. Withers fine, shoulders flat and sloping, chest broad and deep	4
11. Barrel hooped, broad, deep, and well ribbed up	5

ARTICLES.	POINTS.
12. Back straight from the withers to the setting on of the tail...............	5
13. Back broad across the loins............................	3
14. Hips wide apart and fine in the bone.......	3
15. Rump long, broad, and level.............................	3
16. Tail fine, reaching the hocks and hanging at right angles with the back ...	3
17. Hide thin and mellow, covered with fine soft hair....................	4
18. Hide of a yellow color................................	4
19. Legs short, straight and fine, with small hoofs.......................	4
20. Arms full and swelling above the knees........................	3
21. Hind-quarters from the hock to the point of rump long, wide apart, and well filled up..................................	3
22. Hind-legs squarely placed when viewed from behind, and not to cross or sweep in walking.............................	3
23. Nipples to be squarely placed and wide apart.......................	5
24. Growth...............................	4
25. General appearance............................	5
Perfection..................................	100

No prize to be awarded to bulls having less than eighty points. Bulls having obtained seventy-five points shall be allowed to be branded.

RATIO SCALE OF POINTS FOR COWS AND HEIFERS.

ARTICLES.	POINTS.
1. Registered pedigree.............................	5
2. Head small, fine, and tapering.......................	3
3. Cheek small, throat clean.........................	4
4. Muzzle dark and encircled by a light color, with nostrils high and open..	4
5. Horns small, not thick at base, crumpled, yellow, tipped with black......	5
6. Ears small and thin, and of a deep orange color within................	5
7. Eye full and placid.............................	3
8. Neck straight, fine, and lightly placed on the shoulders...............	3
9. Withers fine, shoulders flat and sloping, chest broad and deep...........	4
10. Barrel hooped, broad and deep, being well ribbed up................	5
11. Back straight from withers to the setting on of the tail...............	5
12. Back broad across the loins.........................	3
13. Hips wide apart, and fine in the bone; rump long, broad and level......	5
14. Tail fine, reaching the hocks, and hanging at right angles with the back	3
15. Hide thin and mellow, covered with fine soft hair....................	4
16. Hide of a yellow color.........................	4
17. Legs short, straight, and fine, with small hoofs....................	3

ARTICLES.	POINTS.
18. Arms full and swelling above the knees........	3
19. Hind-quarters from the hock to point of rump long, wide apart, and well filled up..	3
20. Hind-legs squarely placed when viewed from behind, and not to cross or sweep in walking..	3
21. Udder large, not fleshy, running well forward, in line with the belly and well up behind..	5
22. Teats moderately large, yellow, of equal size, wide apart, and squarely placed...	5
23. Milk veins about the udder and abdomen prominent...................	4
24. Growth...	4
25. General appearance...	5
Perfection.....................	100

No prize shall be awarded to cows having less than eighty points.

No prize shall be awarded to heifers having less than seventy-one points.

Cows having obtained seventy-five points and heifers sixty-five shall be allowed to be branded.

The articles Nos. 21 and 23 shall be deducted from the number required for perfection in heifers, as their udders and milk veins cannot be fully developed.

SCALE OF POINTS FOR COWS, ADOPTED BY THE AMERICAN JERSEY CATTLE CLUB, APRIL 21, 1875.

POINTS.	COUNTS.
1. Head small, lean, and rather long................................	2
2. Face dished, broad between the eyes and narrow between the horns.....	1
3. Muzzle dark, and encircled by light color........................	1
4. Eyes full and placid...	1
5. Horns small, crumpled, and amber-colored.........................	3
6. Ears small and thin..	1
7. Neck straight, thin, rather long, with clean throat, and not heavy at the shoulders...	4
8. Shoulders sloping and lean; withers thin; breast neither deficient nor beefy..	3
9. Back level to the setting on of tail, and broad across the loin...........	4
10. Barrel hooped, broad, and deep at the flank.......................	8
11. Hips wide apart, and fine in the bone; rump long and broad..........	4
12. Thighs long, thin, and wide apart, with legs standing square, and not to cross in walking..	4
13. Legs short, small below the knee, with small hoofs..................	3

Points.	Counts.
14. Tail fine, reaching the hocks, with good switch......................	3
15. Hide thin and mellow, with fine soft hair..........................	4
16. Color of hide where the hair is white, on udder and inside of ears, yellow..	5
17. Fore-udder full in form and running well forward.................	8
18. Hind-udder full in form, and well up behind......................	8
19. Udder free from long hair and not fleshy..	5
20. Teats rather large, wide apart, and squarely placed.................	6
21. Milk veins prominent...................	5
22. Escutcheon high and broad, and full on thighs.....................	8
23. Disposition quiet and good-natured...............................	3
24. General appearance rather bony than fleshy.......................	6
Perfection......................................	100

In judging heifers, omit Nos. 17, 18, and 21.

The same scale of points shall be used in judging bulls, omitting Nos. 17, 18, 19, and 21, and making moderate allowance for masculinity.

The American Jersey Cattle Club adopted, February 11th, 1885, a new scale of points, as below :

FOR COWS.

Points.	Counts.
1. Head small and lean, face dished, broad between the eyes and narrow between the horns..	2
2. Eyes full and placid ; horns small, crumpled, and amber-colored........	1
3. Neck thin, rather long, with clean throat, and not heavy at the shoulders	8
4. Back level to the setting on of tail.................................	1
5. Broad across the loin..	6
6. Barrel long, hooped, broad and deep at the flank...................	10
7. Hips wide apart ; rump long and broad..........................	10
8. Legs short...	2
9. Tail fine, reaching the hocks with good switch....................	1
10. Color and mellowness of hide ; inside of ears yellow.................	5
11. Fore-udder full in form and not fleshy...........................	13
12. Hind-udder full in form and well up behind.......................	11
13. Teats rather large, wide apart, and squarely placed..................	10
14. Milk veins prominent..	5
15. Disposition quiet...	5
16. General appearance and apparent constitution.....................	10
Perfection......................................	100

In judging heifers, omit 11, 12, and 14.

FOR BULLS.

The same scale, omitting Nos. 11, 12, and 14, and making due allowance for masculinity ; but when bulls are exhibited with their progeny, in a separate class, add thirty counts for progeny.

SCALE OF POINTS FOR COWS IN MILK.[*]

POINTS.	COUNTS.
1. Weight of milk in twenty-four hours, one count for each pound of yield, 32 lbs.	32.00
2. Total solids by chemical analysis, one count for each percentum, 13.76	13.76
3. Butter fat, three per cent. being standard, add ten per cent. for every one per cent. above, or deduct ten per cent. for every one per cent. below the standard, 5.25	22.50
4. Time since calving, add one count for every ten days—one hundred and twenty-three days	12.30
Total	80.56

The above figures are those of the first prize cow at the Edinburgh Show.

This scale, with an additional point for a butter test on specified rations, would insure a fair test of merit at exhibitions in contests between all breeds of dairy cows.

DIAGRAMS ILLUSTRATING SOME OF THE MORE IMPORTANT POINTS.

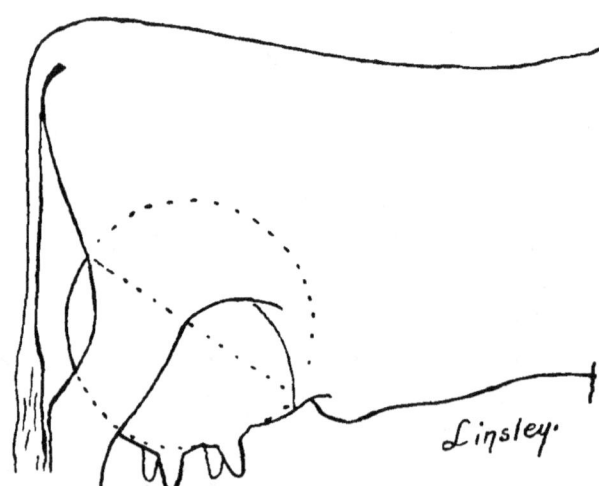

No. 1.—OUTLINE OF UDDER, BARREL, RUMP, AND THIGH OF JERSEY BELLE OF SCITUATE 7828.

[*] Scale devised by James McQueen, judge of Edinburgh Dairy Show, 1885.

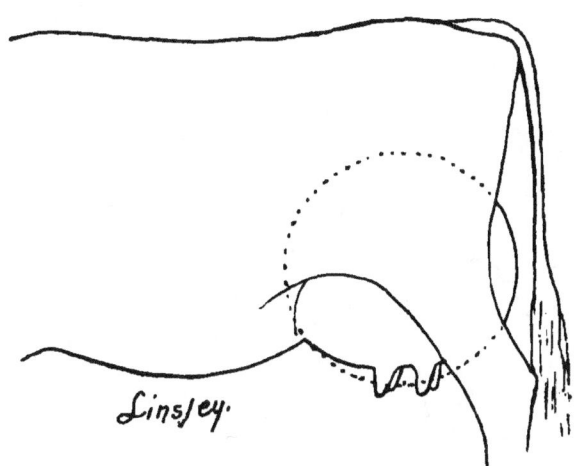

No. 2.—Outline of Udder, Barrel, Rump, and Thigh of Princess 2d 8046.

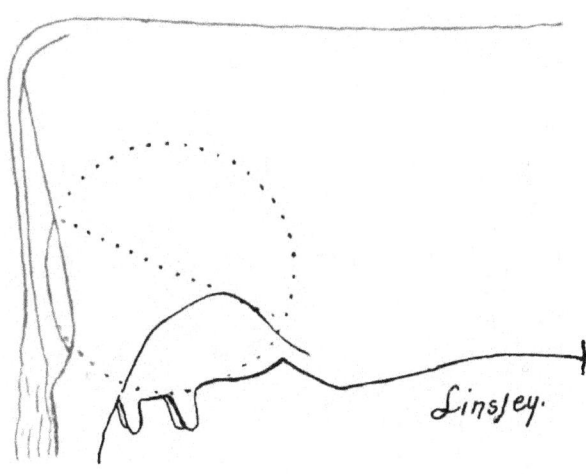

No. 3.—Outline of Udder, Barrel, Rump, and Thigh of Mary Anne of St. Lambert 9770.

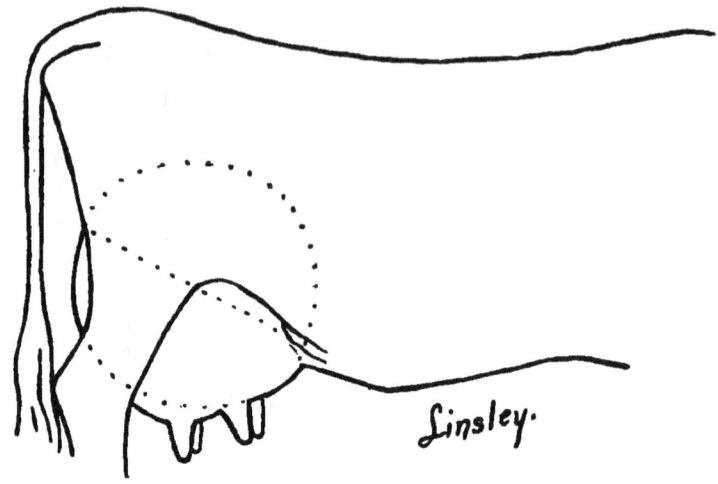

No. 4.—Outline of Udder, Barrel, Rump, and Thigh of Dandelion 2521.

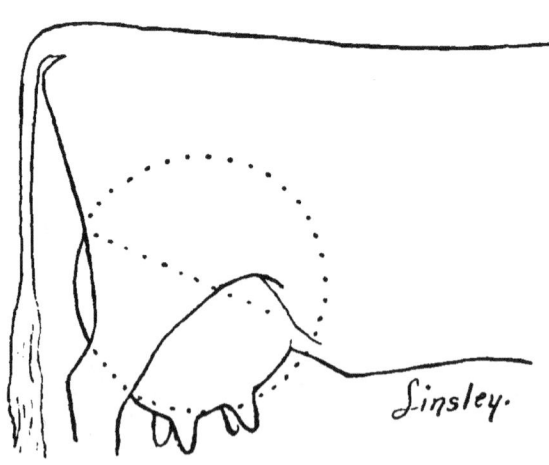

No. 5.—Outline of Udder, Barrel, Rump, and Thigh of Lady Vertumnus
13,217.

THE ESCUTCHEON.

The escutcheon consists of that portion of the surface of an animal which is covered by a reversed growth of hair. It usually includes the udder, the inner surface of the thighs, a portion of the space above the twist, and a part of the surface of the abdomen.

Francis Guenon, the discoverer of this feature in the animal kingdom, by which he rendered his name immortal, was a native of Libourne, France. He was the son of a nurseryman, and had become expert in the art of propagating and grafting fruits. While yet a boy, upon hearing his grandfather say he thought cows might be judged as easily as fruit trees, if we only knew their points, he was ever on the alert to make the saying good, and thereby made the discovery, which, after years of observation, he ingeniously systematized and demonstrated to his own complete satisfaction. This system he afterward disclosed by proving his skill as an expert upon herds of cattle before agricultural societies, and received high recognition, many honors, and medals, and was appointed lecturer on his system in the agricultural schools of France, and also received a pension from the government during his lifetime.

THE GUENON SYSTEM.

The limits of the escutcheon as described by Guenon are from the centre of the lower surface of the udder upward, the inner surface of the thighs and a portion of the perineal region, from the udder to the setting on of the tail. The escutcheon has several regular types, which Guenon classified according to their shape. There are ten of these regular forms, which he described under ten different names, or classes, besides which there are irregular and mixed forms. The ten classes of escutcheon are: 1, flandrine; 2, left flandrine; 3, selvedge; 4, curveline; 5, bicorn; 6, double selvedge; 7, demijohn; 8, square; 9, limousine; 10, carresine. The first class he named because he saw many of them in the province of Flanders, and the cows were great milkers. The second class was left-hand and one-sided; the third class had a narrow strip, like the border on a piece of cloth; the fourth had a curved arch; the fifth had a double top, or two horns; the sixth was an oddity, with two narrow strips; the seventh resembled a wine-jug; the eighth, a carpenter's square; the ninth was common in the province of Limoges, and steeple-shaped; the tenth was level at the top, or horizontal.

In each of these ten classes he made six orders, or sixty distinct forms; also a defective escutcheon, which he called *batard*, or counterfeit. In the first class there are twelve of these counterfeit escutcheons, and in each of the other classes six counterfeits, thus making sixty-six counterfeit escutcheons.

We will first make a brief analysis of the system according to Guenon. In describing his system he used the term "escutcheon," from the shield-like form of the upward growth covering the back part of the udder and thighs; and " Epis" or " feather" for certain peculiar marks on escutcheons by which he designated the various orders. He describes seven different " feathers," five on the surface of the escutcheon and two outside of the escutcheon.

By referring to the diagrams the terms "escutcheon" and " feather" will be fully explained and illustrated.

Stand behind the cow which produces the largest quantity of milk, and you will notice a peculiarity of hair growth which is upward on the udder and above in a broad band to the tail, and outward upon the thighs. Brush the hair in the direction of its growth with your hand, and you will find it softer than satin to the touch. You will notice another peculiar growth, a small oval mark of white hair over each hind teat, where the direction of the hair is downward. This " escutcheon" and this oval, or " oval feather," are shown in Fig. I., p. 59. Fig. II. shows the location of the " buttock feather;" Fig. III., the " babine feather;" Fig. IV., the " vulvous feather;" Fig. V., the " batard feather;" Fig. VI., the " thigh feather;" Fig. VII., the " dart feather."

Fig. I.

THE OVAL FEATHER.

The oval feather is often found on the best escutcheons. If these feathers are small, regular in form, and composed of very fine hair, they are usually an excellent sign; but if large, of irregular shape, and of long coarse hair, they are a mark of inferior quality. This feather should be about two inches long by one inch wide.

Fig. II.

THE BUTTOCK FEATHER.

The buttock feather is on the right and left of the vulva, outside of the escutcheon. Its hair is ascending, and it is usually two to three inches in length by half an inch in width. If smaller than this and of fine hair they are not specially indicative of inferiority; but if larger, and the hair is coarse, they always indicate an earlier cessation of the milk-flow, according to size and coarseness.

Fig. III.

THE BABINE FEATHER.

The babine feather is a narrow streak of down-growing hair within the escutcheon, starting from the side of the vulva—usually upon the left side, but may be upon either or both sides. It is usually two inches long by a quarter inch in width, but may be six inches in length.

I.

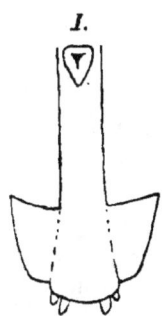

II.

III.

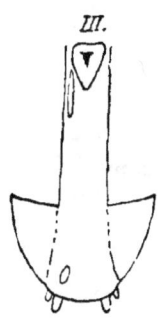

IV.

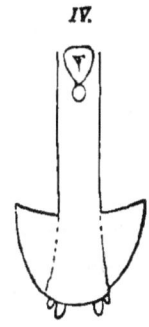

V.

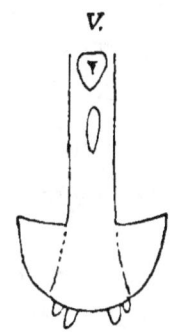

VI.

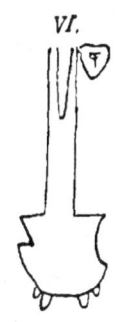

VII.

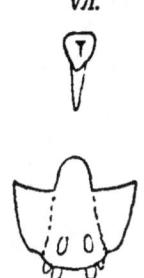

VIII.

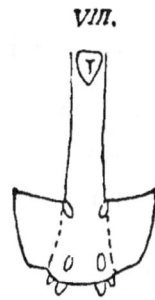

Fig. IV.

THE VULVOUS FEATHER.

The vulvous feather consists of down-growing hair enclosing the lower part of the vulva in a V-shape or forked like a W. It is one inch deep and wide to six inches in depth.

Fig. V.

THE BATARD FEATHER.

The batard feather, or counterfeit oval, is of down-growing white hair in the centre of the escutcheon midway between the udder and vulva, and, according to size and degree of coarseness, it indicates a falling off in milk during pregnancy. It may be six inches by two inches in size, or much smaller.

Fig. VI.

THE THIGH FEATHER.

The thigh feather is an encroachment of ingrowing hair upon the escutcheon of the thigh, and is in the form of a crescent or a triangle, and indicates inferiority, according to size of feather and coarseness of hair.

Fig. VII.

THE DART FEATHER.

The dart feather, also called *epijonctif*, is the result of crossing or compounding a selvedge escutcheon with any of the short escutcheons. It resembles a dart with the point downward, and consists of fine up-growing hair. It is situated beneath the vulva, is an inch wide at the top, and is considered an improvement to the short escutcheons.

The oval feather and the dart feather are good feathers or decorations; the buttock feather, babine feather, vulvous feather, batard feather, and thigh feather are bad feathers, or blots upon the escutcheon.

CLASSES AND ORDERS.

The first class, or flandrine escutcheon, extends from the centre of the four teats upward to the setting on of the tail.

The first order has an oval feather over each hind teat, is full out on the thighs, and has a clean smooth upgrowth fully three inches wide to the root of the tail.

Cow a perpetual milker.

The second order has one oval feather and a babine feather about two inches long on the left or right of vulva. Goes dry two months.

The third order has a vulvous feather one inch to two inches deep, and goes dry three months.

The fourth order has a vulvous feather about five inches long and crescent thigh feather on right thigh. Goes dry four months.

The fifth order has a vulvous feather six inches deep and a triangular thigh feather on right thigh. Goes dry five months.

The sixth order has a vulvous feather eight inches deep and a very small thigh escutcheon invaded by triangular thigh feathers. Goes dry six months.

The size of the escutcheon dwindles in all its parts in each descending order.

THE SECOND CLASS, OR LEFT FLANDRINE ESCUTCHEON.

The first order resembles that of the flandrine, except that it runs up on the left flank, and the right thigh wing is apparently wider than the left wing. There are two oval feathers, and the cow seldom or never goes dry.

The second order has a babine feather on the left, and goes dry two months.

The third order has a babine feather six inches long and crescent thigh feather right thigh. Goes dry three months.

The fourth order has a longer babine feather, a half-moon thigh feather right side, and a triangular thigh feather left side. Goes dry four months.

The fifth order has a coarse flaring escutcheon, a large triangle thigh feather right side. Goes dry five months.

The sixth order has a very small coarse escutcheon, and goes dry six months.

The batard, or counterfeit, has the same marks in all the orders, with the exception of enormous coarse buttock feathers five inches long by three wide. The milk is watery, and falls off rapidly when pregnant.

THIRD CLASS, OR SELVEDGE ESCUTCHEON.

The first order runs fully one inch wide up to the vulva, and is of full width (eighteen inches) on the thighs, with the two oval feathers on the udder. Never goes dry unless forced to do so.

The second order has a left oval feather and a left buttock feather. Goes dry two months.

The third order has two buttock feathers, the left about three inches long. Goes dry three months.

The fourth order has two buttock feathers, the left four inches long, and goes dry four months.

The fifth order has a broken list, the buttock feathers five inches long, and goes dry five months.

The sixth order has a ragged escutcheon; the buttock feathers are six inches long, and she is dry six months.

The batard orders are similarly marked to the six free orders, but have enormous coarse buttock feathers five inches long by three wide, and produce thin milk and are soon dry.

THE FOURTH CLASS, OR CURVELINE ESCUTCHEON.

The first order has the two oval feathers, the escutcheon is eighteen inches wide on the thighs, and ascends in a round arch to within eight inches of the vulva. Never dry unless forced or injured.

The second order has a left oval feather and one small left buttock feather. Dry two months.

The third order has two buttock feathers, the left three inches long, and goes dry three months. There is sometimes a triangular thigh feather right side.

The fourth order has buttock feathers six inches long, right triangle and left crescent thigh feathers, and goes dry four months.

The fifth order has buttock feathers seven inches long and thigh feathers eight by four inches. Goes dry five months.

The sixth order has a very diminutive escutcheon, and goes dry six months.

The batard curveline cows have very large and coarse buttock feathers, and give thin milk, going dry soon.

THE FIFTH CLASS, OR BICORN ESCUTCHEON.

The first order has two oval feathers, a thigh escutcheon eighteen inches wide, and the upper part of the escutcheon terminating in two points, the left higher than the right and within four to eight inches of the vulva. There may be two very small buttock feathers of equal size. Dry one month.

The second order has one left oval feather, and the left buttock feather is two inches long. Goes dry two months.

The third order has buttock feathers three inches long, a triangular thigh feather right side, and goes dry three months.

The fourth order has buttock feathers four inches or longer, a large triangular right-side thigh feather, and goes dry four months.

The fifth order has larger bristly buttock feathers, a larger triangle, and goes dry five months.

The sixth order is a very little bicorn, and there are bristling hairs all over the buttocks. Dry six months.

The bicorn counterfeit has the same marks in each order and two large coarse buttock feathers.

THE SIXTH CLASS, OR DOUBLE SELVEDGE ESCUTCHEON.

The first order has two slender lists running from the bottom of the udder to the tail, with a broad band of descending hair between, reaching to the base of the udder. Thigh wings eighteen inches wide, as in all first-order escutcheons. Dry one month.

The second order has the descending band of hair terminating four inches above hind teats and the thigh escutcheon narrower. Dry two months.

The third order has narrower fillets, descending band stops six inches above teats; thigh wings are still narrower. Dry three months.

The fourth order has coarser hair, descending band twelve inches below vulva, a crescent thigh feather on right thigh. Dry four months.

The fifth order has still coarser hair, the two fillets are ragged, the central band descends to the udder, a triangular thigh feather in both wings. Dry five months.

The sixth order has very small ragged fillets, the right reaching half way up, the thigh wings not discernible. Dry six months.

The batard has the fillets terminating in two large coarse buttock feathers.

THE SEVENTH CLASS, OR DEMIJOHN ESCUTCHEON.

The first order has two oval feathers, and may have two small buttock feathers. The escutcheon is eighteen inches wide on thighs, and the upper part rises like a flandrine, but terminates in a level top four to eight inches below the vulva. Dry one month.

The second order has one left oval feather and buttock feathers two to three inches long. Goes dry two months.

The third order has the left buttock feather about five inches long and a right crescent thigh feather. Goes dry three months.

The fourth order has longer buttock feathers and a triangular right thigh feather. Dry four months.

The fifth order has larger buttock feathers and two large triangular thigh feathers. Dry five months.

The sixth order has a very small demijohn and very large buttock feathers. Dry six months.

The batard has enormous buttock feathers.

THE EIGHTH CLASS, OR SQUARE ESCUTCHEON.

The top of this escutcheon is a narrow list joined to the left corner of the top of a demijohn escutcheon. The first order has two oval feathers. Goes dry one month.

The second order is the only second order which Guenon gave the two oval feathers. There is a buttock feather on the right of the vulva. Goes dry one month.

The third order has a right buttock feather three inches long, a right triangular thigh feather, and left crescent thigh feather, and goes dry three months.

The fourth order has the list ragged in the upper part, the right buttock feather four inches long, the wings very small, with a triangular right side thigh feather. Dry four months.

The fifth order has the fillet still more ragged, the buttock feather five inches long, a triangular thigh feather on each side. Dry five months.

The sixth order is scarcely recognizable. Dry six months. The batard is distinguished by a very large coarse buttock feather on right of vulva, and the fillet on the left of the vulva has bristling hair.

THE NINTH CLASS, OR LIMOUSINE ESCUTCHEON.

The first order has the two oval feathers and wide thigh shield. The upper part terminates four to eight inches below the vulva in a sharp point like a steeple. There may be two small buttock feathers. Goes dry one month.

The second order has one left oval feather and two buttock feathers about three inches in length, the left being the longer, as in all the escutcheons. Goes dry one month.

The third order has the left buttock feather still more elongated; the thigh wings are more contracted. Goes dry three months.

The fourth order has larger buttock feathers; the whole escutcheon is lower and rounded. Goes dry four months.

The fifth order has very long buttock feathers. The wings are small, and each has a triangular thigh feather. Goes dry five months.

The sixth order is so small as to be scarcely distinguishable; the buttock feathers are very long and ragged. Goes dry six months.

The batard, or counterfeit orders, have the same marks in each order, except that the buttock feathers, as in every class, are larger, coarser, and very bristling.

THE TENTH CLASS, OR CARRESINE ESCUTCHEON.

The first order has an escutcheon of full width, but terminated at a line level with the top of the udder. There are two oval feathers and two very small buttock feathers. Goes dry one month.

The second order has one oval feather. The left buttock feather is elongated; the thigh wings are contracted. Goes dry two months.

The third order is still more contracted. The buttock feathers are longer, and there is a triangular thigh feather in the right wing. Goes dry three months.

The fourth order has a triangular gore in each wing. The buttock feathers are long and bristling. Goes dry four months.

The fifth order has very large buttock feathers and very small escutcheon. Goes dry five months.

The sixth order does not rise to the middle of the very small udder. The buttock feathers reach almost down to the udder, and are bristling. Goes dry six months.

The carresine counterfeits have all the marks of the six orders and immense broad buttock feathers in every order.

Batard, or counterfeit cows, in all the classes and orders, only differ from free cows in losing their milk very soon after impregnation.

Guenon arranged cows, according to amount of milk, in three sizes—large, medium, and small. As the Jersey may be properly called a medium-sized cow, the synoptic tables here given are suitable for the Jersey breed.

CHART OF THE GUENON SYSTEM, SHOWING THE DAILY MILK YIELD FOR MEDIUM-SIZED COWS, IN QUARTS.

No. of Class.	CLASS.	First Order.	Second Order.	Third Order.	Fourth Order.	Fifth Order.	Sixth Order.
1.	Flandrine............	20 qts.	16 qts.	12½ qts.	9½ qts.	6 qts.	3 qts.
3.	Selvedge.............	20 "	16 "	12½ "	9½ "	6 "	3 "
4.	Curveline............	20 "	16 "	12½ "	9½ "	6 "	3 "
5.	Bicorn	20 "	16 "	12½ "	9½ "	6 "	3 "
7.	Demijohn............	20 "	16 "	12½ "	9½ "	6 "	3 "
2.	Left Flandrine	18 "	15 "	10½ "	7 "	4 "	2 "
6.	Double Selvedge.......	18 "	15 "	10½ "	7 "	4 "	2 "
8.	Square	18 "	15 "	10½ "	7 "	4 "	2 "
9.	Limousine	16 "	12½ "	9½ "	6 "	3 "	2 "
10.	Carresine	16 "	12½ "	9½ "	6 "	3 "	2 "

It will be seen that the escutcheons are arranged in this table in order of merit.

CHART SHOWING DURATION OF MILKING PERIOD FOR PREGNANT COWS BY NUMBER OF MONTHS EACH ORDER WILL PRODUCE MILK.

No.	CLASSES.	ORDERS.					
		1st.	2d.	3d.	4th.	5th.	6th.
1.	Flandrine................	9	7	6	5	4	3
2.	Left Flandrine..........	9	7	6	5	4	3
3.	Selvedge...............	9	7	6	5	4	3
4.	Curveline.............	8	7	6	5	4	3
5.	Bicorn..............	8	7	6	5	4	3
6.	Double Selvedge........	8	7	6	5	4	3
7.	Demijohn..............	8	7	6	5	4	3
8.	Square................	8	8	6	5	4	3
9.	Limousine.............	8	7	6	5	4	3
10.	Carresine.............	8	7	6	5	4	3

NUMBER OF MONTHS DRY.

1.	Flandrine..............	0	2	3	4	5	6
2.	Left Flandrine..........	0	2	3	4	5	6
3.	Selvedge...............	0	2	3	4	5	6
4.	Curveline	1	2	3	4	5	6
5.	Bicorn................	1	2	3	4	5	6
6.	Double Selvedge.........	1	2	3	4	5	6
7.	Demijohn..............	1	2	3	4	5	6
8.	Square................	1	1	3	4	5	6
9.	Limousine.............	1	2	3	4	5	6
10.	Carresine.............	1	2	3	4	5	6

THE FORE-ESCUTCHEON AND THIGH OVALS.

Guenon did not think it necessary to observe the fore-escutcheon, but many of our best breeders of Jersey cattle regard the fore escutcheon of equal importance with the posterior escutcheon, in judging of the productive and breeding qualities.

The fore-escutcheon consists in the forward growth of hair on the belly of the animal. It is sometimes very large, and extends nearly to the fore-legs, and on the sides it often sweeps over the margin of the curtain, and forms large waves or curls on the sides of the body.

Many great butter cows have a large fore-escutcheon, notably the wonderful cow Mary Anne of St. Lambert 9770, and herewith is shown a diagram of her fore-escutcheon, drawn by the author from memory.

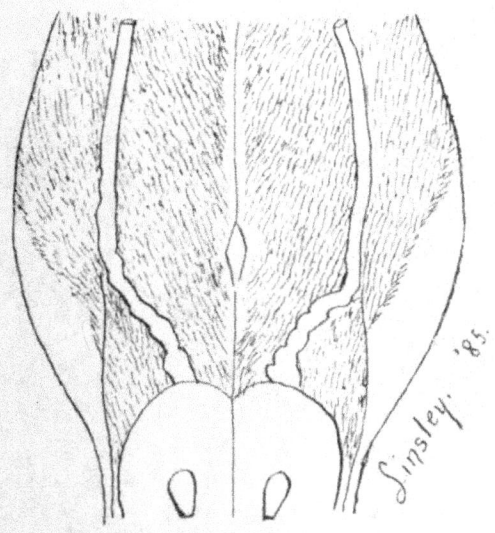

FORE-ESCUTCHEON OF MARY ANNE OF ST. LAMBERT 9770.

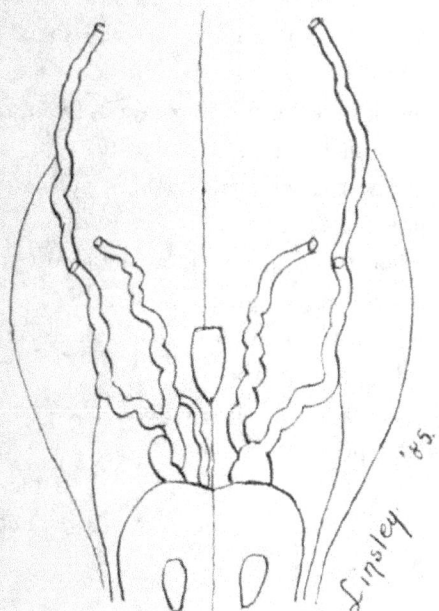

FORE-VEINS OF JERSEY BELLE OF SCITUATE 7828, EXTENDING TO THE SHOULDERS.

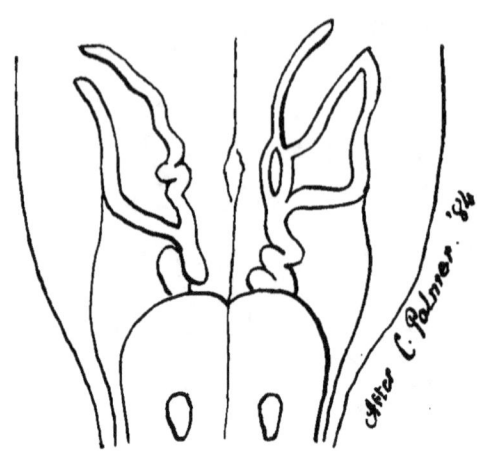

FORE-VEINS OF VALUE 2D 6844.

THIGH OVALS.

Many writers upon Jersey cattle have mentioned a feature that seems to have escaped the notice of Guenon, and that is a feather of an oval shape at the juncture of the thigh wing with the upright portion of the escutcheon. In cows this feather appears at the top of the udder on each side, dipping down from the thigh about two inches. (See Fig. VIII., p. 59, and the escutcheon plates.)

The thigh oval feather is formed of down-growing hair, and is exceedingly fine and soft, and always indicates superior quality in a Jersey of either sex. In the bull it is smaller, and dips down upon the thigh or upon the scrotum. A good thigh oval feather should not be more than three inches in length, but sometimes they extend so far as to coalesce with the lower oval feathers, thus forming a long band of down-growing hair on each hind-quarter of the udder.

The ten plates which follow show an ideal escutcheon of each class. I have added the thigh ovals, with the exception of the double selvedge, of which I have only seen one specimen, and the escutcheon of the bull. The flandrine, selvedge, curveline, bicorn, and demijohn escutcheons are the best classes, and for the Jersey there is no better escutcheon than a perfect curveline.

THE ESCUTCHEON OF THE BULL.

According to Guenon, the escutcheon of the bull starts from the fore-part of the scrotum, extends within and above the hocks, spreads over the hinder surface of the thighs, and in the higher orders of some classes ascends as high as the anus. Those bulls whose escutcheons, in form and dimensions, resemble those of cows of the higher orders, are well adapted to the procreation of offspring of good milking qualities. A bull is well marked and a good breeder when there is no invasion of descending hair into the ascending hair of his escutcheon; and the escutcheon is of large dimensions, in proportion to the size of the animal, and is covered with very fine hair.

"Bulls, like cows, may be arranged in ten classes, of which each class comprises several orders, and every order three sizes. I shall only distinguish three orders in each class, which I shall designate as good, fair, and bad. The same distinctions could be observed as in cows."

The defects in the escutcheon of the bull are coarseness of hair, diminutive size, angular invasions of hair on the thighs, or an oval feather of descending hair an inch wide by two or three inches in length on the inside of the thigh about the middle of the escutcheon and covered with long thick hair.

The fair escutcheon of the bull may be compared with the third and fourth order in cows.

The good escutcheon is equivalent to the first order in cows.

SIGNIFICANCE OF THE SCALE OF POINTS.

The scale of points, including the fore and hind escutcheon, is recommended as a guide in the purchase of animals and as an aid in breeding.

The milk and butter quality, when demonstrated by churn tests, shows what the cow can do, and encourages the breeder to look for the same quality in her male and female progeny.

The escutcheon, according to Guenon, "is the only incontestable characteristic sign that can enable one to discern, by simple inspection, the aptitude for milk production of each animal." Those who decry the escutcheon always like to show a herd of well-escutcheoned animals. The escutcheon when perfect is also one of the chief elements of beauty in a Jersey.

The skin color is an indication that the cow, being richly colored within the ears and on other parts, especially if she retains the color throughout the year (she will show most color when fed on green food), will give a golden tint to her butter in midwinter.

The barrel, if large and capacious, shows that the animal has constitutional vigor and room for the laboratory of digestion and assimilation of food.

The general appearance should be bony and lean, showing that the animal uses the udder and all the lacteal vessels for the special purpose of utilizing all the food elements for the production of milk and cream.

The fore-udder should be full but well rounded rather than square. The angular udder must give place to the spheroid—free from fleshiness, and a true milk-secreting organ.

The hind-udder should project far out from the thighs with a round outline, and well tucked up in the twist. The whole udder should have a spheroidal form, and its supporting ligaments must be so strong as to prevent it from becoming pendent, even in advanced age.

The udder quality should be soft, supple, and elastic, so as to milk empty. There need be only sufficient vascularity to make healthful milk-glands. The udder should have a silken and unctuous touch; the hair very short and fine—a butter udder.

The veins show a capacity for free circulation according to size. The bull should be credited with a "milk" vein, when marking for prizes.

The teats should be just large enough to fill the grasp of a man's hand, and yield the milk upon slight pressure, but never leak. The teats must be kept free from warts and sores by proper treatment. The nipples of the bull, if equal in size and of good length, and set wide apart, indicate the same quality in his heifers.

The high rump is an important point to cultivate; it indicates vigor and less liability to abortion. It also adds much to the symmetry and beauty of the animal.

The thin thigh belongs to the creamer, the round thigh to the beef animal. A curl on each hock is a good sign.

The thin mellow hide is a part of that general make-up which denotes the creamer. The very large pendent navel and loose skin on the belly are associated with a capacious and flexible udder.

Hair as unctuous as vaseline is a prediction of butter in the churn. Avoid dry, wiry hair in every animal, but cultivate the soft, fine coat.

Hips of great breadth indicate great abdominal capacity and room below for an immense udder. Fineness of bone is indicative of fine quality in every tissue of the body.

A level back is an indication of strength, and gives symmetry to the form. An old cow may sway a little below the line.

The double chine is associated with fully developed lacteal and generative organs and first-order escutcheons. When you can lay the fingers in the spaces between the spines, you have a cow with broad hips, large udder, and a very broad escutcheon.

The long, thin neck of the cow is to be matched with the long, well-arched, high-set neck of the bull. A short thick neck indicates fat on the ribs and kidneys.

Large eyes, which for gentleness of expression rival those of the gazelle, are the special feature of loveliness in a Jersey cow. The bull should have a kind but lively eye. The one indicates docility, the other vigor and power.

The lively and playful bull indicates a condition of vigorous health and potency. The gentleness of the cow is manifested by an undisturbed equanimity, a condition of perfect contentment, indifferent to all things except her cud, which she always enjoys when not feeding or sleeping. It is the business of the butter cow to keep the cud in motion.

The tail should be as long, as tapering, and as fine as possible. Such a tail, if tipped with a switch like the tail of a horse, indicates a very well-bred animal.

The shoulders cannot be too oblique or too sloping, and the good butter cow never lays up fat on her shoulder-blades while she is giving a full flow of milk.

The legs should be fine, having flat, hard bones, which with small feet are indicative of good breeding and fine quality.

Thin withers also indicate fineness of breeding, and belong to the wedge-form. Thick withers indicate more lung power and usually a greater feeding capacity; but such animals are liable to become fat, while the thin withers indicate the milking form.

A widening at the crops indicates constitutional stamina and strong vitality.

The deep chest without great breadth indicates a sufficient power of respiration for good health and a form that is compatible with production of milk and cream rather than beef.

The small, lean head, long and tapering, indicates much milk; the short, square head, beef. The arched crown is a beautiful characteristic of the finest Jerseys.

The dished face is attractive and not incompatible with the greatest productiveness.

Breadth between the eyes indicates sagacity and a high degree of bovine intelligence, as well as beauty.

The ash-colored fillet is a striking feature in the Jersey race. If the muzzle is slightly turned up, nostrils wide, the mouth broad, and the masticatory muscles stand out roundly from the muzzle and cheek, it is a good combination of features for business and beauty. A black nose is supposed to be characteristic of the Jersey breed, though not any more essential to purity than a black tongue or a black switch.

The small ear well fringed indicates not only fine breeding but constitutional vigor. The fringe is also a protection from flies.

Horns of translucent amber with black tips are very ornamental, especially if small in size and slender, and if they have a natural crumple, or have been trained to droop or curl about the face. They are as useless as they are ornamental, and have less significance than any other point.

Besides the scale of points, with which one cannot be too familiar, there are several other considerations which need to be remembered in judging of cattle.

1. The race peculiarities.
2. The family traits.
3. The degree of inbreeding.
4. The age.
5. The size.
6. The system of management and care.
7. The health.
8. The variety, quality, and amount of food.
9. The special power to assimilate food.
10. The quality of the cow's milk, cream, and butter.
11. The season of the year, and the weather.
12. The period of gestation.

By familiarity with the animals and with every technicality of these descriptions and points, any one with an eye for a cow can become expert in the selection of the best stock, and while they are yet young calves may apprehend their future excellence.

THE PRINCIPLES OF BREEDING.

Of the ancient methods of cattle-breeding we have little knowledge. The oldest record of skill in the art is found in the Book of Genesis, where Jacob, who was the superintendent of the herd of his father-in-law Laban, the Syrian, after fourteen years of familiarity with Laban's cattle—he had bred cattle, however, all his life, and was past fifty years of age—proposed to take, as his wages, only the spotted cattle. From this it would appear that spotted cattle were then a great rarity, a strange freak of nature not only among sheep, goats, camels, and asses, but bovines. Laban readily assented to the proposition, and Jacob, by consummate skill and selection of the strongest cattle, soon had an immense herd of spotted cattle, notable for their strong constitutions, and, as a sequel, his wages were "changed ten times." This record of breeding, brief as it is, has much that is suggestive to the modern cattle-breeder. Jacob had a plan, adhered to it, and was successful *in changing the colors of cattle; in improving their constitutional vigor, and in overcoming the habit of abortion among his herds.* Some might add that, according to the record, there was divine interposition in his behalf. Well, the record also states that Jacob sought for divine blessing. All modern breeders would also do well to follow his example, and also make confession of the blessing.

The object of the breeder is to produce at will, not by luck or chance, perfect specimens of the race, that shall combine all the qualities desired. Most of the modern breeds of cattle have been developed by a slow, hap-hazard process. Some of the breeds in England have been formed by men of genius after a well-considered plan. The best families of Jerseys have been made by method. The results of the methods practised in England for the last century and in America for a shorter period show conclusively that breeds or races of domestic animals, to be successful and profitable, must combine a few peculiar excellencies which are to be developed to their fullest extent regardless of all other qualities that are incompatible with the object sought.

The dairy breed must have the wedge form and lean general appearance, compatible with a long life devoted to formation of tons of cream in the udder, while the beef animal must have the square form, and make a mountain of marbled meat, rich in osmazome, at or before three years of age. The two types are wholly distinct, and cannot be blended in one breed. As soon think of winning races with the heavy muscles of the cart horse as to win at the churn or cheese vat with a beef breed.

The possibilities of achievement in bringing the Jersey to a high average standard of productiveness have been at least partially shown by the efforts of a few skilful breeders—notably by R. M. Hoe, of New York, O. S. Hubbell, of Connecticut, and Philip Dauncey, in England. Mr. Dauncey began in 1826, and for forty years worked with three distinct objects in view: first, a high average butter yield; second, constitutional vigor; third, coats of uniform style of color, entirely free from patches of white. All these objects were successfully achieved. Mr. Hubbell has accomplished the foundation of a family noted for great yields of butter, beautiful color and symmetry, and remarkable uniformity of excellence. Mr. William Simpson, of New York, is also pursuing a scientific method in breeding.

The great problem that confronts every breeder is that of duplicating at will the animals he has selected as his models.

A thorough knowledge of the history of breeding and the special methods of successful breeders, a taste for the art, and a love for the animals, if combined with a genius for the work, are auspices of great results.

I believe that the laws of breeding may be formulated in such a manner as to insure success to the man of average skill and the requisite education.

ATAVISM.

When the orchardist, by combining the qualities of two excellent fruits, produces, out of many thousand seedlings, one of delicious quality, he very well knows that he cannot reproduce the same or an equally good fruit short of many very tedious experiments, perhaps not in a lifetime, by the process of breeding. The union of two animals produces always a new seedling which varies from the parents more or less widely.

This variation proceeds from the law of heredity, that a seedling represents the sum of the combined qualities of all its ancestors operative at and subsequent to the time of sexual union. The most prepotent force in procreation may revert to some ancestor five, ten, or twenty generations distant, so as to reproduce in all their force of individuality the features of one noted for great merit or marked inferiority. This peculiarity of tracing to a remote ancestor is called *atavism*, and signifies likeness to " an old grandfather."

To avoid the bad influences of atavism, and utilize the good, is the province of the skilful breeder. The orchardist, to avoid the inconvenience and delay resulting from atavism, continually resorts to budding and grafting, perpetuating the identical variety by offshoots. The breeder must assimilate in dealing with animal life as nearly as possible to the process of the orchardist with plant life.

TERMS RELATING TO PURITY OF BLOOD.

A *thoroughbred* or *purebred* animal is one of a race that can be traced back to one common ancestry in both the male and female lines, with close in-and-in breeding for seven or more generations.

A *fullbred* animal is the result of breeding a thoroughbred male to a female of another breed, and successively to her progeny for *six* generations. Thus the progeny of a thoroughbred bull and a native cow gives a female with 50 per cent. of thoroughbred blood, which cow, mated with the same or another thoroughbred bull of the same breed, gives 75 per cent. of the pure blood. The next generation gives $87\frac{1}{2}$ per cent. of pure blood. The fourth generation gives $\frac{15}{16}$, or $93\frac{3}{4}$ per cent., of thoroughbred. The sixth generation gives $\frac{63}{64}$, or $98\frac{7}{16}$ per cent., of pure blood, or a *fullbred*, very nearly.

A *crossbred* animal is the progeny of two thoroughbred animals of different breeds.

A *grade* animal is one that possesses any degree of thoroughbred blood below a fullbred. A low grade has less, and a high grade more than 50 per cent. of thoroughbred.

A *scrub* animal is one whose pedigree has no quality of uniformity or of thorough selection in either the male or female line, *and always gives the highest risk of atavism* toward inferiority in the progeny.

PREPOTENCY.

That peculiar power which is possessed in a very marked degree by a few animals of either sex, of transmitting to their progeny all the striking individual characteristics of the parent, so that the descendants have a uniform resemblance and quality, is called *prepotency*. It is a faculty which implies a special accumulation of vital force in the generative system, in common with all the other departments of the organism, and is not to be confounded with the narrower term, potency, which refers to the physical health of the male generative faculty, regardless of powers of transmission of quality. This element of prepotency may consist in the ability to transmit inferior or mediocre qualities, as well as those of the superlative degree of excellence. The breeder wants animals that overcome atavism by prepotency of the highest order.

Atavism is usually a result of crossing two varieties of the same species, or if it occurs in a thoroughbred family, it is the result of an inharmonious union.

If the breeder could follow the example of the orchardist by budding, the uncertainties of breeding would be neutralized. A seed is but a modified bud; the animal is analogous to the same process of development. The breeder cannot bud from his model, but by a certain formula he can in time produce an animal that shall be nearly identical in blood elements with the selected model.

IN-AND-IN BREEDING.

There is no subject upon which current notions are so wide from the facts as the mating of near kin. A prevalent notion exists that in some mysterious manner the union of the blood of near relations is harmful. All sorts of disasters in man and animals have been attributed to the union of kindred blood. The history of man, and the records relating to the natural history of animals and the science of breeding, show that this current notion is fallacious in the extreme.

According to the Book of Genesis, Eve was identical with Adam. For the first twenty-five hundred years of human history marriage between full brother and sister or half brother and half sister was the recognized order of society. Moses, the greatest man of antiquity—seer, lawgiver, poet, historian, judge, ruler, and leader of a great people just freed from bondage, was the son of his aunt; while his father was also his double uncle. His mother was a daughter of Levi, and his father a double grandson of the same Levi. His brother Aaron and sister Miriam were also highly distinguished for ability. These were the ages of longevity for the human race. From Noah to Moses the average age, for sixteen generations, was nearly three hundred years. To show how close was the consanguinity, I give a chart of the pedigree of Moses. If the lines were all complete, the closeness would probably be still more marked.

I have given this pedigree to illustrate the facts of history, and to show that the closest consanguinity in the human family is not a hindrance to the highest physical and mental perfection. On the contrary, was not this mingling of kindred blood a cause for the great qualities which Moses illustrated? Was the law of marriage which he afterward gave based upon physiological or sanitary necessity, or was it simply relative to a system of social ethics? If the existence of organic diseases

MOSES (Aaron, Miriam).

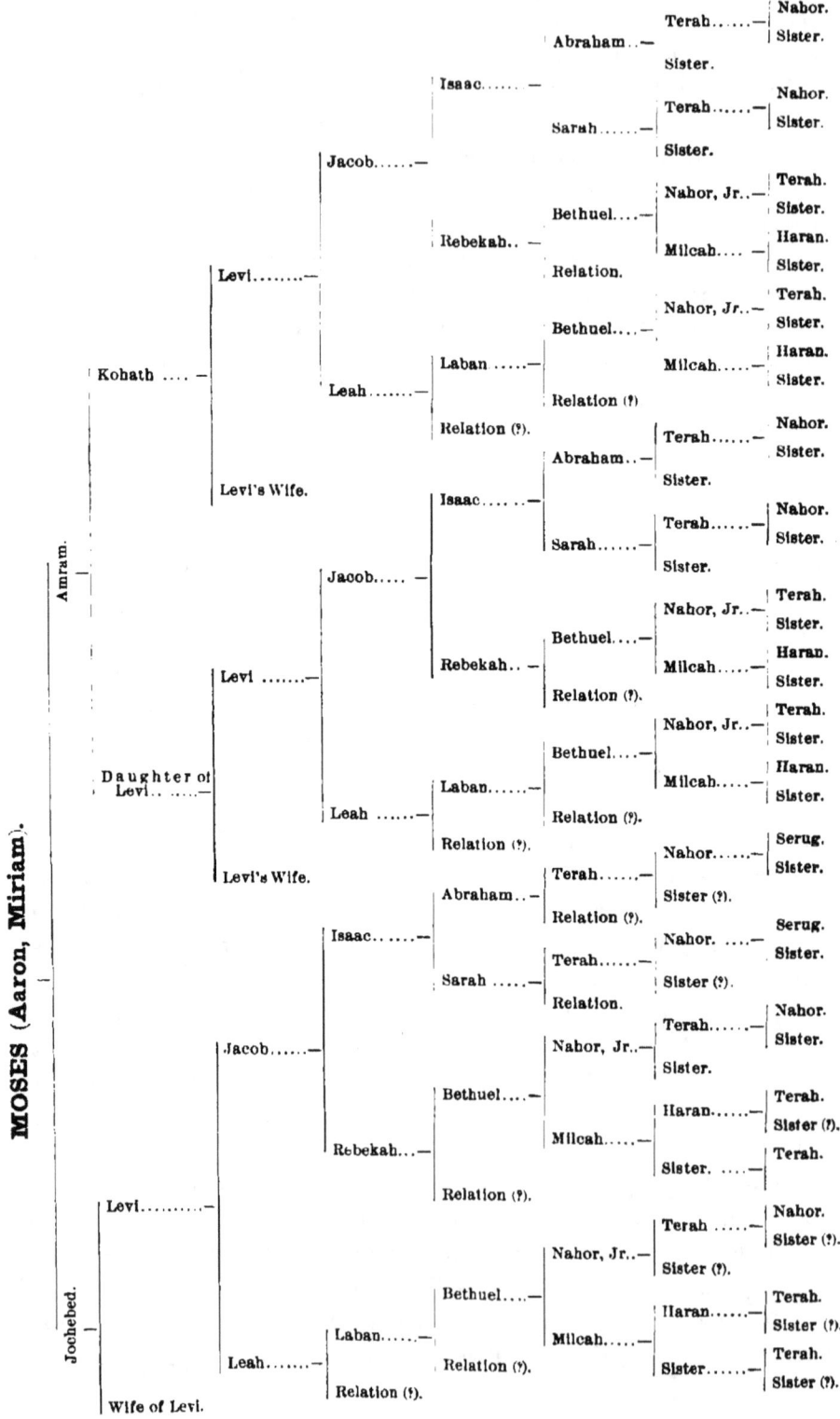

Amram.

Jochebed.

Kohath .... —

Daughter of Levi.. .....—

Levi.......... —

Levi........ —

Levi's Wife.

Levi .... —

Levi's Wife.

Wife of Levi.

Jacob..... —

Leah....... —

Relation (?).

Jacob..... —

Leah ...... —

Relation (?).

Jacob...... —

Leah....... —

Relation (?).

Isaac...... —

Rebekah.. —

Laban .....—

Relation (?)

Isaac..... —

Rebekah.. —

Laban...... —

Relation (?).

Isaac..... —

Rebekah... —

Laban......—

Relation (?).

Abraham..—

Sarah......—

Abraham..—

Sarah......—

Abraham..—

Sarah .....—

Bethuel....—

Relation.

Bethuel....—

Relation (?).

Bethuel....—

Relation (?).

Bethuel....—

Relation (?).

Bethuel.... —

Relation.

Nahor, Jr..—

Bethuel....—

Milcah.....—

Relation (?).

Nahor, Jr..—

Bethuel....—

Milcah.....—

Relation (?).

Terah......—

Sister.

Terah......—

Sister.

Nahor, Jr..—

Milcah....—

Nahor, Jr..—

Milcah......—

Terah......—

Sister.

Terah......—

Sister.

Nahor, Jr..—

Milcah.....—

Nahor, Jr..—

Milcah.....—

Nahor......—

Relation (?).

Nahor. ....—

Relation.

Terah......—

Sister (?).

Terah......—

Sister (?).

Terah......—

Sister (?).

Terah ..... —

Sister (?).

Haran......—

Sister. ....—

Terah —

Sister (?).

Haran......—

Sister......—

Nahor.

Sister.

Nahor.

Sister.

Terah.

Sister.

Haran.

Sister.

Terah.

Sister.

Haran.

Sister.

Nahor.

Sister.

Nahor.

Sister.

Terah.

Sister.

Haran.

Sister.

Terah.

Sister.

Haran.

Sister.

Serug.

Sister.

Serug.

Sister.

Nahor.

Sister.

Terah.

Sister (?).

Terah.

Nahor.

Sister (?).

Terah.

Sister (?).

Terah.

Sister (?).

called for the law, the same organic diseases should have called for a prohibition of marriage without and beyond the prescribed degrees of consanguinity. It is certain that there is no evidence to show that consanguineous union in man or the lower animals ever did or ever can originate disease. On the contrary, we observe that where races diverse in physique and character are bred together the crosses lead to many imperfections. Huth has well said that if organisms are not nearly allied they can rarely be made to interbreed; and that the result of such crosses is an offspring of weedy growth, ill-balanced intellect, often as susceptible to unfavorable circumstances as an unacclimatized animal, and generally sterile; it is impossible that crossing can be considered in any way beneficial except inasmuch as it may relieve a possible hereditary tendency to disease. The Jews, since the Mosaic law, have frequently married cousins, and are the best variety of the human species to illustrate the principle of thoroughbred quality. For ages they have been maligned and persecuted in Eastern countries, and have suffered more hardships than any other people; yet they are possessed of greater viability than any other known race, and can thrive in every variety of climate on the earth, while subject to all the vicissitudes of commercial life. Where other European races would perish, the Jew flourishes and grows rich. Consanguinity is the law among most races of animals in a state of nature, especially those that are polygamous, as cattle, sheep, deer, and antelopes. In herds of wild horses, and also of the wild boar, it is usual to find but one adult male. The elephant, the gorilla, the lion, the ostrich, and many species of birds are polygamous, which always indicates the closest forms of in-and-in breeding perpetuated from the beginning of creation.

IN-AND-IN BREEDING OF ANIMALS—SHEEP.

Doctor Huth, in his great work "The Marriage of Near Kin," in the preparation of which he consulted one hundred and seventy-one different authors, after showing the fallacious character of the meagre statistics purporting to show ill effects attributable to consanguineous marriages, says: "No census could determine whether consanguinity can be a primary cause of disease. For that we must interrogate nature, as she has already been so successfully interrogated on other physiological questions. We must experiment on the lower animals, since we may not experiment on man. Generation varies so little in its essential characteristics from the lowest organisms to the highest, that observations deduced from the breeding of domestic animals may very safely be applied to man. An animal properly bred in-and-in, and a wild animal, is each perfect according to its circumstances. Alter the circumstances, and the animal is at once unfit for its place.

"From the breeder's point of view, in-and-in breeding improves the breed, because it suppresses those qualities which are useless, and develops those which are useful, whether it be for racing, for wool, for the butcher, or for any other purpose; and without in-and-in breeding he cannot alter an animal to suit his purpose.

" Naturally, persons with that preconceived notion which every one is bound to have on this subject who has not studied it are apt to consider any evil result observed in the course of in-and-in breeding as caused by that kind of breeding in animals, without any previous examination whether there may not be other causes to account for it.

" In the study of these cases, therefore, as in others, we must remember that one fact showing the harmlessness of in-and-in breeding is worth a hundred tending to show their harmfulness; since in the former consanguinity is still a factor, but in the latter we are ignorant what other factors may have come into play. Let us now proceed to facts.

" 'M. Allie,' says M. Boudin, 'after a long experience, is of opinion that the system of in-and-in breeding is ruin to sheep. A flock at Petit-Bourg,' he says, ' has diminished greatly in value since it passed into other hands, and this system has been practised.' The observations of Stephens led him to the same conclusion: the progeny, he says, though improved in figure, firmness of bone, etc., are nevertheless delicate-skinned, and therefore liable to the attacks of insects and to inflammation; but this evil is only the result of long-continued in-and-in breeding, and by no means the immediate result.

" M. Aube asserts that sheep will produce a dark kind if bred in-and-in, which he explains as a step on the road to albinoism. While Mr. Giblett, quoted by Walker, asserted that sheep bred in-and-in on Bakewell's principle are fitter for the tallow-chandler than for the kitchen.

" On the other hand, M. Beaudouin gives the following account of a flock of three hundred merinoes bred in-and-in for a period of twenty-two years: the animals originally came from Saxony, were renowned for the purity of their blood, and had only been a few years in the Côte d'Or, when, in 1840, he commenced his observations. At that time, though suffering from no particular disease, the sheep were laboring under general debility, seemingly attributable more to a want of acclimatization than anything else. He began by a little judicious selection, eliminating about 15 per cent. yearly, and the flock soon became remarkably strong and healthy. There was no sign of sterility—altogether, perhaps, the cases of cryptorchis (non-appearance of testicles) and monorchis (single testicle) were not more than 6 per cent., while in the females there were even fewer cases of barrenness. Cases of duplicate organs were about 5 per cent.; and in 1859, a year when these cases were unusually frequent in all the flocks about, there were as many as 7 per cent. in his. The sexes were produced in nearly equal numbers, and cases of miscarriage were not more numerous than among the neighboring flocks. Far from degenerating, they became finer and far more to be depended upon to reproduce their proper type than is ordinary in flocks when crossed. He concludes with the declaration that, in his belief, inbreeding, combined with a moderate amount of selection, has no evil effect. Close interbreeding, says Mr. Darwin, has perhaps been continued longer with sheep than with cattle; but perhaps the nearest relations have not been so frequently matched. Messrs. Brown, during fifty years, have not crossed their excellent flock of Leicesters, nor since the year 1810 has Mr. Barford crossed the Foscote flock. This gentleman asserts that when two nearly related individuals are perfectly sound no degeneracy is produced in their offspring by their union; or, in other words, that there is no danger by in-and-in breeding unless through morbid inheritance. But, on the other hand, he does not pride himself on breeding from the nearest relatives; and I may add that such is not a breeder's object: he does not choose a relative for its relationship, but for its qualities. In France the Naz flock has been bred in-and-in for sixty years, without the introduction of any strange blood. Ferdinand and Louis Fischer started a flock of one hundred ewes of one family and four rams of another; and these families have since been interbred without the admixture of a drop of fresh blood. Mr. Atwood's entire flock, which was so celebrated that it is now scattered by colonization into all the States of the North American Union, originated from a single impregnated ewe; and neither she nor any of her progeny or descendants while in his hands were interbred with any sheep not descended exclusively from Colonel Humphrey's flock, from which she herself came. Mr. Hammond bought a small number of Atwood's flock in 1844, and he has since interbred solely between the descendants of these identical sheep. The Spaniards in their sheep-breeding guard against any admixture between the different cabanas, and they have been bred in-and-in for ages. Hallam says that the fineness of Spanish wool is considered to be owing to an importation of English sheep about the year 1348, and again about 1465, in return for which the Spaniards exported horses. McCulloch says that the Spaniards themselves ascribe their superior breed of sheep to the introduction of a few from England by Catherine of Lancaster in 1394; while elsewhere he says the merino breed is said to have been introduced from Barbary. These importations could not have been very great, and, as it appears, the Spaniards have since bred them in very closely, with the result that they became so valuable that up to the treaty of Basle their exportation was forbidden. By that treaty the French were allowed to buy five thousand merino ewes and as many rams; and from this stock the English sheep, which had also been carefully bred, were improved, while those of France and Germany were almost replaced by them. These sheep, says Mr. Huzard, have been ever since bred in-and-in at Rambouillet, and have never been crossed except by a second importation under the First Consulate. The nearest relatives are generally put together, for the rams are usually put to their own progeny for several generations, and this without any sign of degeneration. The flocks of Tessier, de Sylvestre, Perrault, Girod, and others testify to the same fact. The merino, when introduced

into Germany, was so immensely superior to all the native breeds, that it was everywhere accepted with enthusiasm. In Saxony the greatest attention was paid to
them, chiefly, however, as regards the *quality* of their wool, not as regards the
quantity and quality as well as quality of the meat, as in England.   To this end they
were kept in stables and fed on heating food, such as grain and hay, throughout
the winter.   The result was an unexampled quality of wool; but the animals became
a small and puny race.   In England the breed of sheep was already so good that
men were prejudiced in favor of their own breeds.   Many merinos, therefore, fell
into the hands of men who had no experience in breeding, and they were mismau-
aged; but in the hands of at least one practical breeder they were eminently
successful.   He reports on them: 'Soon after the king's flocks were imported . . .
I purchased a considerable number of sheep from them, and selected from those of the
Negrette blood, as being the largest sheep and carrying the most and softest wool.
These I continued to keep strictly pure, having no other sheep whatever, and I drew
rams from the royal flock, so long as that was kept up, since which I have depended
wholly on my own.   By due attention in breeding, the wool, far from degenerating,
has annually improved in softness and fineness, and these qualities have become
much more uniformly even throughout the fleece; so that I now obtain for the
whole a price beyond what any foreign wool brings in bulk in an unsorted state,
while the fleeces of our own flock are full double the weight of those of the Saxon
sheep.   It is right, however, to state that the staple of my flocks having arrived at a
length beyond that of other merino sheep, has rendered it fit for combing, thus
enhancing the value.   The form of the sheep is also highly improved, while the
disposition to fatten equals that of the Southdown.   The mutton is of the first quality,
and I can readily have for fat wethers the highest price which any mutton brings in
the London market.'   The justly celebrated New Leicester breed of sheep was
entirely created as a distinct breed by this method.   'Taking the native sheep,' says
Macdonald, talking of Bakewell, 'he reduced his size, gave him small offals, induced
him to lay on flesh and fat all along the breech, sides, shoulders, flank, and neck.
He opened his wool, and also reduced it in weight, and a little in length.   He
increased the tendency to lay on fat in proportion to the food consumed, and made
the animal take on fat at least a year or two earlier, thus enabling two or three
animals to be fed where one only was fed before.   Nor was this change fitful or temporary; it was permanent and indelible; and for nearly a century the same breed of
sheep has not only maintained its position, but has been used with more or less of
success to improve nearly every breed in the United Kingdom, and has, moreover,
more or less displaced almost every other breed.'   A correspondent of Walker says:
'I have bred from rams from the same flock in Leicestershire for fourteen years, which
flock has not had a cross since the year 1799.'   Some of the new Leicester breed
appear, however, to deserve the remark of the 'Bond Street Butcher;' for Sir John

Sebright said that Bakewell's principles were followed up too far ; the propensity to get fat has increased so much that their stock has become small in size, delicate, and produces little wool. But another correspondent of Walker points out that a propensity for fat-getting and the production of the finest wool are incompatible ; and it certainly appears from the fact that this breed has supplanted so many others that it cannot have degenerated. Too much fat is always a danger to a breed, for fat is a degeneration of tissue and a cause of sterility ; and although by in-and-in breeding man is able to do a great deal in the way of alteration, he must still follow nature— he cannot go contrary to physiological laws ; he can increase the qualities which he wishes to get chiefly only at the expense of qualities which he is content to do without ; and can no more obtain an animal all fat with every other good quality than he can teach his breed to live without food. We must remember that ill-directed breeding is as bad when there are frequent crosses as when there are none ; that it is selection which is the great improver, when properly directed, and that breeding in-and-in is only advantageous because it fixes the breed and obviates the necessity of crossing from an unimproved breed. Indeed, a careless cross may diminish size, just as careless in-and-in breeding may do so. The Romney Marsh sheep were made smaller in this way ; so were the Teeswater, and so are the mongrels of the merino and Scotch, or the Southdown and Scotch breeds. The sheep of Scotland, says Dr. Copland, are very small, their fleeces fine and soft, their meat delicate and finely flavored. In many parts they have much deteriorated since the introduction of Southdown breeds. Indeed, the sheep themselves seem sometimes to have an antipathy to crosses, for on one of the Faröe isles it was observed that the half wild native black sheep would not readily unite with the imported white sheep. The Shetlanders also tried to improve their native breed of sheep by crosses, and failed signally. So bad are the effects of crossing an improved breed, which must necessarily comprise no very great numbers at first, that some persons keep their animals in different families, and thus while they retain consanguinity, any tendency to disease peculiar to one family from the soil, habit, or what not, is obliterated. On the other hand, so valuable is in-and-in breeding to perpetuate any peculiarity either caused by selection or by what is known as a 'sport,' that nearly all 'created' breeds have been produced in this way, and valuable breeds, such as the Ancon and Mauchamp, would have been entirely lost without it."

IN-AND-IN BREEDING OF CATTLE.

"A majority of the most celebrated breeders and improvers of English cattle, says Mr. Randall, have bred closely in-and-in ; and this was necessary, since an improvement cannot comprise a large number at first. Bakewell was one of these breeders, and his Longhorns were for a considerable time closely interbred, though

Mr. Youatt says that they became delicate, and the propagation of their kind uncertain, a state which seems to have been due to bad management, for Bakewell himself was, as a rule, extremely successful. Knight once in the same season reared two young bulls of which the parents were nearly related; and both proved perfectly impotent, or at least failed to get a single calf; yet the females bred well enough while young. But another correspondent of Walker never found the generative power fail in consequence of in-and-in breeding of cattle; all that is necessary, he says, is to select carefully. The half-wild cattle kept in British parks, at Cadzow Castle, Chillingham, and Chertly are put forward as long-continued in-and-in breeding without any evil results by Culley, Dr. Brown, and Mr. Macdonald. These cattle were parked four or five hundred years ago, and are supposed to be the only remains of the ancient British cattle. Mr. Darwin, however, asserts that, compared to the wild cattle of South America, these are bad breeders; and Dr. Smith says that the Chillingham cattle now produce deviations from the original type of white, with black muzzles and red ears, which deviation he considers a degeneration. It does not follow, however, that this is a degeneration in the ordinary sense of the word; while it must be allowed that selection has not been practised with regard to their breeding, which would prevent any selection on their own part sufficient to allow of the intensification of any particular color, since, though the keepers may shoot these deviations from the original type, this will not prevent it in the first instance. The various colors are there, and it would be contrary to all the teachings of the evolution hypothesis if deviations did not occasionally occur, whether by sports, which would be rare in so in-and-in bred a herd, or by selection among themselves, as explained by Mr. Darwin in his 'Descent of Man.' The fact still remains, however, that these animals have been bred in-and-in for centuries, and still continue to breed without the help of crosses. The South American cattle are all descended from a few brought over from Spain and Portugal; the first by Garay, in 1580, and they have since increased to such astonishing numbers that, even in 1587, there were sixty-four thousand three hundred and fifty skins exported from New Spain. Vast herds of wild cattle are met with in all parts of the country, particularly in the plains of the southern provinces, where they exist in troops of twenty thousand to forty thousand; so that hides, jerked beef, horns, and bones have long formed leading articles of export from Brazil. On the Falkland Isles are herds of magnificent cattle, all descended from a few brought over from La Plata about eighty years ago. They are now breaking up into separate herds of different colors, the white, on the Highlands, breeding earlier than the others. I wish to draw particular attention to this natural segregation, which is also common in horses and sheep, and must be taken in connection with the tendency all polygamous animals seem to have to separate into families. Is this nature's horror of in-and-in breeding? Is this her delight in crosses? Price, the most successful breeder of Hereford cattle on record, until twenty years ago, was a staunch advocate of in-and-in breeding; so were the Collings, Mason, Maynard, Wetherill, Bates, the Booths, Sir C. Knightly, Bakewell, Culley, Ellman, and others. The cow Restless, almost an historical animal, was the result of in-and-in breeding to a degree which would not have been possible to obtain in man, owing to his long childhood. The bull Bolingbroke was put to his half-sister Phœnix, and produced the bull Favorite. Favorite was matched with his dam, and produced the cow Phœnix, a celebrated animal. Favorite was then matched with his daughter, and the produce was the famous bull Comet; then with his daughter's daughter; then with his daughter's daughter's daughter, he being the father in each case. The produce of this last union, a cow, had 93.75 per cent. of Favorite's blood in her, and was bred to the bull Wellington, himself deeply interbred on both sides in the blood of Favorite, of which he had 62.5 per cent. in him. This union produced the cow Clarissa, an admirable animal. Clarissa was bred to the bull Lancaster, who had 68.75 per cent. of Favorite's blood; and this union produced the celebrated cow Restless, a breeding cow of Sir Charles Knightly's herd. The rule of Mr. Bates was always to put the best animals together, regardless of consanguinity. His ' Duchess' family, one of many families thus bred, ceased to breed; but he continued his former course of in-and-in breeding with triumphant success. Mr. Darwin, however, points out that though Bates bred in-and-in for thirteen years, yet during the next seventeen years he thrice crossed his herd, not to improve them, but to increase their fertility; while Nathusius, after a careful study of pedigrees, finds that no breeder has continued in-and-in breeding all his life. But, at all events, many have bred in-and-in far more closely than would be possible in man, for a number of generations longer than the average of human families exist. Mr. Price, whose Herefords were the best in the world in his day, declared he had not gone beyond his own herd for a bull or a cow during forty years. At Earl Ducie's sale, in 1853, a white heifer, only five months old, sold for four hundred guineas; she was the daughter of the bull Fourth Duke of York, who was by Second Duke of York, and her dam was Duchess 59, also by Second Duke of York; consequently the sire and dam were half-brother and sister. Many others which reach high prices are bred on this system. Mr. Gardner gives a most successful case of breeding between son and dam. M. Sanson points out that the Charolaise race of cattle has been greatly improved by in-and-in breeding. At Rambouillet in-and-in breeding was practised among the celebrated cattle of that place—a white hornless breed—with great success, until they were carried off by the cattle epidemic of 1815. M. Huzard also saw at Hohenheim and the royal farm of Holitzchen herds of superior animals, which were always bred in-and-in. In this way, says Mr. Darwin, were in all probability bred the Niata cattle, from one individual sport."

IN-AND-IN BREEDING OF SWINE.

" Breeders are more nearly unanimous on the evils of in-and-in breeding upon pigs, says Mr. Darwin, than perhaps on any other large animal. Mr. Druce says their constitution cannot be preserved without a cross. Lord Weston, the first importer of a Neapolitan boar and sow, bred in-and-in till the breed was in danger of dying out. Mr. J. Wright bred with the same boar from its daughter, granddaughter, great-granddaughter, and so on for seven generations, with the result that the offspring in many cases failed to breed ; in others they produced few that lived, and of the latter many were without instinct to suck, and unable to walk straight. The last two sows were put to other boars, and produced several litters of healthy pigs. The best in external appearance produced during the whole seven generations was one of the last births, the sole one of the litter. She would not breed with her sire, and yet bred from the first trial with a stranger in blood. Nathusius imported a gravid sow from England, and bred closely in-and-in from the progeny for three generations, and with bad results ; yet he esteemed one of the latest sows a good animal, and she bred well with a boar of different blood. On the whole, Mr. Darwin thinks, therefore, that in-and-in breeding does not affect the external form, while it affects the general constitution, the mental powers, and especially the reproductive powers. It must be remembered, however, that pigs are precisely those animals which are cultivated most for their fat, and that fat is very injurious to the health of any animal, and especially in the reproductive powers. Crossing, on the other hand, gives a tendency to reversion, and therefore a relief from fat. Indeed, as I have already explained, facts against the harmlessness of in-and-in breeding have very little value compared with those in its favor, and this is too generally overlooked. These pigs with but little hair on their bodies have by correlation also very bad teeth, and this may be prevented by crossing with hairy breeds. If a breeder, in beginning to breed in-and-in, chose an animal with rather less hair than usual, the progeny would have a tendency to bad teeth, bad digestion, and hence weakness ; and he would naturally conclude, on finding that this weakness was cured by a cross, that it was the in-and-in breeding itself which caused it, and not mere inheritance. Mr. Hobbs divided his stock into three families, and by this device, though he kept the consanguinity, he avoided any chance inheritance of a morbid tendency, and obtained more latitude for selection. Mr. Coate, who won the prize for the best pen of pigs at Smithfield Club Show five times, says : ' Crosses answer well for profit to the farmer, as you get more constitution and quicker growth ; but for me, who sell a greater number of pigs for breeding purposes, I find it will not do, as it requires many years to get anything like purity of blood again.' So Mr. Youatt says : ' A useful pig in these days may easily be bred ; but if you want fixity of type, or, as it is well called, ' character,' you must adopt pure blood.' Red pigs are 'invaluable for giving vigor and constitution to black breeds, when demoralized by over-coddling, over-feeding, and injudicious in-and-in breeding.' "

IN-AND-IN BREEDING OF THE HORSE.

" In Circassia there are six sub-breeds of horses, three of which are asserted, by a native proprietor of rank, almost always to refuse to mingle and cross while living a free life, and will even attack each other. It is a crime punishable by death to forge the mark of pedigree on an animal. The Arabs are equally particular as to their breeds, and their horses are better able to stand a change of climate than are European horses. Mr. N. H. Smith, long a resident among the Arabs, is of opinion that colts bred in-and-in show more blood in their heads, are of better form, and are more fit to start with fewer sweats than are others ; but when the breed is continued incestuously for three or four generations, the animal degenerates. It is difficult to know what is meant by ' breeding incestuously.' Mr. Meynell, it appears, did not think breeding from sire and daughter or son and dam was close in-and-in breeding ; and Mr. Bowly says the term in-and-in breeding ought to be applied only to animals having precisely the same blood, as own brother and sister. Now, breeding from such relationship as this, seeing that the male has only half the blood of the dam, and the female only half the blood of the sire, can scarcely be called pure in-and-in breeding, but may, on the contrary, if carried out with caution, be done with advantage. Our race-horses are derived from a mixture of Persian, Barbary, Arab, and native horses ; but from the first they have been bred closely in-and-in. Rachel, the dam of Highflyer, was the daughter of Blank and grand-daughter of Regulus ; yet both Blank and Regulus were sons of Godolphin. Fox was born under similar conditions of relationship. The dam of Goldfinder was the daughter of Blank and granddaughter of Regulus. The granddam of Brick-hunter was a daughter of Bald-Galloway, who was also the sire of Brick-hunter. The great granddam of Flying-Childers, one of the most famous race-horses, was a daughter of Spanker, while his dam was also the dam of the last. The sire of the Knight of St. George, a winner of the St. Leger, was also his grandsire and great-grandsire. Smith, in his work on breeding for the turf, gives ' once in and once out ' as the rule for breeding ; but 'twice in and once out,' says Mr. Walsh, is more in accordance with the practice of our most successful breeders. The breeder can have no hesitation, continues Mr. Walsh, in coming to the conclusion that in-and-in breeding carried out once or twice is not only not a bad practice, but is likely to be attended with good results. The evidence of repeated success in resort-ing to the practice of in-and-in breeding is too strong to be gainsaid. ' For the race-course,' says Dr. Elam, ' the pure south-eastern breed is adhered to ; but different *stocks* of the same breed, and those brought up in different localities, are selected.' However, by 'crosses' breeders by no means understand the introduction of fresh blood. There are scarcely two thoroughbred horses in the stud-book, says Mr. Walsh, that cannot be traced back to the same stock in one or more lines. An absolute freedom from relationship is not to be found, or, if so, very rarely. Yet continued in-and-in breeding in the closest relationship he does not think advisable—it is apt to develop weak points in the constitution. 'The cautious breeder, therefore, will do well to avoid running this risk, and will strive to obtain what he wants without having recourse to the practice; though, at the same time, he will make up his mind that it is unwise to sacrifice a single point with this view.' Mr. Darwin says that statistics show that nearly one third of our race-horses have proved barren, or have slipped their foals—a fact which he ascribes to their high nurture and close interbreeding. This is very probably the case, since a racing-horse or mare, however delicate it may be, is too valuable not to breed from. Indeed, it is generally a disabled animal—one that has gone lame, and is therefore deprived of exercise and, with this, much of its natural health—which is set apart for breeding. Nor are they chosen for their fertility, but solely for their running powers. In-and-in breeding in horses is carried on at any rate to a very great extent, and with decidedly beneficial effects on the race.

"'Nimrod' concludes a comparison between the thoroughbred and half-bred hunter in these words: 'As for his powers of endurance under equal sufferings, they doubtless would exceed those of the "cocktail;" and being by his nature what is termed a better doer in the stable, he is sooner at his work again than the other. Indeed, there is scarcely a limit to the work of full-bred hunters of good form, constitution, and temper.' Napoleon's celebrated state horses were directly derived, says M. Huzard, from the Arab blood of Count Humiady, who had bred continually from the same two stallions. Indeed, it is the natural state of horses to breed in-and-in."

IN-AND-IN BREEDING OF DEER.

In many of the British deer-parks the deer have been allowed to breed uncrossed for long periods, without any degeneration showing itself or loss of general health. The dark herds of deer in the Forest of Dean, in High Meadow Woods, and in the New Forest, supposed to have been brought by James I. from Norway, have never been known to mingle with the pale-colored herds, although kept together with them—another case showing the rarity of crosses when animals are left to themselves. Dr. Davy mentions the case of a pair of red-deer, who, about the year 1850, were taken from the herd and put into a paddock of twenty or thirty acres adjoining Stornoway Castle, Isle of Lewis; these have multiplied yearly, and numbered, ten years after, twenty-three, not including several which were killed, all descendants of the original pair, and all very much improved in comparison with the deer of the forest. Nevertheless, it is the practice, says Mr. Darwin, to infuse new blood into the fallow deer of the British parks, and this, he says, proves of the greatest benefit in removing the taint of *rickback* and improving their size and appearance. Rickbacked deer are too generally found in many parks, says Mr. Shirley, supposed to be due to weakness, brought on both by breeding in-and-in too much, and also by insufficient food. In other words, we may say that the cause is unknown. The Scotch deer, however, breed naturally in-and-in, and the red deer generally breed between brother and sister for generation after generation, and yet they are, as a rule, perfectly healthy.

IN-AND-IN BREEDING OF FOWLS.

"Sir J. Sebright asserts that his fowls got long in the legs, small in body, and bad breeders from too close in-and-in breeding. Mr. Clark continued to breed in-and-in from his own kind of fighting cocks till they became under the weight required for the best prizes, and lost their pluck. On one cross from Mr. Leighton's they again resumed their former courage and weight. This breeder found that breeding from father and daughter produced a greater loss of weight in the offspring than breeding from the mother and son. Mr. Eyton, of that ilk, says his Dorkings became smaller and less prolific if not occasionally crossed. Mr. Hewitt says the same of Malays, as to size at least. But the fanciers with large stocks can breed from their own stock without this danger, because they keep various families separate for crossing purposes. Mr. Ballance, who breeds in this way, says that breeding in-and-in does not necessarily cause deterioration, 'but all depends upon how this is managed. My plan has been to keep five or six distinct runs, . . . and select the best birds from each run for crossing. I thus secure sufficient crossing to prevent deterioration.' "

Mr. J. S. Rogers, of Paterson, New Jersey, had some Dorking fowls that were inbred for many generations, until they became very diminutive in size. He at once concluded that if any animal could be diminished in size by in-and-in breeding, the converse must be true—they could be bred up in size; and selecting some eggs from a single hen of the large white Brahma breed, he bred in-and-in, always selecting the largest fowls from the descendants of the same hen, but taking care to have several runs of them. He brought them to an exaggerated size, the hens weighing twelve pounds and the cocks as high as fifteen pounds each. This is a good illustration of inbreeding, contrasting neglect and haphazard work with that of careful selection. The key-note of all successful breeding is *intelligent selection*. This, combined with a knowledge of the best formulas for inbreeding, enables the true breeder to accomplish great results in fixing the types of his own selection.

THE " DOWNY FOWLS."

The following instructive lesson was furnished for this work by Mr. J. V. Henry Nott, of Kingston, New York: " When I purchased my farm there were a number of common fowls upon it of no particular breed. We got a number of Plymouth Rock cocks to improve them. After the second year, or when the flock were two thirds Plymouth Rock in blood, we noticed a chicken that looked like a ball of down; and while the rest changed to feathers, she remained downy, and so grew up to henhood, when she proved to be a remarkable mother and layer, raising three broods of chickens in the season, and beginning to lay before the chickens were weaned. So we concluded, as she was a curiosity in appearance, to save her sons and breed them to her, *though none of them were downy.* After three broods one of the chicks turned out downy, and a cock, which we bred to his mother, and their chickens were about half downy and shortwinged, while the rest were common, or feathered.

"We then took the downys, and kept a pair in two separate yards, and when they had chickens took a cock from one yard and pullet from the other, which was breeding cousins [full brother and sister.—Ed.] together, and that is what we are still doing, each generation being a degree of cousinship apart. We are now down to the sixth generation, and the chickens come all downy, but not all shortwinged, or without flight feathers, which is, of course, their great value, though their down is as valuable as goose-down, as far as it goes. But the fact of not being able to fly over a common board fence three feet high makes them the fowl for village people, and to fully establish that improvement we put each new generation in an enclosure with a fence but three feet in height, and keep only those to breed from that cannot get over, without regard to size or appearance.

"Their color is a dark smoky blue, and they are as large and hardy as the Plymouth Rock. Some have single and some double combs."

IN-AND-IN BREEDING OF JERSEY CATTLE.

The Island of Jersey, being but a small tract and isolated from the rest of the world, while its cattle are protected from all foreign contamination, would naturally become a field for the practice of inbreeding cattle. Such inbreeding as has been practised, however, has been mostly accidental and haphazard; yet the pedigrees of imported stock for the past five years show that nearly all meritorious animals trace in several lines to one bull—"Old Noble." Romulus bred to his granddam Musique produced the bull Cetewayo, whose progeny are remarkable for strong constitutions.

Gilderoy, tracing by two or more lines to "Old Noble," was bred to Regina 2d, a granddaughter and great-great-granddaughter of "Old Noble," producing Chrome Skin, a cow that made twenty pounds, thirteen ounces of butter in seven days. Gilderoy bred to Chrome Skin, his daughter, produced Gilderoy 3d, a bull noted for beauty and vigor. This is the breeding practised by Dr. Howe, of Bristol, R. I. There have been many fine illustrations of inbreeding among American bred Jerseys, some by the design of skilful breeders, others that were merely circumstantial. The best model of a Jersey cow ever known—Jersey Belle of Scituate 7828—was produced by breeding Victor 3550 to his own daughter. Victor 3550 was the result of mating full brother and sister. Mr. Simpson's Alphea family has a number of very choice animals, produced by mating full brother and sister, and breeding the progeny to his daughters, granddaughters, and great-granddaughters in double lines.

Young Mercury, whose portrait is shown in this work, and whose escutcheon is used for the illustration on that subject, is the grandson of his sire Mercury and the full brother in blood of his dam Phædra, that made nineteen pounds, thirteen ounces of butter in seven days. Through seven lines he traces to Saturn and Rhea, the sire and dam of the famous cow Alphea. His formula is: Full brother to full sister and their son to his daughters and granddaughters of the same pure Alphea blood. Another noted family, originated by Mr. O. S. Hubbell, of Connecticut, is descended from the noted inbred bull St. Helier. The formula of this family is: sire to daughter, also to granddaughter, and then combine brother and sister; or the grandson of his sire St. Helier is bred to his half sister by St. Helier, and their male progeny is bred to a daughter or granddaughter of St. Helier with most successful results in the production of a choice type of butter Jerseys, that are also remarkable for their uniformity of style and quality. In England Philip Dauncey bred for forty years by coupling half brother and sister and using an occasional outcross from the Island of Jersey. "Pope," Mr. Dauncey's first bull, was purchased in 1826 from Mr. Michael Fowler, by whom another Island bull, "Fowler," was obtained thirty years later. From the combined blood of these two bulls descended the famous bull Riöter 2d 469, imported to America by Col. R. M. Hoe, and also "Stoke Pogis," a bull whose descendants in America have made a great name, the most noted cow being "Mary Anne of St. Lambert," that in her fifth year made eight hundred and sixty-seven pounds, eleven and three quarter ounces of butter, and has an official seven-day test of thirty-six pounds, twelve and one quarter ounces. All the best cows illustrate the success of inbreeding as an essential method of improvement.

INBREEDING AND FECUNDITY.

"Scraps for Breeders," in the London *Live Stock Journal*, contains the following:

"There is probably no opinion more generally accepted among breeders, and taken for granted in every new discussion, than that in-and-in breeding must induce barrenness. That there are grounds for this opinion is certain, for no conclusion obtains wide assent unless it be at least plausible—*i.e.*, consistent with ordinary observation. Yet the first thought, to one reader at least, on turning over the new volume of the 'Shorthorn Herd Book,' was, 'What a lot of twins there are by Booth bulls!' There are not now existing in the kingdom any cattle reared from closer affinities

than those at Warlaby; yet at Warlaby there was in 1883 one pair of twins and a
triple birth; at Killerby there was one pair of twins; at Mr. St. John Ackers' two
pairs; at Lord Polwarth's one pair; another pair at Mr. Talbot Crosbie's; another
pair at Mr. R. Welsted's; while at the Duke of Northumberland's, Mr. Willis's of
Carperby, and at Mr. T. Pear's—whose herds, although not of Booth origin, are very
closely allied, by recent sires, to that strain of blood—there were in each case no
less than three pairs of twins in one season.

"These incidents go far to show that, under proper superintendence, Shorthorns
may yet be very closely bred for concentration of blood, and still remain fecund; and
also that the ordinary allegation against Booth cattle, 'that they are slow breeders,'
is not one which is necessarily true. For in the lot of cows and heifers of which
these herds are composed, and which probably altogether do not much exceed two
hundred and fifty animals, no less than seventeen, or nearly seven per cent., produced
more than one at a birth in 1883. This rate of increase is above that of unpedigreed,
loosely-bred dairies."

It would seem very plausible that the quality of producing twins might be made
a prepotent and permanent trait in any breed, by careful and persistent selection,
though perhaps it would not be so desirable in a dairy race as in beef breeds. The
lack of fecundity in Shorthorns, or any breed, may be induced by allowing individuals of either sex to be kept in a state of obesity that induces fatty degeneration.
It is stated upon good authority that the bull Hubback, from whom the Shorthorn race
was derived, early became impotent, because he was allowed to become very fat,
and consequently his own progeny were very few in number. The quality of the
Jersey breed is such that very little difficulty obtains from a lack of fecundity
through fat.

INFLUENCES DETERMINING SEX.

"In the January number of the *Popular Science Monthly* (1885) there is a
review, by Prof. W. K. Brooks, of an article on the laws which determine sex,
published by Carl Düring in the *Jenaische Zeitschrift*.

"'Each species has acquired, through natural selection, the useful property in
virtue of which any deviation from the average ratio between the sexes is corrected
by an increased number of births of the deficient sex, or a decreased number of the
sex which is in excess.'

"Notice the increased number of male colts as the number of mares put to a
stallion increases.

"Again, notice the increased male births following a war that takes many men
from their homes.

"'A favorable environment causes an excess of female births; an unfavorable
environment an excess of male births.' The female is supposed by Prof. Brooks
to be the conservative element in reproduction, and the male the element through

which new varieties are introduced. Hence, when circumstances unfavorable to the race occur, an increase of males takes place, in order that the race, by altering its habits or structure to some extent, may adapt itself more readily to its surroundings. Assuming a large number of births to be an evidence of favorable surroundings, it has been shown that, in prolific races, the number of females is in excess, and again, in any case, as the number of births increases the ratio of females increases. The birth rate of females is higher in cities than in the country. So much in regard to the human race in general. As regards the individual, Prof. C. M. Hollingsworth puts forth the hypothesis that ' it is a *relative* preponderance of the conditions on which cell division depends which causes the formation of the female or male generative organs and determines the sex of the individual.' The higher plants, he has shown, have female flowers situated in places most favorable for cell growth, and male on places for cell division. The relatively larger plants are female. The sex of a plant can be influenced by placing it in a position favorable or unfavorable for cell growth. It is a fact, arrived at by experiment, that in the higher animals an early impregnation of the ovum results in the birth of a female offspring. It is supposed that in early impregnation an interval elapses before segmentation takes place, and in that time the male element tends to become ' assimilated,' and so, ' by hypothesis, to have its specific capacity or function of exciting cell division to some extent weakened.' In a late impregnation the reverse would occur, and a male offspring be the result."

Many Jersey breeders have made more or less persistent efforts to reduce to practice the theories of biologists in regard to control of sex in offspring. Insufficient data are obtainable upon which to suggest any plan of action or experiment with any reasonable assurance of success.

The Stuyvesant theory of alternating sex in successive periods of heat secured by observing the sex of the last birth, upon the hypothesis that the female gives for each period an ovum of alternating sex, has received some practical attention. Thus far the sexes have been about equally proportioned, the females but slightly preponderating ; and this is doubtless a fixed law of the Creator for the preservation of both sex and species. If the law has been discovered or is discoverable, the knowledge of its application will be of immense advantage to all breeders of cattle.

THE RANKIN THEORY.[*]

"If we take the proper advantage of the fact that the cow has two ovaries, one of which throws off, in her normal condition, an ovum every twenty-one days, which may be impregnated and produce another of her kind, male or female, as is the ovum impregnated. Should a bull calf be the result of the last effort, then the

---

[*] G. T. Rankin, Jersey Bulletin, Sept. 23, 1885.

first ovum passed would be a female, if impregnated, which is the 'Stuyvesant theory.' But what I wish to impress upon the breeder is that in from nine to twelve days after the cow calves she will, through her ovarian system, deposit in the uterus an ovum susceptible of impregnation, although she will show none of the usual symptoms of being in a condition to be served ; but if the last calf should have been a male, then a service within twelve days, if impregnation takes place we may expect a female as the product ; but should she not prove pregnant, as the next ovum would be a male, we must pass the heat, if we want a heifer calf, and breed at what would be called the second observable heat, as the first will not be recognized by any objective signs from the cow, except by her actual exposure to the male. This is my improvement of the 'Stuyvesant theory' of breeding for sex ; and as I have never seen the suggestion of the 'nine-day theory,' I claim it as original, and only ask breeders to report their experience if they think it worthy of trial. No doubt many breeders have tried the 'alternate heat theory,' and been disgusted as well as myself ; but I would like if they would try again, *observing the nine-day caution.*"

The author of this work is not yet satisfied that we have any clew to the law controlling sex, but believes the subject worthy of persistent investigation and thorough systematic experimentation. Possibly physiological research by vivisection, by spaying one ovary, and like experiments, may some time lead to decisive results, which will give us the key of the great secret.

THEORY OF AGE OF THE OVULE.

Still another theory relates to the age of the ovule in determining sex. It has been promulgated and partially investigated by practice, that vitalization of the ovule by the male in the early stage of ovulation, or during the first symptoms of desire on the part of the female, that the resulting offspring will be a female, and conversely if several hours' delay before union of the sexes the ovule undergoes such changes that vitalization by the male then results in a male offspring. It would be well to collect as many observations as possible upon the above theories, singly and in combination. To that end it is commendable in breeders to keep an extended record of all cases in their herds from this time forward, so as to prove or disprove theories.

INFLUENCE OF SEX UPON OFFSPRING.

That the male transmits his peculiarities to the female progeny and the dam yields her characteristics to the male progeny is continually confirmed in nature.

A fine illustration of this axiomatic proposition is given by a correspondent of the *London Field :*

" I put a black-red game cockerel with willow legs to two white game pullets with yellow legs and bills. I have thirty chickens of this parentage. Every cockerel has the shape and yellow legs of the mother, every pullet the type and willow legs of the father. Knowledge of this tendency is capable of rendering good service in many departments thought more highly of than chicken-breeding. Indeed, there is hardly any limit to its usefulness."

SUMMARY OF FACTS ON BREEDING.

1. That man was for twenty-five hundred years under a social system of the closest consanguinity in marriage, the era of the greatest longevity of the human race.

2. That mongrels of the human races, as the Mestizo and the Mulatto, are especially inharmonious mixtures in mind and body, notorious for their depravity and savage-like atavism.

3. "That the effects of crosses in man, animals, and plants are, first, *variability*, which depends, according to Darwin, 'on the reproductive organs being injuriously affected by changed conditions;' and, secondly, on *reversion*, which is generally a change for the worse, as the organism thus reverts to its former unimproved state, and the good effects of natural or artificial selection are thus lost."

4. That in a state of nature, horses, cattle, sheep, deer, elephants, bison, wild boar, and many other animals habitually practise in-and-in breeding, and also selection, the strongest male leading the herd by right of conquest.

5. That among our domesticated animals improvement is made by selection and the closest in-and-in breeding.

6. That fixity of type can only be maintained by perpetual in-and-in breeding.

7. That the qualities of a "sport" or phenomenal animal can only be preserved and perpetuated by close in-and-in breeding.

8. That perfect specimens of any species or breed can be perpetually inbred without any detriment.

9. That no physiologist has ever shown that disease or deformity can be attributable to in-and-in breeding as a cause.

10. That where disastrous results follow the practice of in-and-in breeding, the animals are diseased, and those diseases, like other peculiarities, may be intensified by inbreeding.

11. That a common difficulty is the condition of obesity, which results in sterility and fatty degeneration to those bulls or cows thus kept and overfed.

12. That swine and fowls and all other animals kept for their fat are in an abnormal condition, and consequently difficult to inbreed. Hubback, the bull that founded the Shorthorn breed, early became impotent from fat.

13. That the Jersey race of cattle, being less liable to fatty degeneration than most other races of domestic animals, bears in-and-in breeding well, and such

in-and-in breeding has been the means of developing the most wonderful specimens of productive dairy cows ever known in the world's history.

PLUS INTO PLUS, OR THE TRUE ART OF BREEDING.

> " There is a history . . .
> The which observed, a man may prophesy,
> With a near aim, of the main chance of things
> As yet to come to life ; which in their seeds
> And weak beginnings lie intreasured."—*Shakespeare.*

To all persons who read this book and are desirous of excelling as Jersey breeders, whether they are novices or have had many years' experience, the following directions may be found feasible and practicable for the progressive improvement of their herds.

1. As the bull is the breed, and contributes fully fifty per cent. of the blood value to each of his progeny, he should be the best animal obtainable.

2. The bull should be as largely of the strain, or family, as practicable, from which one wishes to breed.

3. Select the family you wish to breed, and from that family the best bull you can obtain.

4. Select your bull by the new scale of points, from a tested cow.

5. If you cannot find the bull that meets your requirements, have one bred to order, using in the interim of his development the best one that you can obtain.

6. A good bull is one of a thousand ; indeed, you might examine several thousand, and not find a suitable model for your herd.

7. I would designate the bull which possesses all the requirements sought as a *plus* animal ; for if he is properly bred he is likely to be very prepotent, and will transmit to his progeny more than fifty per cent. of blood value ; and in estimating his value he may be marked in the pedigree as (.50+) fifty per cent., *plus.*

8. Similars with similars is the great law for the breeder. St. Helier with St. Helier ; Albert with Albert ; plus with plus ; yellow skin with yellow skin ; butter breeder with butter breeder ; first order escutcheon with first order escutcheon, leaving no element of perfection out of the plan.

9. The bull must not only have an unbroken line of good ancestry for at least six generations without one inferior animal, no weak link in the long chain, but his dam should be extraordinary in all points ; the bull inherits his special qualities from his dam.

10. The cow that fulfils the requirements of the new scale of points may be called a *plus cow,* and is expected, if she be in-and-in bred, to transmit fifty per cent. of her blood value to her progeny. The *plus* cow inherits her special qualities from her sire.

11. To breed a *plus* bull, he should be the product of a formula that would make him the grandson or great-grandson of his dam, and she a twenty-five-pound cow, that he may inherit and intensify her form and character.

12. The *plus* bull should be bred to *plus* cows of the same family as himself. I think a good herd should be so uniform in blood ratios that all should be kept up to a fifty per cent. standard of the family blood. In some cases it would be better to make the animals one hundred per cent. by close in-and-in breeding, for it has been demonstrated that a line of family quality cannot be maintained without purity of blood. If the blood is let loose by crossing it may take years to recover it, or it may be irretrievably lost by such experiments.

13. All animals that fall below the breeder's own standard should be eliminated from the herd.

14. A *plus* cow should be the product of a formula that would make her the granddaughter or great-granddaughter of her sire, and he the son of a twenty-five-pound cow.

15. A good formula for the breeder and worthy of adoption as a motto is the algebraic rule of multiplication.

Plus into *plus* produces *plus*.

Plus into *minus* produces *minus*.

Minus into *plus* produces *minus*.

Minus into *minus* produces *plus*.

The last *plus* is a bad kind for the Jersey breeder to propagate.

16. One cannot be too particular in the breeding of his bull. He must be equally particular as to the cows he breeds.

17. Let the outcrosses, if you make any, not be absolute, but rather let them have at least fifty per cent. of the best blood which characterizes your herd, and the other element should be something that promises to supply a deficiency.

NEGLECTED OPPORTUNITIES.

How many neglected opportunities for doing grand work the history of Jersey breeding sets forth! Look at the portrait of Jersey Belle of Scituate 7828, and conjecture what such a cow would be worth to-day in the hands of a skilful breeder. Study her by the scale of points, her history, and her wonderful record upon moderate feeding. Where shall we see her like again? Where, oh, where were our geniuses— our Collings and Bakewells, our Prices or our Guenons—that they did not see to it that such a wonder of perfection should have been so bred as to leave her form and quality a rich legacy to the Jersey breeders of America, in at least one in-and-in bred *plus* bull that should more than replace her own individuality? Suppose she had been bred to Albert 44 or St. Helier 45, and inbred to her own progeny after the following formula:

HYPOTHETICAL PEDIGREE.

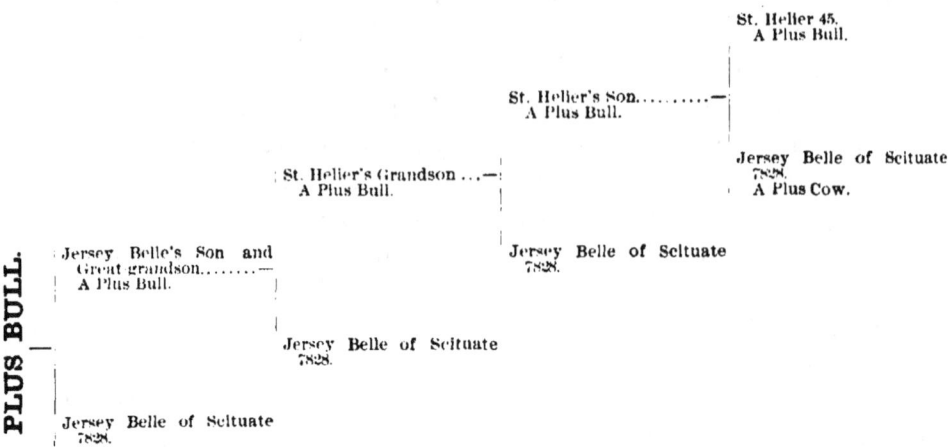

Does any one doubt whether there would have been a "nick" in such a formula? Let him try the experiment with as good a bull as St. Helier and as good a cow as Jersey Belle of Scituate!

Other Jersey bulls as worthy to have their names in such a formula were Albert 44, Landseer 331, Mercury 432, Signal 1170, Top Sawyer 1404, Gilderoy 2107, and Stoke Pogis 3d 2238. Victor 3550, the sire and grandsire of Jersey Belle of Scituate, was certainly worthy of filling such a formula; and if it could have been accomplished, what Jerseys we would now possess for founding herds of superlative excellence!

PART SECOND.

DAIRY FARMING AND MANAGEMENT OF STOCK.

"And he gave it for his opinion, 'That whoever could make two ears of corn or two blades of grass to grow upon a spot of ground where only one grew before, would deserve better of mankind and do more essential service to his country than the whole race of politicians put together.' "—*Swift*.

THE successful Jersey breeder must have a theoretical and also thoroughly practical knowledge of all the principles and requisites of dairy farming.

THE NINE POINTS OF GOOD FARMING.

1. The right selection of soil and location.
2. The right selection of animals, seeds, and plants.
3. The right construction of buildings, machinery, and tools.
4. The right underdrainage.
5. The right making and saving of manure.
6. The right modes of tillage.
7. The suitable rotation of crops.
8. The timely performance of work.
9. The requisite irrigation of all crops.

Volumes have been written upon each of the above-named subjects, and still they prove to be inexhaustible in interest and their importance immeasurable. When the Jersey breeder shall have mastered them all he will have lived long enough to become famous. All Jersey breeders who have good farms are very fortunate. In the selection of a farm for dairy purposes, one must first seek for fertility ; buy a farm that is of the richest soil, or soil that can be made rich ; secondly, it must be in a healthful region ; thirdly, near enough, but not too near, to the best of neighbors ; and fourthly, contiguous to a good market.

The farm should be stocked with the best strain of Jersey cattle and have buildings constructed suitable to their use. The crops grown must be the best assortment for the comfort and health of the occupants of the farm, and afford a sufficient variety of economical and wholesome food for all the stock. Every slough, swamp, or unprofitable acre of wet land must be thoroughly underdrained in the best manner by the use of the best quality of glazed pipe and collar drain tile, thus improving the fertility and healthfulness of the farm.

The farm should continually grow richer by the saving and properly utilizing all the manurial elements, and by turning under green crops to make vegetable mould. The tillage should be done with the most effective and labor-saving implements and always thoroughly and appropriately qualified according to the needs of each crop and the condition of the soil. Tillage enables plants to digest and assimilate manures, as a thorough mastication prepares food for the animal economy. The crops must be so arranged in order of rotation as to utilize the various elements of cumulative fertility and allow of a restoration of those that are deficient or exhausted. If work is always done at the right day and hour much needless expenditure of vital force and money will be saved, and the farm will become like a well-regulated workshop, where every employé knows his place and fulfils the expectations of his employer.

It needs a good deal of careful planning to keep the machinery in smooth running order. To drain at the right time ; to manure at the right time, in the right way ; to plow at the right time ; to pulverize thoroughly at the right time ; to cultivate and harrow and till at the right time and all times ; to plant at the right time ; to reap at the right time ; to turn on the water from the brook, fountain, or reservoir just at the right time, and save a crop from the drouth or a pasture from scorching ; to raise big crops and keep down weeds at the right time—to do everything in the easiest and most expeditious manner and make it pay in money returns, is the province of good farming.

THE SOIL.*

BY DR. AUGUST VOELCKER.

"On examining the various soils of this or any other country, they will be found to consist generally :

" 1. Of larger or smaller stones, gravel or sand.

" 2. Of a more friable, lighter mass, crumbling to powder when squeezed between the fingers, and rendering water muddy.

" 3. Of vegetable and animal remains (organic matter).

" On further examination of the several portions obtained by means of washings, we find :

" 1. That the sand, gravel, and fragments of stones vary according to the nature of the rocks from which they are derived. Quartz-sand, in one case, will be observed as the predominating constituent ; in another this portion of the soil consists principally of a calcareous sand ; and, in a third, a simple inspection will enable us to recognize fragments of granite, feldspar, mica, and other minerals.

---

* Morton's Encyclopædia of Agriculture.

"2. In the impalpable powder, the chemist will readily distinguish principally fine clay, free silica, free alumina, more or less oxide of iron, lime, magnesia, potash, soda, traces of manganese, and phosphoric, sulphuric, and carbonic acids, with more or less organic matter.

" 3. The watery solution of the soil, evaporated to dryness, leaves behind an inconsiderable residue, generally colored brown by organic matters, which may be driven off by heat. In the combustible or organic portion of this residue the presence of ammonia, of humic, ulmic, crenic and apocrenic acids (substances known under the more familiar names of soluble humus), and frequently traces of nitric acid, will be readily detected. In the incombustible portion, potash, soda, lime, magnesia, phosphoric, sulphuric and silicic acid, chlorine, and occasionally oxide of iron and manganese, are present.

" All cultivated soils present a great similarity in composition : they all contain the above chemical constituents. This similarity becomes still more apparent after burning, when nearly all soils will assume a red color, which is due to the presence of the oxide of iron.

" At first sight this might be regarded as opposed to the great diversity of soils ; but if we examine the relative proportions in which the several constituents are mixed together, the state of combination in which they occur, and the manner in which the different soils are formed, we shall find that diversity is compatible with a certain similarity in elementary composition.

" In all fertile and arable soils organic matters, more or less decomposed, varying in quantity from one half of one per cent. to twelve per cent., occur ; and as in good garden mould the proportion of such organic matters frequently amounts to twenty-four per cent. of its own weight, and seldom is less than ten to twelve per cent., it was believed that the amount of organic matters in soils determined their relative degree of fertility. This, however, is a great mistake, for there are soils containing only two per cent. of organic substances which are, notwithstanding, greatly superior to others containing six or eight per cent. ; and, again, in peaty or boggy soils, belonging to the worst description, sixty or seventy per cent. are by no means uncommon. In soils celebrated as good wheat soils we have found not more than three to three and one half per cent. of organic matter ; while in far less productive land we have found as much as ten to twelve per cent. That no reliance can be placed on the amount of organic matter in soils, as indicating their productive powers, is also clearly seen in the following determinations made by Dr. Anderson, in some of the best wheat soils from different parts of Scotland.

LOCALITY.	Organic Matter in Soil.	Organic Matter in Subsoil.
Mid Lothian Wheat Soil	10.19	4.83
East Lothian Wheat Soil	6.32	5.85
Perthshire Wheat Soil	8.55	6.82
Morayshire Wheat Soil	4.54	3.76
Morayshire (different specimen)	3.47	
Berwickshire Wheat Soil	6.67	

"The organic matter in the soil is due, for the greater part, to the vegetable remains of former crops, and partly to animal matters, derived from the decay of insects or the excrementitious substances contained in manure. The vegetable and animal remains, under the influence of water, air, and heat, gradually decay, producing a brownish or black powdery substance, or rather a mixture of substances, which is known to practical men under the name of humus, or vegetable mould. There are principally two kinds of humus—brown and black; the former is contained in large quantities in the brown variety of peat; the latter, the result of further decomposition of the brown, is found in black peat.

"Brown and black humus have a very complex composition, which is changing every day as the decay of the vegetable remains in them proceeds. During this decay a number of peculiar organic acids are formed, as, for instance, ulmic, humic, crenic, apocrenic, and geic acids. These acids resemble each other very much in their general aspect, as well as in their composition. Humus plays an important part in the process of the nutrition of plants, but its functions cannot be explained by one action only, for it is evidently subservient to the luxuriant growth of plants in more than one way.

"Thus it exercises a beneficial action in condensing ammonia, as well as moisture, from the atmosphere, and likewise by furnishing a continual source of carbonic acid, arising from its decomposition. Again, the vegetable remains in humus always contain a certain amount of inorganic matters, but the latter are not soluble in the fresh roots, stems, and other parts of plants, and only become available to vegetation during their gradual decay and conversion into humus.

"Notwithstanding a general similarity in the composition of arable soils, the appearance and general character of many soils, in every country, present striking differences, which cannot fail to strike the attention of every superficial observer.

"The forms and proportions in which the chemical elements usually constituting soils are mixed together, in different localities, explain, in some measure, though by no means fully, the various appearances and agricultural capabilities which they possess. These forms and proportions themselves depend on the causes and circumstances under which they originated.

"The manner in which some soils are formed will not be long doubted by any one who has observed the appearance of large rocky masses, the clefts and crevices they present, the bare surface of their smoother and harder parts, the growth of mosses and smaller plants on the more softened portions, the accumulations of gravel, smaller fragments of minerals, and fine mud, with their luxuriant vegetation at the foot of these rocks, and in the valleys of mountainous districts.

" These soils evidently have originated in the degradation and decomposition of the solid rocks in their immediate neighborhood, especially of those which occupy the surrounding eminences. But as rocks differ much in composition, the soils which are formed on their decomposition must necessarily present, in many cases, great differences equally with the rocks themselves ; and the study of the latter will therefore be of considerable interest to the cultivator of the soil. In other instances, however, the nature of the soils, in a given locality, partakes nothing of the characters of the rocks in the immediate neighborhood, nor even of those on which they rest. The causes which are instrumental in the formation of soils fully explain this apparent anomaly ; and we shall, for this reason, draw attention to the various causes which give rise to the formation of arable soils. In some instances we can trace the changes rocks undergo in the course of time, step by step, and refer them to their true causes ; in others only the ultimate products of decomposition are well described, and their primary causes less clearly understood. This much is sure, that the causes which operate in the formation of soils are various and often complicated. Some of them may be referred to chemical forces and agencies ; others, which are based on purely mechanical principles, we shall distinguish as mechanical causes ; and a few partake of the nature of both—they act partly chemically, partly mechanically.

" I. Chemical causes of the degradation and disintegration of rocks.

" 1. One of the principal agencies in effecting a gradual disintegration of solid rocks is the atmospheric oxygen. In the course of the formation of oxides the compact texture of the rock is broken up, and the whole mass of the rock gradually crumbles down.

" 2. A second and no less powerful chemical agency in the formation of soils is the carbonic acid of the atmosphere, carried down by the rain. The affinity of carbonic acid for different mineral compounds varies greatly. Limestones are easily attacked by rain-water, while pure quartz and sandstones are scarcely acted upon by rain-water.

" Under the influence of carbonic acid and water, feldspar, granite, and other minerals consisting of silicate of alumina and an alkaline silicate are decomposed into alkaline silicates, which in turn give rise to silica and carbonate of potash or soda, and into silicate of alumina or pure clay.

" 3. In the formation of soils from solid rocks the lower orders of plants and animals take an active share. The seeds of lichens and mosses floating in the air attach themselves to the roughened and partially decomposed surfaces of rocks, and finding here sufficient food, germinate and throw out roots, which penetrate the little crevices in the rocks like wedges. These widening and multiplying crevices hasten the final disintegration of the rock. Mosses and lichens likewise retain the atmospheric water and keep the surface of the rock moist for a longer time, giving in this manner rain-water a better chance of exercising its dissolving powers on the constituents of the rocks. Insects and other animals of the lower orders collect and feed on the lichens and mosses, and both insects and plants in due time die, decay, and leave all the mineral matter which they have originally obtained from the rock behind, mixed with vegetable and animal remains or humus. A thin layer of a more fertile soil is thus formed, on which plants of a higher order may spring up ; in the course of time these die, and enrich and increase the soil.

" II. Mechanical causes acting on the formation of soils.

" Generally the first stage in the disintegration of rocks can be referred to a chemical force. The described chemical agencies, however, are often associated with mechanical ones, or followed by purely mechanical causes, which produce great changes in the appearance of rocks, and contribute much to the rapid formation and the peculiarity of some soils.

" 1. One purely mechanical agency is the force of gravitation. When the force of gravity preponderates over cohesion, the rock so influenced contributes to fill up the valley below with disintegrated fragments. According to the nature of the rock, vegetation springs up on these débris more or less luxuriantly, often very rapidly.

" 2. The finer portions of broken rocks are easily moved by the winds.

" 3. Water exercises a powerful influence in changing rocks in a mechanical way.

" By freezing it expands and bursts the rock. The rains continually wash off particles and carry them to lower levels.

" The finer deposits form the alluvial soils of our river-banks. The vast mass of materials deposited at the mouths of large rivers alters the condition of the soils along the banks of the deltas from a naturally sterile into a most rich and fertile one.

" 4. The sea likewise plays an active part in changing the character of the land near the shore and in giving rise to new soils.

" 5. Vegetable remains, and especially animal remains, contribute much to the formation of some soils. Vast numbers of infusoriæ, near the mouths of rivers where salt and fresh waters mingle, die daily, mix with the mud, and are deposited along the banks, and thus alluvial soils of the utmost degree of fertility are formed.

CLASSES OF SOILS.

"Soils in general consist of a mechanical mixture of the following four ingredients:

"1. Silica, silicious sand, and gravel.

"2. Clay.

"3. Lime.

"4. Animal and vegetable remains (humus).

"There are few soils which consist of only one or two of these four substances; most contain them all, but the relative proportion of each in different soils varies considerably.

"A simple classification of soils, accordingly, may be founded on the preponderance of one of these four chief constituents:

"Soils may be conveniently classified as follows:

"1. *Sandy soils*, containing above eighty per cent. of silicious sand.

"2. *Calcareous soils*, containing above twenty per cent. of lime.

"3. *Clay soils*, containing above fifty per cent. of clay.

"4. *Vegetable moulds* (humus soils), containing more than six per cent. of organic matters or humus.

"5. *Marly soils*, or soils in which the proportion of lime is more than five, but does not exceed twenty per cent. of the whole weight of the dry soil, and that of clay is more than twenty, but less than fifty per cent.

"6. *Loamy soils*, or soils in which the proportion of clay likewise varies from twenty to fifty per cent., but which at the same time contain less than five per cent. of lime.

CHARACTERISTICS OF SOILS.

"1. *Sandy Soils.*—They are generally of a loose, friable, open, dry character, and for that reason are more easily and less expensively cultivated than any other description of soils.

"Many consist almost entirely of silicious sand and gravel, with but little alumina and calcareous matters. Such soils are almost absolutely barren, and in general termed *hungry* soils, from their tendency to absorb manures without any corresponding benefit to the land. Others contain a large proportion of alumina and lime, which render them more compact and always more fertile.

"On these richer kinds of sandy soils, beans, peas, and spring wheat succeed well; and as turnips are frequently grown with advantage on them, they are called also turnip soils.

"Sandy soils are capable of improvement.

"Clay, marl, chalk, and many other substances counteract the loose texture and porosity, and may with advantage be applied to them.

" 2. *Calcareous Soils.*—As the physical characters of calcareous soils depend chiefly on the relative proportions of lime and the other constituents which enter into the composition of this class of soils, it is impossible to give a short general characteristic. While some are deep, dry, loose, and friable in their nature, and as productive as some soils resting on the lower chalk formation, others are stony, poor thin soils, producing but a scanty vegetation. Beans, peas, and clover are grown with advantage on this class of soils.

" They are subdivided into calcareous clays, loams, and sands, according to the proportion of clay and silica.

" 3. *Clay Soils.*—The properties of clay soils are diametrically opposed to those of sandy soils. Stiffness, impenetrability, great power of absorbing and retaining moisture, and great adhesiveness characterize this class of soils. They are consequently cold, stiff soils, which are expensive and difficult to cultivate. When properly cultivated some are turned into highly fertile soils. Their mechanical structure may be corrected by drainage, burning, bulky manures, and the addition of lime, ashes, and sand.

" 4. *Vegetable Moulds.*—Any soil containing more than six per cent. of organic matter, whatever else its composition may be, is called a vegetable mould. Soils of the most opposite physical characters may be thus grouped in this class. They are clayey, loamy, or sandy. Many are highly fertile; others are more or less unproductive, but capable of improvement; and others again contain so large a preponderance of organic matter that they are called *peaty* or *boggy*.

" 5. *Marly Soils.*—Marly soils resemble more or less in their characters calcareous and clay soils.

" They are always less retentive, less impervious than clay soils, but generally not so open and porous as many calcareous soils. On the whole, marly soils belong to the better, more productive, and generous soils.

" A sandy marl is a marly soil in which a large proportion of clay is replaced by silicious sand. Clay marl, on the contrary, is a marly soil in which clay preponderates.

" 6. *Loamy Soils.*—The term loam is reserved to all soils which contain the four chief constituents—silicious sand, clay, lime, and vegetable and animal remains—in a fine state of subdivision, intimate mixture, and in such relative proportions that the quantity of lime does not exceed five per cent. nor that of clay fifty per cent.

" Loamy soils, next to the richer garden moulds, belong to the very best soils. They are easily cultivated, and yield abundant crops of almost any kind. Many alluvial deposits that are celebrated for fertility belong to this class.

" Sandy loam, clay loam, marly loam, are terms applied to soils wherein sand, clay, or marl appear more prominently than in others."

ANALYSIS OF LOAMY SOILS.

BY DR. ANDERSON.

CONSTITUENTS.	Soil.	Subsoil.
Silica	63.19+	61.63+
Alumina	14.04+	14.24+
Organic Matter	8.55+	6.82+
Peroxide of Iron	4.87+	6.23+
Potash	2.80+	2.17+
Water	2.70	4.57
Soda	1.43	1.04
Lime	0.83	1.27
Phosphoric Acid	0.24	0.26
Sulphuric Acid	0.09	0.03
Carbonic Acid	0.05	....
Chlorine	0.009	0.02
	100.00	100.00

"It is a natural inference to expect in unproductive or barren soils a deficiency or total absence of one or more of those constituents which are highly conducive to the luxuriant growth of plants. A chemical examination in such cases must prove of utility to the practical man, inasmuch as it not only is calculated to point out the cause of infertility, but also to suggest an efficient means to raise its productive powers. In many other cases, in the majority of instances, the barrenness cannot be traced to the deficiency or total absence of an important soil-constituent nor to the existence in the soil of a substance injurious to vegetation. The fault may be one, not of the existence, but of the accessibility of the requisite ingredients for the crop. All the substances needed by the plant may be present, and in sufficient quantity; the soil, considered as a storehouse, may be full; and the infertility complained of may simply be the want of the key.

"This is the case of a soil locked up in stagnant water, which only needs drainage to prove the fertility which one would expect from its analysis. But independently of this, as a general rule, even a minute chemical analysis, in which only the proportions of the several constituents are indicated, is of comparatively little, and often of no practical utility to the individual who has had a reproductive soil analyzed, with the view to have a remedy suggested by the analytical data for bringing it into a better state of cultivation. . . .

"A point of great practical importance is the state of division in which the constituent parts of soils are mixed together; and as a chemical analysis gives no information in this respect, the necessity for submitting the soil to a mechanical examination becomes apparent.

" Such an examination enables us to ascertain whether its mechanical condition is such as to render its cultivation economical or expensive, and at the same time llows us to recognize the nature of the stones which are found in the soil. An acquaintance with the composition of the stones affords a good criterion as to its probable state of productiveness, and in many cases suggests the propriety of leaving the stones on the land or of removing them.

" The property of absorbing water, either in the form of vapor or in the state of dew from the atmosphere, has a material influence upon the productive characters of soils, and contributes to explain the superiority of one soil over another. Intimately connected with the preceding property is the power of soils absorbing fertilizing gases from the atmosphere. Generally speaking, those soils which absorb a larger amount of moisture from the air than others are also the better absorbers for carbonic acid and ammonia.

" This property, though dependent in a great measure on the porosity or the state of division of the various constituent parts of the soil, is still more intimately connected with its chemical constitution."

RIGHT SEEDS, PLANTS, AND ANIMALS.

Under this second point of good farming might be arranged the discussion of the raising and improvement in quality of all kinds of seeds and all the grasses and clovers and the root crops used upon dairy farms. But the scope of this work will hardly admit of such an extended discussion. It is well, however, to advise that, as far as practicable, farmers raise their own, and patronize those dealers only who have a reputation for careful selection as to purity and quality of all kinds of seed.

In regard to the selection of stock for the dairy farm, no one who has taken pains to inform himself in regard to the merits of the various races could hesitate to give his choice to the Jersey as pre-eminently the best of all breeds of dairy cattle.

QUANTITY OF SEED REQUIRED TO PLANT AN ACRE.

Barley, in drills	1	bushel.
Barley, broadcast	2½	bushels.
Beet, in drills 2½ feet	9	pounds.
Cabbage, sown in frames	4	ounces.
Carrot, in drills 2½ feet	4	pounds.
Clover, Lucerne (Alfalfa)	10	"
Clover, Alsike	6	"
Clover, Large Red	16	"
Clover, Large Red, with Timothy	10	"

Corn, Sweet	10	quarts.
Corn, Field	8	"
Grass, Timothy, with Large Red Clover	8	"
Grass, Orchard	64	"
Grass, Italian Rye	20	"
Grass, Mixed (twenty varieties with Clover)	40	"
Mangold, in drills 2½ feet	9	pounds.
Millet, broadcast	50	"
Oats, in drills	¾	bushel.
Oats, broadcast	1½	bushels.
Parsnip, in drills 2½ feet	5	pounds.
Peas, in drills	2	bushels.
Rye, in drills	1	bushel.
Rye, broadcast	1½	bushels.
Turnips, in drills 2 feet	3	pounds.
Wheat, in drills (*best conditions*)	¾	bushel.
Wheat, broadcast	1½	bushels.

THE BARN.

The barn is a storehouse for fodder, and should never be used for a stable. The contamination of sweet hay, grain, and roots by the putrescible exhalations and vapors of a stable is neither conducive to the health of the animals nor compatible with the highest excellence of quality for the butter and cream.

In a barn that costs $1000 a man can keep $50,000 worth of Jerseys; but if from any cause the building takes fire, he is sure to lose his herd. The risk is too great.

The most economical form of barn is the octagon. A fifty-foot octagon, suitable for a fifty-acre farm and a storage of fodder sufficient for fifty head of Jerseys, can be built at a cost of about $700. The barn may be of lumber, but preferably a concrete wall, with a lumber framed roof covered with slate. The walls may be twelve to eighteen inches thick and twenty-four to twenty-eight feet high; the rafters thirty-four feet long; the roof lighted by a cupola. If the barn can be located in a side hill of sufficient height, a bridge or an earth driveway can be constructed so as to drive in at a gable door and dump all loads from a floor resting on the top of the walls, thus saving a vast amount of labor in unloading hay, grain, and roots. Where this plan is impracticable, the next best thing for hay is the horse-fork, which has a free swing in such a barn, there being no cross-ties or beams to obstruct its working.

The walls should be made of water-lime cement, sand, gravel, and small fragments of broken stone.

ELEVATION FOR OCTAGON BARN, WITH SIDE-HILL DRIVEWAY AND GABLE ENTRANCE.

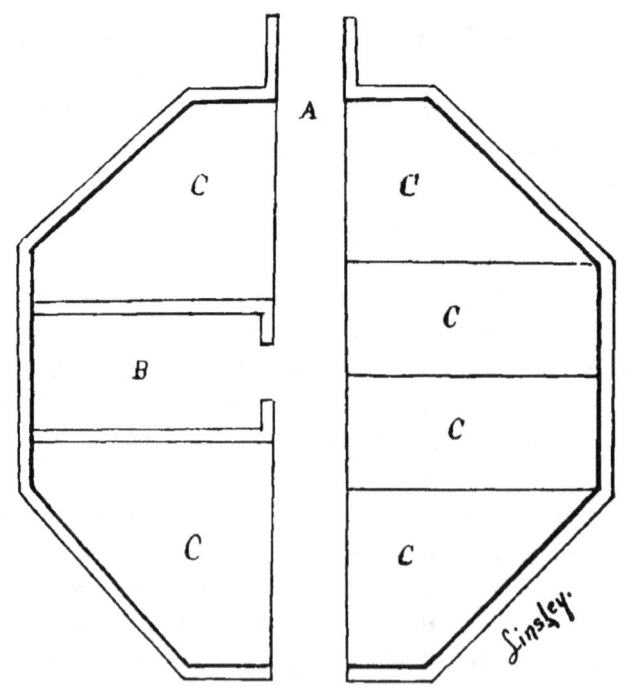

GROUND PLAN OF OCTAGON BARN.

A, Driveway.   *B,* Root Cellar.   *C, C, C,* Compartments for Hay and Corn Stover.
Upper story may be used for storing grain crops.

FORMULA FOR CONCRETE.

Sharp sand, 4 parts ; water-lime, 1 part. *Mix thoroughly before wetting.* Mix very wet; add gravel, 4 parts ; mix very wet, and work over thoroughly four times. Add small broken stone, 4 parts.

Put into the box or form a layer of an inch of the mortar and then a layer of stone, always taking care to have the stone in the centre and a layer of mortar making fully two inches of the outer portion of the wall. The mortar should be tamped in, so as to make it solid. Let it dry forty-eight hours for each tier of one foot in height. If care is observed the building will be better in quality than stone or brick, as it makes a very dry wall. Sills placed on the top of such a concrete wall are liable to rot from being coated with lime. This can be prevented by spreading a layer of gas-tar or asphalt on the top of the wall. No moisture should be allowed to come in contact with a concrete wall until it has become hard ; then it will be water-tight. There should be a drain cut lower than the foundation wall to carry off any water that might come against it from the hill-side. Fill in the space above the drain, which should be of good pipe, with small stone as high as the bank in which the excavation is made.

The boxes or forms for a wall one foot thick should be made of plank fourteen inches wide, one and a half inches thick, and of the right length. The standards or posts may be three by four scantling a little exceeding the height of the wall. These posts are set fifteen inches apart, with the planks on the inside. The standards are held in place by nailing thin pieces of board across. These remain in the wall. The planks on the outer side of the octagon must of course be longer than the inner by the thickness of the wall. The boxes need a clamp to prevent their springing between the standards, and it is well to have the plank lined with tin or zinc to prevent their becoming flexible from the excess of moisture while the wall is drying. The clamps may be made of hard wood two feet long, with a two-inch hole at each end, and fifteen inches apart. A strong pin two feet long is set in each hole so as to protrude ten inches, and these pins will just fit over the outside of the plank box, and a brace driven between the upper ends will make them clasp the box. Two or more of these are needed for each form. Door and window frames have jambs the width of the wall's thickness, and must be put in place at the proper time and plumbed the same as the standards. The usual cost of the concrete wall is about ten cents a cubic foot. It will be a little more when the walls are high. Walls twenty-four feet high give a capacity of hay storage eighty per cent. greater than when but sixteen feet, because of the closer packing of the deeper mow. The floor of such a barn may be of concrete and laid directly upon the earth. The root bins should be walled in with concrete partitions on the hill-side of the basement. The grain bins are to be placed in the attic. From these the grain is drawn down through a cloth spout into bags or barrels, as needed.

ELEVATION FOR SIDE-HILL BARN, WITH DRIVEWAY ENTERING AT GABLE.

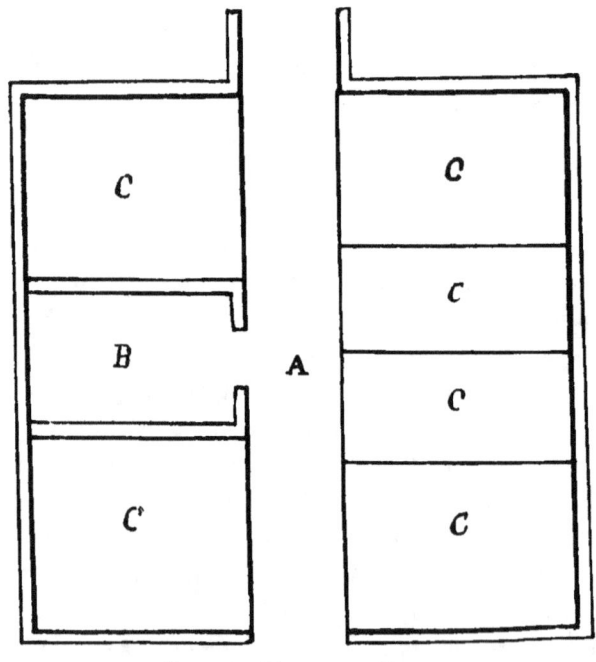

GROUND PLAN OF BARN.

A, Driveway.　　　B, Root Cellar.　　　C, C, C, Compartments for Hay and Corn Stover.
Upper story used for grain crops.

The corn crib, for storing maize in the ear, is best made of slats with spreading top, and set upon posts capped with flaring tin pans as a guard against rats.

The STABLE is the most important of farm buildings, and needs as much care and forethought in planning as the dwelling-house of the owner. The stable is the home of the Jerseys, and should be devoted solely to their comfort and health. Neither horses, sheep, swine, fowls, nor dogs should be allowed to occupy the same building with Jerseys. The reasons for such exclusiveness should be obvious to every butter-maker, as the object of the true breeder is not only to develop the Jerseys to the highest perfection, but to produce the best quality of dairy products.

The requisites for a Jersey stable are : (1) a fire-proof building of the parallel form, built of material that shall render it cool in summer and warm in winter, and free from dampness or frost on the floor and wall ; (2) to afford a full supply of sunlight ; (3) perfect ventilation ; (4) convenient facilities for feeding and watering ; (5) comfortable stalls and fastenings ; (6) the best means of cleanliness ; (7) the manure storage to be conducted in a separate structure.

As far as walls are needed for a stable, water-lime concrete, made according to the formula for the barn, is the best material. Such a wall has a peculiar porosity of an infinite number of very minute air cells, which render it almost the equivalent of a double wall of brick or stone as a non-conductor of heat and moisture. The wall must be so built as to have a stratum of the concrete mortar on the external and internal surfaces of about two inches thickness, and the centre of the wall a stratum of broken stone mixed with the water-lime cement and mortar, the whole to be thoroughly tamped and well dried in each successive tier of building. The roof should be of slate. The stable needs a dry floor. This should be made of concrete. First, a film of coal-tar upon the levelled hard earth ; second, a layer of soft mortar and gravel, which, after drying forty-eight hours, may be topped with a layer of several inches (four to six) of small fragments of broken stone ; third, a layer of gravel or sand which must be thoroughly rolled and worked into the stones ; fourth, a layer of concrete mortar three inches deep, well tamped and left forty-eight hours to dry. The ground must first be marked out according to the plan of the stable, making excavations for the water troughs in front of the cattle and the manure gutters behind the platforms.

As the cattle may be allowed to drink frequently, the gutter beneath the manger need be but shallow—five or six inches deep—with fall sufficient to empty when plug is removed at the lower end. The manure gutters are to be cleaned three times daily, and may be thirty inches wide by six inches deep, with sufficient fall to discharge the urine. This will be treated of under "Cleanliness."

SUNLIGHT.

The direction of the building, if from north to south, is best, as the windows then receive the early rays of the sun; those who prefer the noonday sun can have an east and west plan; but any arrangement whereby a large amount of sunshine may fall upon the cattle is desirable, and that which provides most is best. One large window for each animal, or at the least for every two animals, is a plan to be commended—windows that come within a foot of the floor and the same distance from the top of a ten-foot wall. The windows should also be double, with an air-space of six to eight inches between, thus saving warmth in winter and rendering the stable cool in summer. But the sunlight is as needful to the thrift of animals as it is to plant life, and the breeder who gives his cattle sun baths in winter will soon learn its vitalizing effects.

VENTILATION.

As in the human dwelling, so the stable should have a perfect system of ventilation; for cattle have the same lung diseases or a similar loss of vitality when deprived of oxygen as the human race. The air should not only be kept as near a state of purity as possible, but as far as practicable at a healthful temperature. Cattle thrive well and make their best growth at a temperature of about 60° Fahrenheit. "A fight with flies and poverty" at 90° or with foul air and poverty at 10° below zero may suit the fancy of some theorists who hold that "roughing it" is the correct system of disciplining cattle into endurance; but humane treatment only will be found profitable with Jerseys or any other breed of cattle.

STABLE.

a, Air-Receiver. *b,* Ventilator.

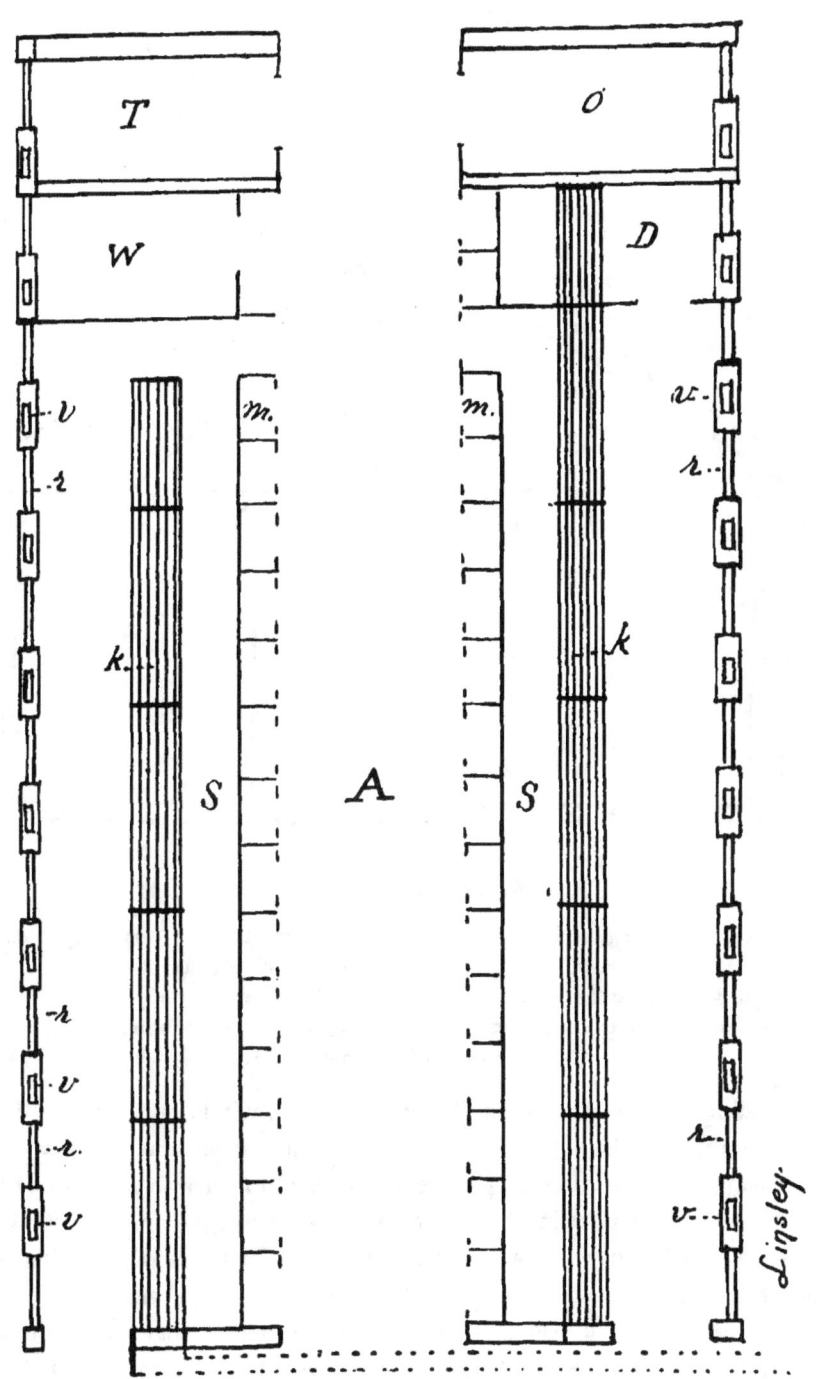

GROUND PLAN OF STABLE.

T, Tempering Room.   *W,* Wash Room.   *O,* Office.

D, Bull Stalls.   *S,* Cow Stalls.   *k,* Grating over Gutters.

A, Feeding-way between cattle which may have railway-track for feeding-car.

m, Mangers.   *v,* Ventilating Flues.   *r,* Windows.

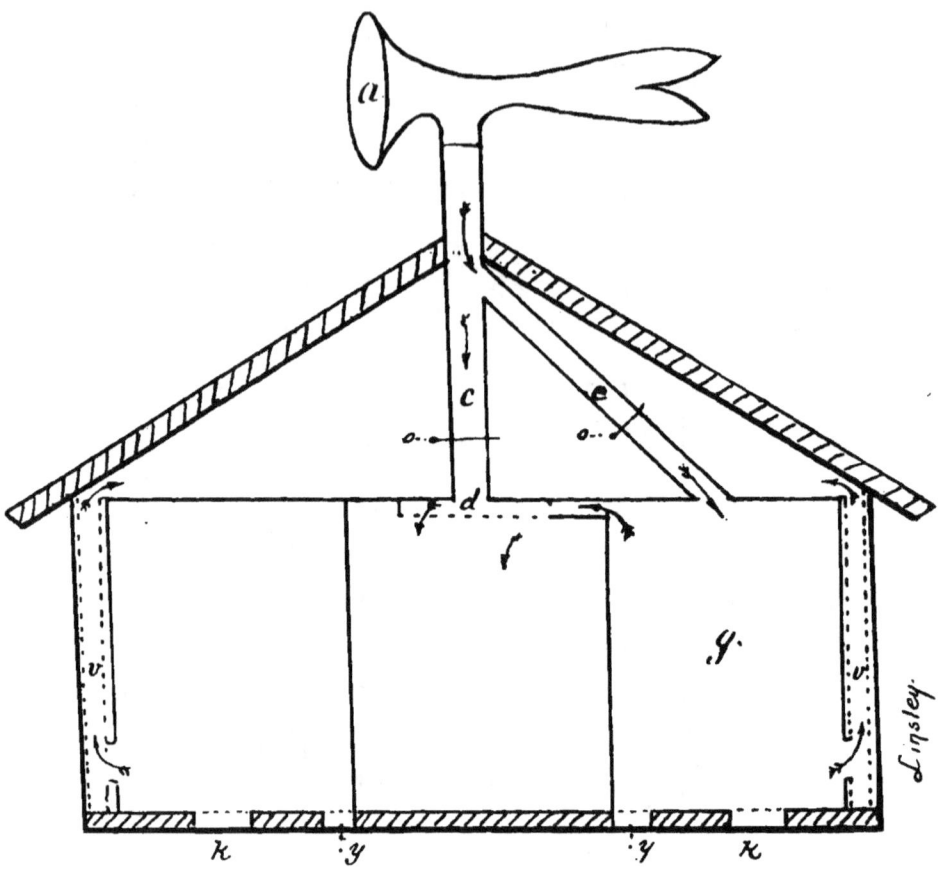

SYSTEM OF VENTILATION.

a, Supply Funnel facing the wind.	*g*, Warm Air from Tempering Room.
c, Supply Pipe for Summer.	*v*, Foul-air Exhaust Flues.
d, Air Box perforated at the bottom.	*o*, Dampers.
e, Winter Supply Pipe to Tempering Room.	*y*, Channel for Drinking-Water.
k, Channel for Manure Gutter.	

Buildings need means of ventilation other than windows or doors. Least of all can we count upon the natural ventilation resulting from the porosity of building materials. There should be ample provision made to furnish a supply of fresh cool air sufficient for comfort in the heat of August, or to keep up a constant and pure flow at 55° to 60° when the outside temperature is far below the freezing-point. We must provide for the removal of the exhalations of lungs and skin and the evaporated particles from dung and urine; for all putrescible matters that are dangerous sources of tubercular lung disease, or that in any way tend to lower animal vitality.

The fresh air is best introduced from above and in the faces of the animals. The exhaust or outward flow should always be from the base of the wall behind the cattle. One large flue, with an opening a foot square, between each pair of windows will conduct the foul air into a ventilating loft in the attic. Large ventilators on the ridge disperse the current outwardly. The outflow may be enforced by heat or regulated by an air-tight blower. The thermometer is essential to show the temperature of the stable, and the plan should be to pass a river of pure air perpetually through the stable as near the temperature of 65° as it is practicable to produce. The pure air must not be shut out because of pinching cold, but some artificial means of heat must temper the air before it is introduced. During the severe winter weather the most convenient method of tempering would be the "ventilating stove," which combines all the elements of stove, furnace, and open fireplace. Where a steam-engine is used the heat may be utilized. The subterranean system of the deep earth duct may be found of great advantage in some large herds. By referring to plate the elements of ventilation are illustrated. The air is best introduced by a long duct at the top of the stable between the two rows of cattle. This duct may be of pine smoothly planed within or of galvanized iron, the bottom of the duct to be closely perforated with half-inch holes. Such a system will not only prevent tubercular disease, but insure normal health and full constitutional vigor.

THE WATER SUPPLY.

Very fortunate is the farmer who can turn the water of a pure spring from the hillside into his dwelling and stable. Cows require more water than any other stock. They drink enormous quantities when in a full flow of milk. The experiments of Prof. Horsfall and of M. Dancel illustrate the necessity of an abundant supply. The former "found that cows, when giving only twenty pounds of milk per day, drank forty pounds of water more than fattening cattle of the same weight." The latter says that " by inducing cows to drink more water, the quantity of milk yielded by them can be increased many quarts, without injuring the quality." By moistening their fodder and adding a little salt the milk was increased from nine and twelve quarts on dry fodder to twelve and fourteen quarts daily. The amount a cow drinks is a criterion of her milking powers, a cow that drinks fifty quarts of water daily giving eighteen to twenty-three quarts of milk. The water should be pure and about 65° in winter and summer. It should run through a gutter in front of the stalls. The gutter is to be covered by a hinged lid, which forms the floor of the manger, when closed. Some breeders prefer a trough which is raised or lowered like a dumb-waiter. Water should always be pure and perpetually supplied in the manger gutter, winter and summer.

CAPACITY OF TANKS AND CISTERNS.

Two feet in diameter and ten inches deep holds.................... 19 gallons.

Three feet in diameter and ten inches deep holds.................. 44 "

Four feet in diameter and ten inches deep holds .... 78 gallons.

Five feet in diameter and ten inches deep holds................ ... 122   "

Six feet in diameter and ten inches deep holds.................... 176   "

Seven feet in diameter and ten inches deep holds.................. 239   "

Eight feet in diameter and ten inches deep holds.................. 313   "

Nine feet in diameter and ten inches deep holds.................. 396   "

Ten feet in diameter and ten inches deep holds..................... 489   "

Eleven feet in diameter and ten inches deep holds... 592   "

Twelve feet in diameter and ten inches deep holds.................. 705   "

Thirteen feet in diameter and ten inches deep holds............... 827   "

Fourteen feet in diameter and ten inches deep holds............... 959   "

Fifteen feet in diameter and ten inches deep holds...............1101   "

Twenty feet in diameter and ten inches deep holds................1958   "

Repeat the quantity for each ten inches in depth.

A good cow requires from twelve to fifteen gallons of water daily.

A herd of fifty good Jerseys require seven hundred and fifty gallons daily.

A tank six feet in diameter and five feet deep holds ten hundred and fifty-six gallons—an ample supply for fifty cows, if it is kept filled or replenished each day.

THE STALLS.

It is desirable to have as little wood in the stable fittings as possible, and *no paint on any part of the interior of the stable.* The cattle stand in two rows, facing inward. The concrete floor, covered with suitable bedding, makes a good arrangement for a stable bottom. This floor must be level, from front to rear, and have the same slope and incline as the gutter. The gutter ought to be covered with an iron grating (Stewart's). The part of the stalls between the manger and gutter may be three feet six inches wide. The gutter may be from thirty to thirty-six inches wide. The grating consists of flat wrought-iron bars three eighths by one and one half inches, riveted to an iron frame, and hinged so as to be turned up when cleaning the stable. The cattle stand with the fore-feet on the bedding, and the hind-feet reach the first and second or third and fourth bars, so that all the dung and urine fall into the gutter. The space allowed for each cow should be about three feet six inches in width. A pavement of brick saturated with boiling asphalt is an excellent stable flooring.

CLEANLINESS IN THE STABLE.

A first-class breeding establishment should always be in a condition for visitors to see, especially in regard to cleanliness, which is essential to the health of the animals, the purity of dairy products, and the morality of the workmen. The stable should be cleaned regularly three or more times a day. The attempts to store manure in a cellar beneath the stable or to deodorize it in a deep gutter are not commendable. The receptacle for the manure should be a separate building devoted to the collection of all the excrement from all the stables, cattle, horses, sheep, swine, and fowl houses ; to the drainage of the dwelling, including the contents of the water-closets, washtubs, sinks, and the kitchen garbage. This manure factory may be a large concrete water-tight reservoir, roofed over to keep out rain and flies. This vat will require a great quantity of water from the roofs of buildings or a reservoir, and in winter if there is not rain enough snow must be applied, so that the contents may be kept in a condition of moderate fermentation, and it can be applied as needed to fields and crops by the manure-spreader. Some might make a step in advance and liquefy the whole mass, to be applied to the soil by a sprinkling cart or by irrigating pipes.

BEDDING.

All bedding should be short. Marsh grass, salt hay, straw, the refuse hay and corn stover from the mangers should be run through the cutter to a length of one or two inches. This may be used alone or mixed with "peat moss," and being kept clean by the Stewart grating, will last a long time, and make a comfortable bed. Cocoa matting has been used for bedding cows. The peat moss of commerce is excellent bedding.

THE FASTENINGS.

Some form of stanchion may be used if one wishes the advantage of cleanliness to be secured. There is a rotary stanchion, which promises to be just what is needed. Whether the best contrivance for fastening has yet been devised remains to be proven.

MANGERS.

The hinged lid over the water gutter forms, when closed, the floor of the manger. All the mangers require side partitions to keep each cow's mess isolated, but the front may be open. This renders feeding from a car convenient as it is moved through the stable, and the manger is conveniently kept clean by being brushed out daily.

CHEAP STABLING FOR COWS.

"Lay out, for twenty-five cows, a space one hundred feet long by fourteen feet wide. Set cedar or chestnut posts, six feet apart, nine feet high for the front, and seven feet high for the rear. Set a row of posts four feet high, four feet apart, and four feet from the rear row. Board up with twelve-foot hemlock boards laid horizontally all these three rows. Close in the ends. Put on rafters spiked to the posts, so that the roof boards will fit quite close to the plates. A 2×4 scantling nailed to the top board and spiked to the posts will make a sufficient plate.   Lay the roof boards of sixteen-foot hemlock from front to rear ; the roof will have two feet slope.   Cover the roof joints with three-inch strips well nailed.

"The roof boards rest upon three boards nailed to the rafters three and one half feet apart.   If strong boards are selected the roof will be firm.

"Make a feed trough along the inside of the inner partition two and one half feet from the ground ; leave out one board—the third ; hinge this to the lower board and with cords, so as to make a falling door at an angle of forty-five degrees for a 'shoot' to the feed trough.   Fasten the cow to the post by a strap, or use a stanchion.   Give each cow five feet of space, and make a plank gutter fourteen inches wide, leaving three feet space behind.   Make a concrete floor at a cost of fifty cents per cow.   Cost of stable, $125."—*N. J. H., New York Tribune.*

THE CALF STALLS.

In the plan for stable shown herewith the attic or second story is appropriated as the most suitable place for the keeping and rearing of calves.

A convenient arrangement is to have two rows of box stalls or pens, each stall four by eight feet, to be occupied by a single calf, as this prevents annoyance of sucking each other.

A passage-way between the rows of stalls serves for a cart to carry the milk for feeding.   The stalls must be well lighted ; indeed, a glass house would be the best for calves in this respect.

Upon the front of each stall, beneath a feeding door, fix a band of hoop iron of size and shape to hold the pail securely while the calf is drinking, or, better still, place the sucking feeder within the stall.   When the milk has been warmed to the temperature of 102° by the thermometer, add the requisite quantity of prepared rennet, and set the pails into the receptacles in front of the stalls, and open the feeding doors.   These doors may be nine by thirteen inches in size, and should swing so as to clear the top of the feeding pail.   The calves readily learn to drink from a feeding pail thus placed, but one can use the sucking apparatus instead of the pail if he prefers that method.

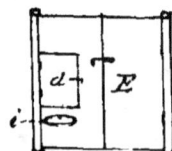

FRONT OF CALF STALL.

E, Door.   *d,* Feeding Door.   *i,* Ring for Pail.

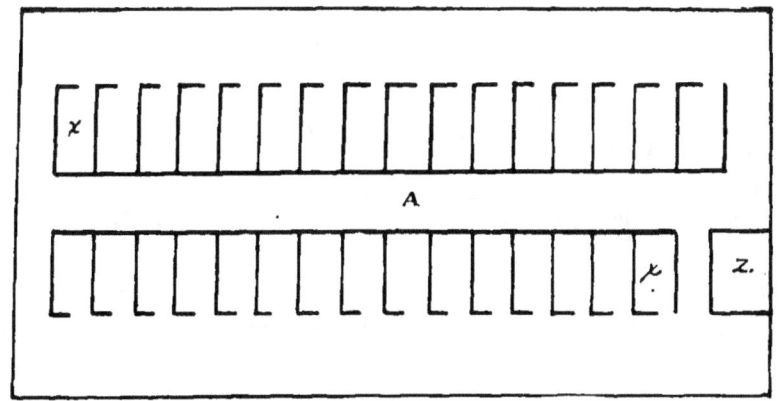

ATTIC-FLOOR CALF STABLE.

A, Passage-way.   *r,* Stalls.   *z,* Elevator.

SELECTION OF IMPLEMENTS.

An important element in the education of every farmer is the cultivation of the faculty for judging and selecting labor-saving tools and machines.   Probably there is no department of agriculture upon which so much of the final success depends as the right selection and use of implements.   When one instrument will do double the work of another of the same cost it is a matter of economy to know it and make it available.   The farmer needs the best of everything, and always the peculiar implement suited to his own special circumstances.

The plow should have all the qualities of the best invention, the lightest draught, the best material, and also be specially suited to the quality of soil and the situation of the land.

The harrow should combine the qualities of pulverizer, smoother, and cultivator.

The drill should be adapted to all kinds of seed, and perfect in its mechanism.

Cultivators should be adjustable for smooth, shallow, surface pulverization, or for the needs of special crops.

LIST OF APPARATUS FOR A DAIRY FARM OF ONE HUNDRED ACRES.

One sulky plow.

One swivel plow.

One iron frame steel-tooth harrow; pulverizer, smoother, and cultivator.

One two-horse drill, with force-feed grass seeder.

One two-wheel cultivator, for root crops.

Six steel-prong hoes.

One twelve-foot poly-section roller.

One mowing machine.

One wheel horse rake.

One reaper and binder.

One hay tedder.

One thresher.

One fan mill.

One corn sheller.

One improved grinding mill.

One hay cutter.

One cutter and crusher for corn stover.

One root cutter (Clark's).

One bone mill.

One motor—two to four horse-power, with attachments suitable for all machines to be used.

One hay loader.

One power hay fork.

One dumping hay cart.

One hundred hay caps.

One root cart.

Three steel hay forks.

Three manure forks.

Three shovels.

One post-hole digger.

One power manure lifter.

One manure cart or spreader.

One sprinkling cart for liquid manure.

LIST OF APPARATUS FOR BUTTER DAIRY USING THE CREAM OF FIFTY COWS.

One three-can Stoddard creamery for testing cows.

One largest size dairy creamery, Stoddard.

Six Perfect Milk Pails.

One No. 1 Stoddard churn for testing cows.

One No. 6 Stoddard churn.

One centrifugal butter worker.

One lever butter worker.

Two fifty-gallon cream tempering vats.

One weighing scale.

One butter salting scale.

Two dairy pails.

One half-gallon dipper.

Two butter ladles.

Two dairy thermometers—eight-inch nickel.

One cream strainer.

One buttermilk strainer.

One self-gauging butter printer.

One butter tray

LIST OF APPARATUS FOR CHEESE DAIRY USING THE WHOLE MILK OF FIFTY COWS.

Six Perfect Milk Pails.

One two-hundred-gallon self-heating vat.

One eight-blade metallic-head curd knife, perpendicular.

One six-inch by twenty-inch curd knife, horizontal.

One curd scoop.

One curd pail.

One weighing scale.

One dairy thermometer.

One whey strainer.

One syphon.

One gallon dipper.

One curd mill, rotary disk with cutting blades.

Two moulding presses for ten three-pound cheeses.

Tin-foil for cheese wrappers.

FARM IMPLEMENTS OF SPECIAL MERIT.

As an illustration of the points of excellence in farm implements, the author has deemed it expedient to show several inventions of special merit by the following series of cuts and a mention of their salient points of superiority and utility.

CHAMPION GRAIN AND FERTILIZER DRILL.*

* Gere, Truman, Platt & Co., Owego, N. Y.

DEVICE FOR CHANGING FEED.   SPRING HOE.

Points of Excellence.

1. It has force-feed grain distributers.
2. It has force-feed grass-seed distributers.
3. It has force-feed fertilizer distributers.
4. You can sow grass seed equally well in front or behind the hoes.
5. It has a special device for dropping and fertilizing corn.
6. It has a cold-rolled steel axle, which has three times the strength of the iron axle.
7. It is well balanced and of light draught.
8. Its frame is braced with heavy castings at the four corners.
9. By a very simple device, change of feed is made by changing speed of distributers, but no loose or detached pinions are used.

CLARK'S ROOT CUTTER.

10. It is easily handled by the team and operator.

11. It can easily be set to sow accurately any desired quantity of grain, grass seed and fertilizers.

12. It is made of the best materials and the best workmanship.

CLARK'S ROOT CUTTER.*

(THREE SIZES.)

Description.

This cutter is built with a heavy oak frame, well bolted together; is stanch and strong, neatly finished and handsomely ornamented.

The cutting apparatus consists of a cylinder of steel knives, shaped like a chisel gouge, so arranged on a wrought-iron shaft that they are perfectly secure; no chance of becoming loose or breaking.

Operation.

It works with rapidity and ease, cutting the roots into thin, narrow pieces, which are thoroughly crushed and fitted for easy mastication by the animal. Nos. 1 and 2 are for hand use; No. 3 for power.

Points of Excellence.

It is compact and portable.

It is strong and durable.

It is effective and uniform in its work.

It is worked with ease, a boy cutting forty bushels, and the power cutter one hundred bushels an hour.

It facilitates mastication and digestion, and promotes health.

It precludes all danger from choking.

It will reduce apples, beets, carrots, mangolds, pumpkins, parsnips, and turnips to the proper condition for use.

It is an indispensable apparatus in every stable for economy of labor and the safety and health of the animals.

---

* Manufactured for R. H. Allen Company, 189 and 191 Water Street, New York.

DAIRY IMPLEMENTS OF SPECIAL MERIT.

THE PERFECT MILK PAIL.*

Description.

The pail is made of the best tin plate, and will bear a weight of three hundred pounds; holds fourteen quarts; has a concave lid; a broad funnel upon the spout; a rubber tube renders the spout flexible, and there is a strainer at the lower end of the spout.

Points of Excellence.

It prevents the entrance of dirt or dandruff, and excludes foul air.

It forms an easy seat for the milker.

It enables the milker to do rapid work.

It strains the milk.

Its funnel is adjustable to low or high cows.

It can be used without the rubber tube.

It is very durable.

It secures clean milk, sweet cream, and better butter and cheese than can be obtained without it.

It is indispensable for comfort, cleanliness, and consummate quality of product in every dairy.

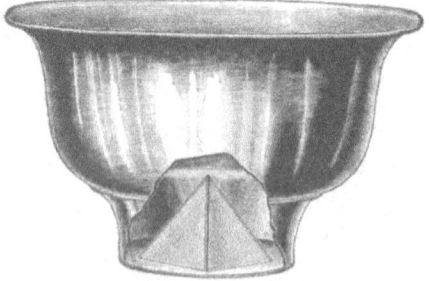

MORE'S PYRAMIDAL STRAINER.

* R. H. Allen Company, 189 and 191 Water Street, New York, General Agents.

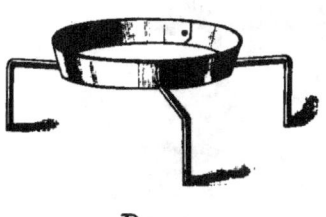

REST.

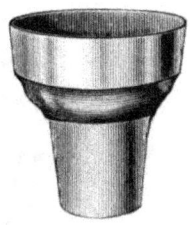

FUNNEL.

MORE'S PYRAMIDAL STRAINER.*

Points of Excellence.

It is durable, being stamped from heavy tin and retinned.

It cleans milk rapidly, and will not clog when regularly cleaned at using.

It gives a large straining surface.

It uses finer cloth than a flat strainer.

Its form allows the milk to fall on its apex, and the sediment settles at the base.

It is a perfect milk strainer, and can be used with the rest and funnel upon any sized pan or can.

THE STODDARD CREAMERY AND REFRIGERATOR,† WITH PATENT SKIMMING ATTACHMENT.

(ELEVEN SIZES.)

Points of Excellence.

The cream is drawn off the milk through an adjustable tube passing down through the milk and bottom of the can through the faucet. The milk is afterward drawn through the faucet.

No watching for cream line or cream flakes. No cream wasted.

There is no sediment drawn with the cream, as is the case when the milk is drawn from *under* the cream and the cream afterward drawn out or poured from the can. When the milk is drawn from *under* the cream the sediment is not drawn out with the milk, or but a small part of it, but runs out when the last of the contents of the can is discharged, which is the cream.

The skimming is done quicker than by any other method, which is a great advantage in the cream-gathering system.

Milk or cream can be drawn out at any time.

---

* Moseley & Stoddard Manufacturing Co., Poultney, Vt.

† Moseley & Stoddard Manufacturing Co., Poultney, Vt., or their agents.

THE STODDARD CREAMERY AND REFRIGERATOR, WITH PATENT SKIMMING ATTACHMENT.

Ventilation of milk, removing animal odors, and saving of ice.

All faucets are brass, nickel-plated, with ground joints.

Cans are easily removed if necessary for any purpose, and are interchangeable.

The construction of tank affords space for large pieces of ice.

It has a perfect refrigerator.

The walls of the creamery are thick and built refrigerator style; has lining of two thicknesses of heavy paper and *double air space*, thus effecting a great saving of ice.

No other creamery so thoroughly constructed in this respect.

It is made in two styles, with and without the separate refrigerator compartment. The refrigerator is built in one end of the creamery, is lined with zinc, and has slate shelves. It has no connection with the milk receptacle. The door to refrigerator is in the end of creamery. It is *dry* and *very cool* when ice is used for cooling the milk.

Can be used with running water at a temperature of 55° F.

It is an indispensable convenience for the purpose of testing butter cows.

It is the most economical in price, considering material, workmanship, convenience, and utility.

It is fully protected by letters patent.

THE STODDARD CHURN.*

Points of Excellence.

Its form gives thorough mixture of cream, with degree of concussion desired.

There are no dashers or floats to injure grain of butter.

Simple in construction and very durable.

It is especially adapted to the granular system of making butter; glass indicator shows when to stop the churn.

Being air-tight, it never leaks.

It is readily ventilated, allowing gas to escape.

Butter can be rinsed in the churn and thoroughly drained.

It can be used with any regulated motor.

It makes the best quality of butter.

It is an indispensable convenience in making butter tests of cows.

It is fully protected by letters patent.

* Moseley & Stoddard Manufacturing Co., Poultney, Vt., or their agents.

DRAINAGE OF LAND.[*]

" There is no subject which of late years has attracted more attention or excited more discussion among those interested in the cultivation of the soil than the drainage of land. It requires no great research to discover the reasons for the interest which this subject has attracted to itself. Whether we look at the vast amount of capital annually expended on it or at the great improvement in the agriculture of the country which it is gradually effecting, we cannot fail to perceive it to be a subject of the greatest importance. Were any extraneous evidence of its importance required, it might be found in the recent vote of Parliament, by which the enormous sum of two millions sterling was set apart to be loaned out for its encouragement and extension ; and in the avidity with which the whole of that large sum was applied for and absorbed in the course of a few months.

" Notwithstanding that the benefits to be derived from draining are now so well known and appreciated—so much so, indeed, that most agriculturists, if asked what they considered the first requisite toward good farming, would, without hesitation, answer, thorough drainage of the land—still, the careful observer, casting his eye over the surface of the United Kingdom, cannot fail to be struck with surprise at *the vast extent of available surface* which is rendered partially or absolutely valueless to the community by the presence of an excess of water.

" The great extent of land which is still undrained excites the more surprise when we reflect that experience has shown that even very unpromising portions yield large and remunerating returns for the outlay.

" The beneficial effects which result from complete drainage of land may be classed under two heads — mechanical and chemical. The mechanical division includes the improved efficiency of all those laborious operations carried on for the purpose of pulverizing and cleaning the soil, such as plowing, harrowing, and weeding. It also includes the saving in time and labor in carrying out the general business of the agriculturist, as well as the saving of that portion of seed which is destroyed in wet soil from mechanical causes.

" The chemical division is a copious one, and embraces more than our philosophy even dreamed of twenty years ago. It includes all that great class of phenomena relating to the improved fertilizing powers of manures and alteratives, as we may, in certain cases, denominate lime, marl, clay, etc.; the improvement of climate ; the raising of the temperature of the soil ; the acceleration of the period of the harvest ; the decomposition of substances in the soil injurious to vegetation ; the improvement in the nutritive value of herbage, and other crops ; and, in consequence of all these, improved races of animals, including even man himself.

---

[*] Encyclopædia of Agriculture, by J. C. Morton, Edinburgh. Extract from article by John Girdwood.

MECHANICAL ADVANTAGES.

"Let us first, then, consider the mechanical advantages. Every one at all acquainted with the conduct of agricultural operations must be aware of the great difficulties which a wet state of the soil throws in the way of performing these operations with propriety, despatch, or economy of labor. The great object of all the operations of tillage is, along with the removal of weeds, to reduce the soil to a finely divided state, through every part of which the fine filamentary roots of plants may spread themselves, in order to obtain supplies not only of moisture and air, but of those substances of which they are partly composed, and the due preparation of which is one of the most important functions of all mechanical operations of the soil.

"The tempering of mortar or clay affords a very apt simile for any operations undertaken on land in a wet state, and furnishes a very true analogy as to the results. It will, therefore, be evident that, so far from furthering the object in view, plowing, or other working of land when wet, will have the directly contrary effect of rendering it more stiff and close; and instead of producing a finely divided and porous state of the soil, so indispensable to the healthy and vigorous growth of crops, will leave it, when dry, a hardened mass, in which useful plants will find it difficult to obtain even the most scanty subsistence.

"In such a climate as that of Britain, where there is generally a great deal of rain and very little evaporation during the greater portion of the period in which the preparation of the soil must go on, and where wet, undrained land, once thoroughly moistened, hardly dries until the searching breezes of spring begin to act upon it, it is a matter of no small difficulty to find a season when operations may be carried on with propriety upon the land.

"In order to meet this difficulty, it is the custom on wet land farms to maintain an extra force, both of men and horses, in order to seize such favorable opportunities for working the land as may present themselves; to take advantage of a good 'tid,' as such an opportunity is sometimes called; and to complete, within a few weeks of early autumn and late spring, those operations which the cultivator of dry or drained land may carry on at his convenience during the greater part of winter. The latter is enabled to effect the tillage of his land in a careful manner at absolutely less cost than that for which it can be slurred over in the most imperfect way by his less fortunate or less improving neighbor, who, notwithstanding all the haste he can make, is frequently 'caught out,' and compelled to leave unsown, fields which have been prepared and manured, and to substitute, at a more propitious season, some less valuable crop for that which he intended.

"It is not in the operations of tillage alone that extra labor is demanded from men and horses on wet land; the carting on of manure, the carting off of produce— in fact, all operations whatever carried on upon its surface are alike impeded.

"The saving effected by drainage in the number of horses required on a farm has been variously computed by different authorities; but it seems to be a very reasonable calculation to estimate at one in four, or twenty-five per cent.; while even with the smaller number the preparation of the soil is effected in a more complete manner.

"The power of laying land flat with safety is one of the important advantages which draining confers.

"The narrow, high-backed ridge, which wet, undrained land requires, is but too often accompanied by a bare and sterile furrow, hardly replacing the seed bestowed upon it; whereas, after thorough drainage, by gradually levelling down the ridges, every part may be made to yield alike, and to present the appearance of a garden clothed with equal and uniform luxuriance.

"There are, perhaps, few occasions on which the value of efficient drainage addresses itself more powerfully to the mind of the farmer than at the season of the year when the preparation for the root crops goes on. Upon farms where large breadths of potatoes or turnips are grown, with what anxiety does the cultivator of undrained land watch every cloud! well knowing that a few days of rain may destroy the results of weeks of laborious exertion; and that on the very eve of commencing to ridge his fields the effects of the various plowings, harrowings, and rollings, which have cost him so much care and expense, may be annihilated. The turnip has with truth been called 'the root of good husbandry.' It may be likened to a miner; for it explores the soil, and brings up from it much valuable material, in a state fit to be converted into beef and mutton, while the refuse of that conversion forms food for new tribes of plants. Like a miner, however, it cannot work unless the mine be kept 'water-free.' It cannot be called 'the root of good husbandry' when it barely replaces the manure which may have been supplied to it. It is only when it yields fair crops that it is so; and in order to obtain these, the first requisite is to have the land freed from stagnant water.

"It is found that, coincident with drainage, an important alteration takes place in the texture of tenacious soils, by which their nature is so modified as to permit of the most perfect pulverization, without any very great expenditure of labor.

"If the extension and improvement of the cultivation of the manure-making crops were the only advantages of thorough draining, it might with truth be asserted that these would amply repay the country and individuals for the outlay; for without an abundance of root-crops there can be no very large manure heaps; and without the latter well-filled barnyards cannot be obtained.

"It is found, however, that *all* the cultivated crops are benefited by the drainage of the soil, in some cases to such an extent as to repay the outlay in a single crop.

"The advantages resulting to the grain crops are not confined to increased luxuriance and bulk. The ear is found to be better filled, and that with a weightier and more valuable grain ; the harvest, too, is generally found to be hastened, which is no unimportant consideration, especially in the later districts of the country.

"There are few cases in which the value of drainage is more strikingly illustrated than in the case of wet grass lands. The first effect of a judicious and thorough system of drainage on such lands is the speedy disappearance of rushes and the coarse subaquatic grasses, and the substitution of a rich sward of sweeter and more nutritious herbage, which not only maintains a larger number of animals, but maintains them in superior health and condition. There are no more effectual means for the extirpation of that most destructive disease, the rot in sheep, than removing the superfluous water in the soil. So efficient, indeed, has this been found, that on farms where rot annually destroyed large numbers of them not a single instance of the disease has occurred since the land has been drained.

"Paradoxical as it may appear to the inexperienced, the drainage of watered meadows, where the soil is retentive, is a most valuable and profitable improvement, and has, in many cases, at once doubled the crop both of hay and aftermath.

"Drainage has a most important effect in preventing land from burning in dry seasons and in preserving a certain degree of moisture in the soil. This arises wholly from the more perfect division of the soil which takes place after land is drained, and not from drains forming reservoirs of moisture, as some have asserted. Soil has the power of absorbing much moisture from the air ; and this power, as might be expected, is increased in proportion to the surface exposed. This peculiar property of soils did not escape the notice of the illustrious Davy, who, in speaking of this subject, says : 'The power of the soil to absorb water, by cohesive attraction, depends in a great measure upon the state of division of its parts ; the more divided they are the greater is their absorbing power.' And again : 'The power of the soils to absorb water from the air is much connected with fertility. When this power is great, the plant is supplied with moisture in dry seasons ; and the effect of evaporation in the day is counteracted by the absorption of aqueous vapor from the atmosphere by the interior parts of the soil during the day, and by both the exterior and interior during the night. The stiff clays, approaching to pipe clays in their nature, which take up the greatest quantity of water, when it is poured upon them in a fluid form, are not the soils which absorb most moisture from the atmosphere in dry weather. They cake, and present only a small surface to the air ; and the vegetation upon them is generally burnt up as readily as upon sands.' *

"There needs no apology for transcribing this passage from the great pioneer of scientific agriculture. It explains, in a clear and forcible manner, one of the most important advantages of the thorough comminution of the soil, which draining so greatly tends to promote. It will, no doubt, be eminently suggestive to the

* Elements of Agricultural Chemistry, by Sir Humphry Davy.

thoughtful agriculturist, and it will explain some apparent anomalies; among others, the reason that horse-hoe work among turnips, in a dry season, has very much the same effect as successive showers of rain.

" Liebig and others have shown that rain and snow generally contain substances in the highest degree useful to plants, and that soils have the power of abstracting these substances from the rain which passes through them. It has been further shown by chemists that various injurious substances are washed out of the soil where a perfect system of drainage is in operation, or are so changed in their nature as to become innocuous.

" Various experiments have shown that rain, when percolating through the soil, has a strong influence in raising the temperature of the latter. The causes of this will be readily understood when we reflect that rain, at those periods of the year when vegetation is in progress, generally possesses a temperature considerably above that of the soil. In passing through the soil each successive portion of the rain gives off part of its excess of heat, until a mean temperature is established. This may be termed a positive cause of increased temperature; but there is also a negative cause, tending to the same end, in the great decrease of evaporation from drained soils. A great amount of evaporation is constantly taking place from the surface of soil saturated with water, and the temperature of the soil is consequently lowered; whereas, when the amount of moisture does not greatly exceed that for which the soil has a natural affinity, but little evaporation takes place, and that portion of the solar heat which would be dissipated in evaporating this water is applied to raising the temperature of the soil itself.

" Mr. Parkes has detailed a set of very valuable experiments in this important branch of the philosophy of drainage, in which he compared the temperatures of drained and undrained portions of bog. He found the temperature of the undrained portion to remain steadily at 46°, at all depths, from one to thirty feet; and at seven inches from the surface the temperature remained at 47° during the experiments. During the same period the temperature of the drained portion was $48\frac{1}{4}°$ at two feet seven inches below the surface; and at seven inches reached as high as 66° during a thunder-storm; while on a mean of thirty-five observations the temperature at the latter depth was 10° higher than at the same depth in the undrained portion of bog.*

" The sources from which excessive moisture in the soil is derived may be classified under two general heads: (1) *Springs* rising to the surface, and pouring out their waters over the adjacent land, or saturating the soil and subsoil at those points which they approach, without directly discharging on the surface. (2) *Rain* stagnating in the soil and subsoil. To these might be added such occasional and

---

* Parkes' Philosophy and Art of Land Drainage.

accidental sources as the overflowing of rivers and streams presents; but, as in such cases the water is generally carried off on the subsiding of the flood by open ditches and water-courses, that portion remaining in the soil may be considered in the same light as if derived from a heavy fall of rain.

SPRINGS.

" These owe their formation to certain peculiarities in the crust of the earth, by means of which rain falling on more elevated ground is collected and poured forth in a perennial discharge at a lower level, and frequently at very distant places.

" The crust of the earth is composed of numerous strata, or layers, lying one over the other, sometimes in a nearly horizontal position, but more frequently in one more or less inclined, or dipping, as it is termed, to the horizon. Some of these strata, such as gravel and sand, are highly porous and absorbent, and readily permit the passage of water; while others, such as clay and some rocks, are nearly or altogether impervious.

" When rain falls upon a tract of country part of it flows over the surface, and makes its escape by the numerous natural and artificial courses which may exist, while another portion is absorbed by the soil and the porous strata which lie under it. Again, rocks lying under the surface are sometimes so full of fissures that although they themselves are impervious to water, yet so completely do these fissures carry off the rain, that in some parts of the county of Durham they render the sinking of wells useless, and make it necessary for the farmers to drive their cattle many miles to water.

" It sometimes happens that these fissures penetrate to enormous depths, and are of great width, and filled with sand or clay.

" These are termed *faults* by miners, and some which we examined, at a distance of four hundred yards from the surface, were from five to fifteen yards in width.

" These faults, when of clay, are generally the cause of springs appearing at the surface; they arrest the progress of the water in some porous strata, and compel it to find an exit by passing to the surface between the clay and the faces of the ruptured strata. When the fault is of sand or gravel the opposite effect takes place, if it communicates with any porous stratum; and water which may have been flowing over the surface on reaching it is at once absorbed.

RAIN.

" This, as it is the most universal, so it is the most important source of an injurious excess of moisture in land. Before the introduction of thorough draining rain-water in excess was hardly looked upon as an evil with which the drainer could deal. Land was divided into two classes—wet and dry. No one contemplated the possibility of converting all the wet lands into dry lands by artificial means; the attention of drainers was, therefore, attracted merely to the removal of springs. These rendered certain parts of most districts of country useless for agricultural purposes. They arrested the course of the plow, and thus demanded attention and remedy. We accordingly find that the early methodical applications of draining were mainly directed to the removal of spring water, to rendering the 'springy' and boggy ground equal to that among which it lay, and not to the amelioration of the whole body of the soil, as is now the case.

"The quantity of rain which falls varies most materially in the different latitudes of the world; thus, according to Humboldt, one hundred and forty-one inches annually fall in Cuba, while only twenty inches fall in Paris. The quantity of rain varies, too, very much in different localities of the same country.

"But in estimating the quantity of rain which requires to be provided for in draining operations, it is not merely necessary to take the average annual fall into account; provision must be made to meet the greatest fall which is likely to take place in a limited period. To carry the system which was required to accommodate a fall of fifty inches of rain annually to a part of the country where the fall was only twenty inches might fairly lay the drainer open to a charge of wastefulness.

"Pipes of an inch bore, and laid at wide intervals, have been highly recommended by several drainers of great experience, without any caution as to their application in wet localities; and yet our own experience convinces us that larger pipes, laid at little more than one half the distance recommended, are barely adequate to perform the work required of them in some districts of country.

"The whole of the rain which falls is not carried off by drainage, but a large proportion of it is carried into the atmosphere by evaporation. The experiments of Mr. Dickinson show that of the rain which fell rather more than one half was evaporated, leaving rather less than one half to be carried off by drainage.

DRAINS.

"The drains used may be divided into two classes—*open* and *covered*. These again may each be subdivided into drains intended merely to act as water-courses, and drains which, in addition to acting as water-courses, are also intended to carry off the surplus water from the land through which they pass.

OPEN DRAINS.

"The rudest forms of open drains are the deep furrows, lying between narrow, high-backed ridges, which are still to be found in some parts of the country, with their accompanying water-furrows for discharging their streams.

"These are only meant to carry off the surplus water after the soil is completely saturated; and this they effect by carrying along with it all the best portions of the soil and of the manure which may have been spread upon its surface, as the turbid waters discharged from fields so treated abundantly testify. These require no other remark here than a recommendation to substitute for them some of the more perfect forms of drain to which we shall have to advert.

OPEN DRAIN WATER-COURSES.

"The ordinary ditch is the common form of this kind of drain, and though rude, it is one which cannot altogether be dispensed with, although where a perfect and complete system of drainage has been effected few indeed of them ought to be found. They are constant sources of annoyance, from their sides crumbling in; and they are constant sources of expense, not only by occupying much valuable space, but also by requiring a thorough annual scouring, wherever any pretensions to good farming are made. They are also fruitful nurseries of weeds. Open ditches occupy an important place in the early stages of draining bogs; but after the bog has become consolidated, the greater portion of them may be dispensed with, and their places supplied by large covered drains.

"In forming open drains in loose soil the sides should generally slope at an angle of 45°, which is the smallest angle at which earth, if it be at all crumbly, will retain its position; indeed, in deep cuttings, such as railways present, a fall of one in two, or a slope of 27½°, is that which is generally preferred. Where the soil is excessively stiff, as strong clay, or where the sides of the drain are to be lined with masonry, or where the channel is cut in rock or marl, the slope may be less than 45°.

"The depth, dimensions, and direction of open water-courses must be determined by the purposes they are to serve. Sharp turns are to be avoided, more especially where the fall is rapid or the quantity of water is great; for the banks are generally hollowed out by the force of the current where such turns occur; and thus it may happen that at the very time when a free channel is of the utmost consequence in discharging some flood of rain, a stoppage or impediment may be created by the fall of a portion of the undermined bank. Where sudden and steep descents are required in the course of open drains, it is a good practice to line the sides with rough masonry, and to pave the bottom, so as to prevent that hollowing out which sometimes converts a moderate-sized ditch in a few years into a gully or ravine of formidable dimensions.

"Sometimes in large drainage works, where outfalls require to be formed for extensive districts, calculations of a complicated nature have to be made, to show the size of main drains which will be required to void the waters which may be expected to flow into them. Sir John Leslie has given the following rule for ascertaining the velocity due to declivity and depth of current in streams: 'Multiply the mean hydraulic depth* of the river by the declivity per mile, both in feet, and extract the square root of the product; the result, diminished by one-sixteenth part, will be the mean velocity of the river in miles per hour.'

" Having thus found the velocity due to any proposed channel, the quantity of water which it can discharge will be found, in cubic feet, per minute, by multiplying the area of the transverse section of the stream by its mean velocity in feet per minute.

" Thus, a channel or water-course six feet wide and two feet deep, and having a fall of only five feet per mile, will discharge 2464 cubic feet, or 15,400 gallons per minute; and would, therefore, be capable of discharging a fall of two inches of rain in twenty-four hours from 488 acres of land, supposing the whole quantity to be carried off by it.

COVERED DRAINS.

" We now come to the consideration of the more important description of drainage—the removal of water by means of covered drains.

" For hand-made drains the tools consist of a set of spades—generally three of different sizes—gradually diminishing in width to suit the different parts of the drain.

" For soils free from stones these spades work better when curved; but for stony lands the flat form is preferable. For taking out the last narrow spit, to form the seat for the draining pipe, long narrow spades are used, called bottoming tools. There are also scoops of various widths, furnished with long handles, and rounded or flattened in the soles, according as they are required to finish the bottom of the drain, for the reception of stones, a horse-shoe tile and sole, or a draining pipe. In the formation of large and deep drains a shovel, very much bent at the neck, or having a great 'lift,' as it is termed, is very useful for finishing the bottom. Where the subsoil is stony or hard, a hand pick or foot pick is required to loosen it before it can be shovelled out.

" For the purpose of laying pipes in narrow deep drains, an instrument called a pipe-layer has been invented, and is indispensable; for the narrowness of the drains prevents a man from standing in them, nor is he able to reach the bottom with his hand, without much trouble; while, with the assistance of the pipe-layer, he does not require to go into the drain at all, and can proceed with great expedition.

" In addition to the tools named, a drain-gauge is a necessary instrument.

---

* The mean hydraulic depth is the depth which would obtain if a stream were made to flow in a new channel, the breadth of which was equal to the sum of its present mean breadth and twice its mean depth. Thus a stream having a mean breadth of six feet and a mean depth of two feet would give a mean hydraulic depth of $1\frac{1}{3}$ feet.

Different sizes of these are required for the different widths and depths of drains. Where the fall of the ground to be drained is very slight, the workmen should be provided with a level. The best form of this instrument, because that which they most readily understand, is the ordinary mason's level, but made of large dimensions, and having a stem of such height as to show the 'bob' above the drain, when the level is applied at the bottom.

TILE DRAINS.

"Of all the materials which have yet been brought forward for forming the conduits of drains, none are so well fitted for the purpose as tiles or pipes of burnt clay. Draining tiles, especially those in the form of pipes, possess all the qualities which are required in the formation of drains. They are cheap, durable, and portable. They afford a free ingress to water, while they effectually exclude vermin, or earth, and other materials, which often destroy the less perfect forms of drains. They afford a ready passage to the water which enters them, and a moderate amount of superintendence will insure their being properly laid in the drains, while the expedition with which a great extent of draining operations can be executed where they are used permits an extension of this improvement, which could not be even thought of if stones or other weighty materials had to be employed.

DURABILITY.

"There seems to be hardly any limit to the durability of a well-burned pipe in itself; and if they are carefully laid the drains formed of them should be equally durable with themselves.

"Their structure prevents the entrance of gross matters, by which they might become choked, while no amount of violence to which they are likely to be subjected can injure them. Cases have, no doubt, been discovered in which tile drains have become choked by a ferruginous deposit of peroxide of iron, which, having entered in the form of a solution of the protoxide, becomes deposited on contact with the air in the drain. It is of very rare occurrence, and is equally or even more injurious to stone drains.

"The entrance of roots into pipes is almost the only other accident they are liable to; but it is believed that a complete remedy for this, so far as cultivated plants are concerned, can be pointed out, and will be considered under the 'Depth of Drains.' No drain should pass near to water-loving trees, as no crevice, however small, is proof against the entrance of their roots. An instance of stoppage came under our notice where the roots of the common willow had so completely filled the pipe of a drain for thirty feet that not the smallest quantity of water could pass through it. The ash tree is also very destructive to drains.

"A ton of two-inch pipes will furnish forty-eight rods of drain, while a ton of broken stone will only form two rods.

"*Pipe Tiles* have been made of a great variety of shapes, but experience has convinced us that there is no form so good as the cylinder. The cylinder can hardly be placed improperly, if the trench be finished with a semi-cylindrical scoop, as it at once finds its place in the centre of the cavity.

"In clay soils the trench should be cut of a convenient width for the operations of the workman, to within nine inches or a foot of the total depth; the bottoming tool is then employed to take out the remaining portion, in the form of a narrow spit, of just sufficient size to admit the pipe. By this means no more work is done in cutting than is required, while the fitting of the pipes to each other is secured.

"Where a sudden or steep descent occurs in the course of a drain, or where there is a running sand or boggy place, pipes of one size should either be entirely sheathed in larger ones, breaking bond with them, or they should be furnished with collars. These collars are short sections of pipe of such size as to fit upon smaller ones.

"The question of the size of pipes proper for drains is not entirely dependent upon the quantity of rain to be discharged. An important consideration is the probable effect of a slight displacement upon the drain. Some have advocated the use of one-inch pipes; but in some situations these would prove quite insufficient to discharge the quantity required of them. But, apart from the question of capacity, it must be obvious that a very slight sinking of an inch pipe or a slight inaccuracy in placing it would entirely destroy the drain. It is true that this objection is of no force where collars are used; but the cost of collars and trouble and loss of time in fitting them are so great that a larger size of pipe is a cheaper alternative. The smallest size of pipe that can be employed with safety seems to be that of one and a half inch diameter in the bore. In wet districts the two-inch size is to be preferred, although the one and a half inch size may, with propriety and economy, be used in the first one hundred yards of each drain, or throughout, when the drains are under that length, as in such cases the accumulated quantity of water is not great.

"The same rule which governs the flow of water in streams also governs the flow in covered drains, theoretically speaking; but the great inequalities and asperities in pipes occasioned by imperfect joinings and otherwise reduce the results in practice in some cases nearly fifty per cent.

DRAINING OF SPRINGS.

"The drawing off of the pent-up waters, which are the sources of springs, is a department of draining which requires, for its successful practice, a considerable knowledge of the different varieties of stratification which occur, and is, probably, for that reason too little practised. When the theory of springs is understood, and a knowledge of the strata obtained, the judicious application of a few simple drains, made to communicate with the watery layers, will often dry swamps of great extent, where large sums of money, expended in forming furrow drains in the swamp itself, would leave it but little improved. It is to the application of both kinds of drainage where required that we are to look for the best results; and the judicious drainer will well consider and avail himself of those means which are capable of producing the maximum effect required, without staying to consider whose or what system it is which he employs.

"In endeavoring to drain springs, the point to be sought for is to furnish outlets sufficiently numerous to discharge all the water from the porous bed, at the lowest point to which it reaches in the land to be drained.

"Elkington had the merit of reducing the draining of springs to a system, and the rules which he laid down were so simple and complete that their authority has remained undiminished. Johnstone thus describes the method: 'Draining, according to his (Elkington) principles, depends upon three points: (1) Upon finding out the *main spring*, or cause of the mischief, without which nothing effectual can be done. (2) Upon taking the level of that spring, and ascertaining its *subterranean bearings*, a measure never practised by any till Elkington discovered the advantage to be derived from it; for if the drain is cut a yard below *the line of the spring*, you can never reach the water which issues from it, and by ascertaining that line by means of levelling, you can cut off springs effectually, and, consequently, drain the land in the cheapest and most eligible manner. (3) By making use of the auger to reach or tap the spring where the depth of the drain is not sufficient for the purpose.'

"The term 'main spring' in the passage just quoted refers to what is sometimes called the *true* spring, in contradistinction to those termed *false* springs. The true springs seldom cease to flow, whereas, in dry seasons, the false springs sometimes intermit for considerable periods of time. The true spring is the natural outlet of the enclosed water which gives rise to it, whereas the false springs are occasioned by the backing up of a large quantity of water from the insufficiency of the outlet, till it flows forth at some higher level, in which case they appear above the true springs; or they owe their existence to water which, after having issued from the true spring, has soaked into the soil, and has again appeared where some obstruction forces it to the surface. In the latter case the false are below the true springs.

"Having ascertained the line of the true springs, the next step is to cut a drain sufficiently deep to reach the watery stratum at a short distance below the line of the springs.

"If, upon experiment, it turns out that the superincumbent impervious layer is considerably more than five or six feet in thickness, it will be proper, instead of incurring the great expense of forming an enormously deep drain, to cut a drain four feet in depth only, and then to sink small wells down to the watery bed at intervals along the course and a little to one side.

"These wells are to be filled with small stones, so as to afford a ready passage for the water to rise up through them to the drain. The conduit may be formed of draining pipes or bricks or stones, as may be most convenient, taking care, however, that the culvert is securely formed, and that the floor of the drain is protected by tile soles or slate or some other material, to prevent the hollowing action of the great flow of water which may be expected to proceed along it. The small stones should be continued to the height of ten or twelve inches above the culvert, so as to furnish a free passage for the water into the chinks and joinings of the culvert. When the watery stratum lies at a depth exceeding eight feet, it is usual, instead of sinking the small wells just described, to make use of an auger, or boring rod, in order to reach the reservoir of water. The auger-hole, like the well, ought also to be sunk a little to one side of the drain, so that the discharge from them may not interrupt the course of the drain, by rising at right angles with the flow of water in it, and so as to guard as much as possible against the choking of the culvert by any bodies which may ascend through the well or auger-holes, such as sand, which is sometimes discharged in large quantity on the first tapping of the spring.

"In the plan the feeder is carried horizontally along the line of springs, so as to communicate with the tail of the watery stratum along the whole of its course. From the horizontal drain or feeder there must be carried a main drain, to convey the water to the nearest brook or water-course; or where there is a large extent of horizontal drain or a great quantity of water, it may be necessary to make several mains. These should always be laid out in the line of the fall of the land, so as to discharge the water as quickly as possible; for it must always be borne in mind that a drain, with a depth of water constantly flowing therein, in a direction transverse to the slope, cannot fail to supply spongy soils with more water than is consistent with a healthy state.

"It often happens that instead of a line of springs there is but one, arising from some chink or fissure communicating with a watery stratum at a considerable depth. Such are the 'well eyes,' 'piping springs,' and 'quags,' as they are called, which one meets with so often in moorland tracts. The proper course to pursue with these is to cut a drain of three or four feet in depth from the point of discharge toward them in a direct line up the ascent, deepening the drain considerably as it approaches the spring. In such cases it is always necessary to reach the orifice of the spring, if the watery stratum itself is out of reach; for the constant flow of water keeps up a rank growth of peat and subaquatic plants, which act like saturated sponges, and retain the water even although the drains be within a few feet of them.

"When the main orifice of the spring has been reached, and its waters are confined to the channel of the drain, any subsidiary outlets will be more readily discovered, if such exist.

"These must be reached by short drains, branching off the main one described.

"Valleys between rising grounds or hollows in an undulating country are sometimes kept in a marshy state by springs, which are fed from the higher ground in the vicinity. In this case a deep drain, carried along the lowest ground, and either reaching the watery stratum directly, or by means of wells or auger-holes, will generally dry a very large extent, if not the whole, of the swamp.

"It frequently happens that although the surface of a district is wet there exists below it, at some distance, a layer so dry and porous as to be capable of absorbing any quantity of water which may gain access to it. Well-sinkers sometimes meet with such strata, which at once absorb all the water they may have met with in the upper strata. This peculiarity has been taken advantage of in draining, and copious springs may be made to disappear by simply boring an auger-hole into such a stratum, where it is known to exist, and turning the water of the spring into it.

ABSORPTION AND RETENTION IN SOILS.

"All porous bodies have the power of attracting or absorbing liquids in a greater or less degree, by virtue of a particular property which they possess, which has been denominated capillary attraction, from the minute tubes in which its influence is exhibited. Capillary attraction acts more rapidly in some soils than in others; thus we find that in pure clays it exhibits its influence but slowly; in agricultural clays, into the composition of which some of the more porous earths enter, its action is more rapid; while gravel, sand, or peat—which latter may be likened to a vegetable sponge—speedily absorb as much water as they can hold on being brought in contact with it. This power of attraction also manifests itself on the surface of bodies, and may then be called the attraction of adhesion. Soils, in common with all other bodies, possess this property, and in a greater or less degree, according to the aggregate surface which the particles of a given bulk present.

"Thus clay may, by means of kneading, be made to contain so large a quantity of water as that at last it may almost be supposed to be divided into infinitesimally thin layers, having each a film of water adhering to it on either side.

"Such soils, again, as sand or chalk, the particles of which are coarser, exert a less degree of adhesive attraction for water. Professor Schübler, of Tübingen, found that sand was capable of holding twenty-five per cent., loamy soil forty per cent., clay loam fifty per cent., and pure clay seventy per cent. of their own weights of water, when the water was merely poured upon them in a dry state, till it began to drop. Sir Humphry Davy found that the power of attraction for water generally proved an index to the agricultural value of soils. This sort of attraction, however, depends upon other causes besides the adhesion to which we have been alluding. The power of attraction which certain substances exhibit for the *vapor* of water is more akin to the force which enables certain porous bodies to absorb and retain many times their volume of the different gases, as charcoal or ammonia, of which it is said to absorb ninety times its own bulk. And as finely-divided mineral matter, as well as vegetable matter in a state of decay, at the same time that they possess this power in a high degree, are also indications of fertility, according to the proportion in which a soil contains them, so the relationship observed by Davy between fertility in soils and their affinity for aqueous vapor admits of easy and satisfactory explanation. Clay soils are called impervious soils, because in their natural state they resist the passage of water through them. They are also called retentive soils, because if water does gain access to them their power of adhesion enables them to retain a large quantity of it for a great length of time. These are properties which have a very injurious effect on all agricultural operations, and their removal is one of the results which the scientific drainer seeks to effect.

" We have it in our power to increase for a time the permeability of clay soils by mechanical means.

" By pulverizing them when dry, we so separate their parts as to afford a ready passage to water.

" Natural causes also have a like tendency. The summer drouth causes numerous cracks and fissures, which admit the rains to all parts of the soil. This temporary permeability on undrained clay lands is, however, found to be an evil; for by means of it the rain is enabled to penetrate and saturate the soil, in autumn, to a considerable depth; while their great adhesive power retains it to an extent which reduces the soil to a state of quagmire during the winter months. Accordingly, we find that the clay-land farmer is by no means ambitious to pulverize his soil very finely when it is undrained. He prefers a rough clod on his wheat land, which has to contend with the watery influences of the winter months; and he very properly eschews all attempts at subsoiling in the wet months of the year, or anything which may bring into play the water-retaining powers of his soil. When clay is properly and thoroughly drained, however, a new element is brought into operation by the constant supply of air to the soil. By its means the permeability is increased, while the adhesiveness, if not removed, is at least prevented from exercising any other than a beneficial influence.

" The water-resisting power of soil which has become slightly dry is familiar to every farmer, although many may not be aware of the cause.

" When a piece of damp land is plowed it is very apt to get 'soured' if rain falls immediately after it is turned over; whereas, if it gets somewhat dry before the rain falls it is but little injured. This effect is entirely owing to the air, which takes the place of the evaporated moisture, and acts like a waterproof garment in warding off the rain.

"When rain falls upon the surface of soils which rest upon an impervious or very slightly pervious substratum it is gradually diffused through all the porous and absorbent portions by capillary attraction, assisted in clays by the cracks and fissures they may contain.

"If the fall continues the soil becomes saturated, and the excess then forms pools, or makes its escape by flowing over the surface to any neighboring water-course which may exist.

"When the rain ceases to fall those parts of the surface which are higher than the rest gradually become drier, because the water being no longer poured upon them, the law of gravitation produces its natural results. Now, we cannot raise the soil, but we can lower the impervious or saturated bed on which it rests, and so increase the depth of porous soil.

"If we cut a trench or drain into the subsoil, we immediately disarrange the hydrostatic relations which exist in its neighborhood in a greater or less degree, according to its depth. The capillary force which retained the water in the soil to a height of a few inches is no longer able to sustain it when the height is increased to feet, and a portion descends into the drain, leaving the upper part of the surface comparatively dry. Now, the unequal pressure of different heights of water in the land immediately compels the portion of soil next to that from which the water has been drawn to yield up a portion of its excess to it, obtaining, in its turn, a portion from that farther off, and so on through the whole mass of the surface soil; but as fast as it is supplied the drain draws it off, so that in a short time the level of the water in the whole mass is lowered. This is the action which is indicated by the term *drawing*, which is so often applied to drains, probably in many cases without any very definite idea of its meaning.

"All soils, too, but especially those containing clay, possess the property of expanding when wetted, and contracting when dried; so that after the drain has removed a portion of the water a considerable contraction takes place, especially in a dry season; but as the ends of the field cannot approach each other to suit the contraction, both soil and subsoil are torn asunder, and divided into small portions by a network of cracks and fissures, the sum of which represents the amount of lateral contraction throughout the field. This circumstance is familiar to every one, and most persons who are conversant with strong land are aware that in some seasons the fissures extend to a great depth.

"These phenomena are of the utmost consequence in draining land; indeed, it may well be doubted whether without such properties in the soil or subsoil we could drain our clay lands at all. It is worthy of remark here that as on stiff soils the cracking action is strongest, nature seems to second the efforts of man, and compensates the want of porosity in clays by the more powerful development of a property which, under skilful treatment, renders them almost as easy to drain as the more porous soils. The tendency of draining is to increase and guide the course of this cracking action.

"The main fissures all commence at the drain, and spread from it in almost straight lines into the subsoil, forming so many minor drains or feeders, all leading to the conduit.

"These main fissures have numerous small ones diverging from them, so that the whole mass of earth is divided and subdivided into the most minute portions. The main fissures are at first small, but gradually enlarge as the dryness increases, and at the same time lengthen out, so that when a very dry season happens they may be traced the whole way between the drains.

"When the fissures are once formed the falling of loose earth into them and the growing action of the water which passes through them prevent them from ever closing so perfectly as to hinder the passage of water, while each successive summer produces new fissures, till the whole body of the subsoil is pervaded by a perfect network of them, which gradually alters the very nature of both soil and subsoil, and in connection with judicious and liberal manuring has the effect of converting poor cold clays into something not very different from a good clay loam.

DEPTH OF COVERED DRAINS.

"Such drains have a twofold office to perform. They have to collect the superfluous water from the soil, and then to carry it off in a certain fixed course. They must, therefore, afford free access to water at all points, and, at the same time, prevent that which they have collected from leaving them by any other way but by their own channels. They must also be covered to such a depth as not to interfere with the working of the land. Let us fix the minimum depth of this covering of soil.

"Modern agriculture, practical as well as theoretical, has shown that 'to have large crops we must have a deep soil.' The soil is a great storehouse of materials of which plants are composed; but these require a certain amount of preparation before they become fitting food for our crops. That preparation is effected by exposing them to the action of the elements, through the operation of tillage. Plants have the peculiar property of being able to adapt themselves to almost any amount of food which may be presented to them. Take turnips, for example; these will be found varying from the size of a pigeon's egg to that of a man's head, or larger according to the amount of food with which they have been supplied. It is, therefore, an object of first importance that a large quantity of what chemists call the *inorganic* constituents of plants be constantly in course of preparation in a soil deeply stirred by the subsoiler or trench plow. If, then, it is considered probable, or even possible, that subsoiling and trench plowing may become general, it is imperative that drains be so put in as not to interfere with or be injured by such operations.

"Subsoiling as hitherto practised has reached a depth of eighteen inches, but it is highly probable that future experience may demand a still greater depth. If, however, we take the depth at eighteen inches, we cannot with safety place the *upper* part of the drain nearer to the surface than this depth at least.

"But, further, if such an instrument as a subsoiler was to pass close to the top of a drain, it could not fail to injure or destroy it; and even although an inch or two of soil were to intervene between the instrument and the top of the drain, still the shaking and crushing which take place would, in all probability, materially injure it. It must, therefore, be concluded that from four to six inches of soil must be left undisturbed between the *top* of the drain and the subsoiler, so as to insure the safety of the former. If to this we add the depth of the subsoiling operations, we obtain data showing that the top of a drain should never be nearer the surface than twenty-two to twenty-four inches.

"Where a precise rule cannot be laid down, it is best to keep on the safe side; we must therefore assume that there should be at least twenty-four inches of soil *above* every drain.

RESULTS REQUIRED BY DEPTH OF DRAINS.

"There is hardly any subject connected with agriculture which has excited such an amount of controversy as the proper depth for drains.

"A careful consideration of the very numerous recorded opinions published during the last few years only leaves the conviction that there is no settled rule as to the depth of drains *best* adapted to all soils and all circumstances.

"We say *best* adapted, for we believe that there is a depth and distance of drain which will effectually remove the *surface water* from all soils; but whether that might be the most economical and most judicious mode of proceeding in particular cases can only be settled by a thorough investigation of the particular case to which it is to be applied.

"Cases are sometimes adduced as successful examples of deep thorough draining where the drains are placed at great intervals, and in some cases ten feet in depth. These have no claim to be considered as examples of thorough draining at all. They are merely successful examples of Elkington's principles to the removal of springs, by furnishing outlets to the water in the stratum from which they arise. The proper function of thorough draining is the removal of rain-water, which would otherwise lodge and stagnate in retentive soils.

"Gravitation is not the only agent to be considered in thorough draining.

"The first consideration to which we must address ourselves, in fixing the depth of drains, is the depth of soil which is required to be laid dry. There is a limit to the depth of drained soil required for the purposes of cultivation; and any extra expenditure in drying soil at greater depths than will yield a return must be regarded as waste.

"It is probable that, as a general rule, the roots of our cultivated plants do not penetrate to a greater depth than two feet, or two feet six inches, even in soil fitted for their reception. That they descend to the former, at least, of these depths, has been put beyond dispute, by the roots of mangold having been found in a drain, the top of which was two feet from the surface; but there is no case recorded in which the roots of cultivated plants have been found in three-feet drains. This is, no doubt, only negative testimony; still, the spirit of inquiry in regard to draining would probably have discovered such a case if it had occurred. It will not be a very forced conclusion if we take it for granted that thirty inches represents the ordinary depth to which the plants of agriculture are likely to penetrate.

"It is further necessary that water should not be allowed so near the surface as to create any chilling effect on the vegetation. It is also desirable that it should not be so near as to be capable of injuring the surface, by ascending in too great quantities, by means of capillary attraction. A depth of from three to four feet seems to be as great a depth of drained soil as can be required.

"But there are still other considerations which must influence us in fixing the proper depth for drains in particular cases. In the case of a porous soil and subsoil saturated with water, in consequence of resting upon an impervious stratum, it will be proper, if practicable, to sink entirely through the porous stratum, and to form the conduit of the drain in the impervious layer. By this plan of proceeding a very limited number of drains may be made to dry a great extent of surface; for it may be laid down as a general rule in regard to *very porous soils*, that the deeper the drain the further it will draw.

"It sometimes happens in clayey soil that at no great distance from the surface there is a watery stratum composed of porous materials.

"If this stratum is not more than six or seven feet from the surface, it may be turned to excellent account in draining; for by cutting a smaller number of drains down to it than would have been required in ordinary cases, it will not only be emptied of its own water, but will be converted into one extensive natural drain under the whole surface, of which the drains which are cut will form the outlets. In the drainage of *shallow* peat bogs it is always desirable to cut through the peat to the solid stratum on which it rests, for the very unstable nature of the peat renders it a very bad foundation on which to form a drain.

"It appears, then, that drains ought to vary in depth according to each particular case to which they are to be applied.

"Drains in porous soils may be deep and wide apart, because the water will readily flow to them from all parts, and the greater the depth, the more powerfully will the capillary attraction of the soil be neutralized.

"In clay soils, again, the drain has not only to carry away the water, but to aid in maintaining the artificial porosity of the soil, by means of which the water is to gain admission to it. This it cannot effect if placed at a depth to which the shrinkage of the soil does not extend; and it must not be forgotten that this shrinking action is much greater in certain parts of the country than in others, and in some seasons than in others.

"The comparatively slight benefit derived in many cases from drains in clay during the first season after their formation, more especially if that has been a wet one, is sufficient confirmation of this view of the matter.

"Our own experience and observation, combined with the experience of others, have convinced us that no drain should be put in at a less depth than three feet, where this is practicable.

"That clay-land farmers will be found to advocate the use of much shallower drains, and will point to the water standing above such *deep* drains (as they style them) as a conclusive proof of their inefficiency, is no doubt true; but an examination of the shallowest drains, where the land has been stirred or trod upon when wet, will exhibit the same appearance. One inch of wet and worked clay will prevent water from passing through, so long as it is kept wet, as effectually as a yard will do.

"The true remedy is to refrain from working such land when it is too moist; any stirring of it in that state is only undoing all that the summer drouth has effected in rendering it porous.

"Taking three feet as a minimum depth for drains, three and a half and four feet will be found safe and efficient depths at which to place them, where there are no peculiar circumstances demanding special depths to suit them. What the nature of these circumstances is has been stated in a general way; the limits to which we are necessarily circumscribed in such a work as this prevent our referring to them more in detail. Neither our nomenclature of soils nor our knowledge of the laws which govern the flow of water through them is at this time sufficiently exact to permit us to frame rules to be implicitly followed.

FREQUENCY OF DRAINS.

"The distances apart at which drains ought to be placed is a subject of great importance, and one on which much difference of opinion exists.

"Smith contends that drains should be placed at very short intervals. He says: 'In laying off the drains, the first object for consideration is the nature of the subsoil. If it consists of a strong stiffy 'till' or a dead sandy clay, then the distance from drain to drain should not exceed from ten to fifteen feet; if a lighter and more porous subsoil, a distance of from eighteen to twenty-four feet will be close enough; and in very open subsoils forty feet distance may be sufficient.'

"On the other hand, Parkes, who represents the deep and distant drain system, says: 'It consists with my own practice, at the present time, that drains are being

executed at depths of from four to six feet, according to soil and outfall, and at
distances varying from twenty-four to sixty-six feet, complete efficiency being the
end studied, and the proof of such efficiency being that, after a due period given for
bringing about drainage action in soils unused to it, the water should not stand
higher, or much higher, in a hole dug in the middle between a pair of drains than
the level of those drains.'

" The distance, like the depth of drains, must be governed by a variety of circumstances, all of which demand strict and careful investigation before proceeding
to set off any system of drainage. The most important of these considerations is
the nature of the subsoil, and the effects which the removal of stagnant water will
produce upon it. If the subsoil be very porous, or, although not porous in itself, if
it rests upon a porous substratum, from which the drains are calculated to remove
the water, the parallel drains may be deep and placed at considerable intervals. On
the other hand, where the subsoil is impervious the drains must be placed at much
shorter intervals.

" In estimating the imperviousness of subsoils, it is not only necessary to have a
due consideration of their nature before drainage ; the effects which drainage will
produce upon them must also be taken into account. In some soils, as we have seen,
a great degree of artificial porosity will be produced by draining; on these the
drains may, with propriety, be at wider intervals than on soils in which this cracking
action is less powerfully developed.

" The subsoils upon which draining acts to a shorter distance perhaps than any
others are those clay subsoils, containing a large quantity of imbedded stones, which
characterize a large portion of the surface of the carboniferous and Cambrian forma-
tions. They are often so completely indurated as to be almost impervious to water,
and when cut into are almost dry, even although the surface soil which rests upon
them may be at the same time of the consistence of soft soap. The great portion of
their mass, which consists of inexpansible materials, prevents the production of that
artificial porosity which plays such an important part in the draining of the purer
clays. Subsoiling as an adjunct to drainage on such soils proves of the greatest
value.

" In planning the draining of clay soils, climate must also be allowed its due
effect.

" We have seen that a drain may pass very near a spring without drawing off its
waters, because the perennial supply of water prevents the formation of fissures by
shrinkage.

" Our own experience over a considerable range of soils and climate, collated
with the experience of a very large number of careful and unprejudiced observers,
has convinced us that the extreme distances named both by Smith and Parkes are
to be avoided.

"There can hardly occur any instance in which drains require to be placed at such close intervals as ten feet. There may be isolated spots in a field into which it may be necessary to extend a branch, to draw off some minor spring; but, as a general rule, it may be held that draining at ten feet apart is a waste of labor and materials. On the other hand, we think sixty-six feet an extreme and unsafe distance for thorough draining.

"A scale of distances ranging from eighteen to forty feet will be found to suit almost any case which may occur, while it will not incur the charge of waste of means on the one hand, or inefficiency on the other. We have found a distance of twenty-four feet, with a depth of from three and a half to four feet, produce very perfect results on soils of considerable tenacity, in districts subject to more than the average fall of rain in the British Islands. These will be found safe examples to follow under similar circumstances; and where there is nothing in the formation of the subsoil calling for a particular arrangement to meet it, these intervals and depths will generally be found perfectly successful.

DIRECTION AND DECLIVITY OF DRAINS.

"As the law of gravitation, when permitted to act by either natural or artificial porosity, is that which governs the descent of water into drains, the chief object to be considered in laying out drains is the placing of them in such a position as will bring this principle to bear most fully upon them, in reference to the land on which they are intended to act.

"Where land is altogether level, all parts of the surface will be in the same relative position as to height above any drain which may be cut into it. In such a case, therefore, as in the flat alluvial tracts which border some rivers, and are to be met with in various districts, the choice of direction for the drains ought to depend, in a great measure, on the convenience of outfall. It is a matter of no consequence whether the drains run in the line of the ridges, at right angles with or diagonally to them. The main consideration necessary to be attended to is how they may be most conveniently disposed in reference to the main drain or place of discharge.

"Where, however, the land is possessed of any degree of slope, other considerations must guide the drainer.

"Where the slope is very slight, the necessity for selecting the line in which it is greatest for the direction of the drains, in order to obtain a flow in them, will be admitted by all. This rule ought also to obtain *in all cases* of sloping land, though for different reasons.

"There are reasons for selecting the line of the greatest fall for the direction of the drains, which are applicable to all lands alike.

"The most important of these is, *that the line of the greatest fall is the only line in which a drain is relatively lower than the land on either side of it.*

LAYING OUT DRAINS.

" Before proceeding to lay out drains, the depth, frequency, and kind of drain to be used must be fixed upon. In deciding upon these points an experimental examination of the subsoil should take place, where its nature is not already known.

" Pits of three, four, five, or six feet should be dug, and these questions decided upon the principles already explained, according to the indications which the pits afford. In deciding on the frequency of the drains, it is worthy of reiteration that extreme distances ought in every case to be avoided. A due regard should be paid to economy of labor and materials, but the object of the drainer ought rather to be to effect a perfect drainage than to convert extensive works into an experimental trial of the effect produced by drains at wide intervals.

" The same remarks apply to the size of pipes to be used. One-inch pipes ought *never* to be used without collars, and the locality must determine whether they may be used at all. In some districts, if placed at twenty-four feet apart, they would not void one-third part of the water required of them.

THE PLACE OF OUTFALL.

" This should always afford a free and clear outlet to the drains, and must of necessity be at the lowest point of the land to be drained. It will often be found necessary to cut across other land, in order to obtain a proper outfall ; but this is an expense which should readily be gone to, where drains require it ; for draining without a proper and clear outfall is only a waste of money. The position of the proper points of outfall should be determined by means of levelling instruments ; and wherever there is a considerable extent of work to be done a competent surveyor should be employed to fix these, as well as some other points which we shall have to advert to, if not to lay out the whole works ; for it cannot be too strongly enforced that there is no more worthless economy than that which entrusts the planning of operations involving an outlay of hundreds or thousands to the rule of thumb proceedings of some laborer whose sole qualification is derived from the fact of his having helped to *cut* some hundreds of rods of drains.

POSITIONS OF THE MINOR DRAINS.

" There is a very simple mode of laying out the minor drains, which will apply to most cases, or, indeed, to all, although in some its application may be more difficult. The surface of each field must be regarded as being made up of one or more planes, as the case may be, for each of which the drains should be laid out separately. Level lines are to be set out a little below the upper edge of each of these planes ; and the drains must then be made to cross these lines at right angles. By this means the drains will run in the line of the greatest slope, no matter how distorted the surface of the field may be.

"When the furrows happen to coincide with the line of drains, it may sometimes be proper to take advantage of them, in order to lessen the cost of cutting; but where either their distances or direction do not coincide with those which are ascertained to be the proper ones, they should without hesitation be disregarded.

THE POSITION AND SIZE OF THE MAIN DRAINS.

"All the minor drains should be made to discharge into mains or submains, and not directly into an open ditch or water-course. There are many reasons for this.

"Grass and weeds, and débris of various kinds, collect in open ditches, and are apt to choke up the mouths of drains, and thus greatly injure them, especially if the fall is slight; but when many drains are collected into one main, the run of water at its mouth becomes so strong as to clear away and overcome these obstacles, if through negligence they are allowed to accumulate. It is also much more easy for the farmer to look after the working of a few main drain mouths than to have a large number of small drains requiring examination from time to time.

"There is also the further advantage of there being less risk of damage from roots in the fences entering the drains, and the entrance of vermin can more readily be guarded against.

"The mains should intercept all the minor drains, at eighteen or twenty feet distance from the fences to which they tend, and conduct the accumulated waters toward the place of outfall. There must also be submains in all the hollows. As a general rule, there should be a main to receive the waters from every five acres, as a great current is apt to injure them.

"The rule of Leslie which has been given will serve to determine the necessary sizes of mains required, by deducting in round numbers twenty-five per cent. of the gross discharging power, on account of friction, and some other phenomena connected with the discharge of water from pipes. Main drains may be conveniently formed of one or more large pipe tiles. Main drains should be three inches deeper than the minor drains, so as to give the latter a drip, and prevent any damming up from sand. The minor drains should enter the mains with a curve, in the direction of the current of the mains; and when they enter on both sides of the mains they should not be exactly opposite to each other, as such an arrangement is apt to produce stoppage of the full flow in the mains.

EXCAVATIONS.

"In excavating the trenches for drains, the first operation should be to cut the main, beginning at the place of outlet. The width must be carefully and neatly set out with a line, as indeed that of the whole of the drains should be. The earth should be thrown on the lower side of the main. The minor drains are next to be cut, commencing with those farthest from the outlet. The object of this is that as fast as each drain is cut it may be laid with pipes, or other material, and covered in, as well as the pieces of main between it and the next mouth or joining; for nothing is more improper than having a great extent of drains open at one time, as a moderate degree of frost will cause much expense and trouble from the crumbling in of the sides. Two workmen generally work together in each trench, commencing at the lower end, that they may not be incommoded with water. As soon as they have completed it, it should be carefully inspected by the overseer of the work, after which the laying of materials should be immediately proceeded with. Where pipes or tiles are used, these should previously be laid ready along the sides of the trenches, taking care, however, in laying horseshoe tiles down from the carts, to place them on their backs; for if their edges are placed in contact with the earth, a very slight degree of frost causes it to adhere to them with such tenacity that they cannot be used until a thaw sets in.

" The laying of materials should commence at the upper end of the drains, so that all mud may be cleared away without the risk of its entering the conduits.

" As a general rule, the laying of materials should be performed by a trustworthy person paid by the day, for on the perfection of this operation the value of the drain in a great measure depends. The joinings at the mains should be made either by means of tiles made for the purpose, by having a hole cut in the side before being burned, or by neatly chipping out a small piece by a smart blow in the proper direction. All faulty tiles should be rejected, as holes in drains are fruitful sources of injury.

" When the materials for the conduit have been placed in the trench, the earth may either be returned upon them by manual labor or by the plow.

" Where the plow is to be used, the earth must be placed on the right and left of each alternate trench, so that the plow may make a full bout by passing up one drain and down another. The horses walk on either side of the trench, and a wide swingle-tree must be used. Each drain will require from four to six furrows to complete the turning in. Where labor is not too high, the spade or drag hoe will generally be found nearly as cheap methods of filling in as the plow; and with the latter there is a great risk of accident to the horses."

The Theory of Drainage.*

BY LYON PLAYFAIR.

" The theory of drainage, an operation in agriculture of almost equal importance to that of plowing, is, in reality, very simple, although it depends upon several physical and chemical conditions in themselves very distinct. The mechanical

---

* Morton's Encyclopædia of Agriculture.

conditions effected in thorough drainage require: (1) That all the rain which falls on the surface should quickly sink to the level of the drain, and be carried off; (2) that, in thus sinking, the finely divided portions of the soil should not be carried away, but that the water should be filtered before entering the drain; (3) that the depth of the drain should be sufficient to carry off underground water, and produce amelioration in the soil to sufficient depth. Keeping these mechanical conditions in view, the two principal effects produced, and which require explanation, are the following: (1) The increased temperature of the soil, by which crops mature upon it with greater rapidity; (2) its increased fertility and better adaptation for all kinds of cultivated crops.

"These two main improvements require separate consideration.

"When water stagnates in a soil, air is at the same time excluded, and the necessary amelioration of the organic and inorganic ingredients cannot be effected. In all cultivated soils decaying matter has a positively injurious action, even on its mineral ingredients, by reducing the higher state of the oxidation of the iron generally present into the lower and injurious condition. In soils permeable to air, this evil is at once counteracted by a fresh absorption of oxygen from the atmosphere; but in soils in which water stagnates this remedial process does not exist.

"The heat of the sun falling upon wet land does not exercise its genial influence in promoting the growth of plants, but expends it in evaporation of the stagnant water. In doing this much of the sensible heat is rendered latent, or, in other words, is deprived of its warming properties. Water, in being converted into steam, absorbs or renders latent an enormous amount of heat, which is of course robbed from the soil; for it otherwise would be used in the more profitable manner of maturing the crops growing upon it. Some idea may be entertained of the amount of heat absorbed and rendered useless to the plants growing on the soil, if a special case be taken for illustration. It is found that porous chalk soils evaporate only one half the fall of rain, the rest infiltrating and running off as springs and streams, or being afterward found as wells.

"This, therefore, is a case very favorable to a wet soil, which would in reality allow a very much smaller quantity of rain to pass it; nevertheless, the porous land would require an expenditure of nearly twelve hundred weight of coal per day to evaporate artificially one half the rain which falls on an acre during the year. In other words, more than two hundred and nineteen tons of coal annually would be required for every acre of undrained land, so as to allow the free use of the sun's rays for the legitimate purpose of growing and maturing the crops cultivated upon it.

"It will not, therefore, be surprising that undrained soils are, in the language of the farmer, ' *cold.*'

"But in addition to the heat abstracted by the evaporation of water in undrained soils, other physical properties combine in reducing their temperature.

"One of these is the low conducting power of water. When the sun's rays infringe upon the surface of a watery soil, it raises the temperature of the water; but the heated water, being lighter than the cold water beneath, remains on the surface, and the heat cannot penetrate into the interior of the soil. But at night the very reverse action ensues; for the water, rendered cold at the top, descends, by an interchange with the hotter water beneath, which, in its turn, being cooled again, sinks; and thus the whole soil becomes quickly reduced to the same temperature as the external air, and the roots of the plant frequently suffer from being thus chilled. Water radiates its own heat freely into space, and hence a watery soil is quickly cooled in a cold night by the heat which the water distributes into the colder atmosphere.

"All these evils tend to reduce the temperature of undrained soils, and to render them less fitted for the growth of cultivated crops, which, in general, require a genial warmth.

"When soils are drained to a sufficient depth, the condition of the soil, with regard to temperature, is entirely altered. The redundant water does not now stagnate on it, but is immediately carried off. The aqueous moisture of the atmosphere, condensed into raindrops, is of a higher temperature than the air itself. This arises from the circumstance that when vapor becomes liquid it renders sensible that latent heat which it had absorbed to keep it in the gaseous state. The rain, therefore, in its passage communicates its own natural heat in addition to the higher heat of the soil's surface, and quickly percolating through it, and being removed by drainage, it does not require an additional amount for the purpose of evaporation. The similar warming action of rain on a drained soil is also exerted by dew in the coldness of the night.

"Soils and the plants upon them radiate heat into the atmosphere, from which is deposited water in the form of dew, as soon as their temperature is lower than that of the surrounding air. But the dew deposited upon the cold surface still preserves the latent heat, rendered sensible by its condensation, and this heat prevents the extreme chilling which would otherwise take place. The texture and porosity of drained soil soon change by chemical actions, so that they become more absorptive for moisture during dry weather. In fact, such soils do attract a large quantity of the aqueous vapor always present in the air, even in the driest weather, and thus prevent the parching of plants from the heat of the sun in the absence of moisture.

"These very obvious improvements in the condition of soils, depending upon their relation to heat and moisture, have practically the effect of an amelioration in the climate of a district.

"The sun's rays now produce their full effect on the soil and on the crops, without being robbed of their heat by the stagnant water of the soil, unable to effect its escape except by evaporation. The chemical effects of drainage, in promoting increased fertility, are not less striking. Rain-water always contains in solution air, carbonic acid, and ammonia. The first two ingredients are among the most powerful disintegrators of a soil. The oxygen of the air and the carbonic acid being both in a highly condensed form, by being dissolved exert very powerful affinities on the ingredients of the soil. The oxygen attacks and oxidizes the iron; the carbonic acid, seizing the lime and potash and other alkaline ingredients of the soil, produces further disintegration, and renders available the locked-up ingredients of this magazine of nutriment. Before these can be used by plants they must be rendered soluble; and this is only effected by the free and renewed access of rain and air. The ready passage of both of these, therefore, enables the soil to yield up its concealed nutriment. The soil thus acted upon becomes soon changed to a certain extent in its mechanical as well as its chemical character. The particles of the soil being comminuted, are rendered more absorptive of the gaseous foods of plants—carbonic acid and ammonia. The porous soil thus becomes richer in organic food at the same time that it is made to yield its nutritive mineral riches to the plants growing upon it. The peculiar chemical action exerted by the surface of soils for fixing ammonia and other soluble ingredients in water becomes more powerfully exerted.

"The water being removed from beneath the roots of plants by an adequately deep drainage, prevents the depression of temperature in the manner described. But, at the same time, it opens a new magazine of nutriment, by enabling the air and carbonic acid to reach the lower parts of the soil, and to ameliorate its injurious ingredients, while it liberates those which are useful.

"The plant has, therefore, a wider range in which it may seek its food, and is thus enabled to extend its roots in search of nutritive matter, which it formerly refused to do in a cold wet soil, in which the constituents were unfit for its healthy growth.

"Hence it is apparent that drainage is a most powerful agent in agriculture. By it the temperature and therefore the climate of soils is elevated; their porosity for moisture, though not for *wet*, is increased; their disintegration is effected, and nutritive, soluble materials are liberated; the organic gaseous food of plants is furnished by absorptive action in greater quantity than before; and the injurious organic and mineral ingredients of the soil are so far altered as to be positively beneficial to vegetation. With such advantages it is not surprising that drainage has become an essential operation in agriculture."

DRAINAGE IN AMERICA.

As compared with Great Britain, the climate of America presents a greater variety, but with a lower average rainfall, a less humid atmosphere, and greater rapidity of evaporation, because of the very much larger prevalence of sunshine at all seasons of the year. Nevertheless, all that has been said in favor of drainage and its results applies with equal force to the greater part of the United States and Canada, while many portions of the country require in connection with thorough underdrainage an equally elaborate system of irrigation either with river water or the sewage of towns, in order to derive the full benefit of the system of drainage. The testimony of those who have had long experience in drainage, where the work was thoroughly well done (and all else is absolute waste of money and labor), assert that with drainage alone the crops are so largely increased in quantity and improved in quality that the expense of the original cost is repaid in from two to three years. It is believed by many who have given the subject careful study that in America great advantages will accrue if the systems of drainage and irrigation shall be combined for all soils where practicable.

Great improvement has been made in recent years in the quality of pipe tile. The best are those that are hard-baked and glazed.

The plows and machines for use in soils which are free from stone promise to greatly lessen the cost. Vast areas of the best land are at present saturated a large part of the year, or entirely and perennially drowned. These are also a prolific source of fevers, ill-health, and poverty, to the majority of the inhabitants of neighboring districts.

Drainage is the only remedy.

Among the various ditching machines, there are several that promise to give aid in reducing the cost of excavation, and also greatly facilitating the speed of the work.

VALUE OF MANURE.

" No cattle, no dung ; no dung, no crop."—Flemish Adage.

Most farmers keep themselves in a state bordering on impoverishment by a neglect to save manures. To allow the liquids to percolate through a porous soil beneath the barn and stable yards is to lose the greater part of manurial value. According to the best authorities, the urine is of more value than the solid excrement, being ordinarily of double value, and under high feed is quadruple the value of the dung of equal weight. According to experiments of German chemists, fully ninety-five per cent. of all the valuable fertilizing elements digested were recovered in the liquid excrement. The undigested elements are passed as solid excrement. The feed was barley meal.

NITROGEN STORED UP AND VOIDED FOR 100 CONSUMED.

ANIMALS.	Stored up as Increase.	Voided as Solid Excrement.	Voided as Liquid Excrement.	In Total Excrement.
Sheep	4.3	16.7	79.0	95.7
Oxen	3.9	22.6	73.5	96.1
Pigs	14.7	21.0	64.3	85.3

ASH CONSTITUENTS STORED UP AND VOIDED FOR 100 CONSUMED.

ANIMALS.	Stored up as Increase.	Voided in Total Excrement.
Sheep	3.8	96.2
Oxen	2.3	97.7
Pigs	4.5	95.5

The combined excrements are rich in both nitrogen and mineral constituents. Two thousand pounds of the solid would contain fourteen pounds, and of the liquid twenty-eight pounds, on light feed; but a rich food would give nearly double for the liquid, or more than fifty pounds. The great essential to an improved agriculture is the saving of all manure, and then properly applying it to the soil when and where needed. It is simply a question of prosperity with manure, or poverty without manure.

TABLE OF MANURE VALUES.

ARTICLE.	Estimated Value of Manure from 2000 lbs.
1. Linseed Cake	$19.54
2. Peas	13.65
3. Clover Hay	9.65
4. Oats	7.40
5. Wheat	7.08
6. Maize	6.76
7. Meadow Hay	6.43
8. Barley	6.27
9. Oat Straw	2.90
10. Wheat Straw	2.68

Article.	Estimated Value of Manure from 2000 lbs.
11. Barley Straw	$2.26
12. Potatoes	1.51
13. Mangolds	1.08
14. Carrots	.86

The above is a part of the table from Stewart's work, " Feeding Animals," which is made up from the estimates of Sir J. B. Lawes.

TABLES SHOWING AMOUNT OF NITROGEN, POTASH, AND PHOSPHORIC ACID IN 1000 POUNDS, AND THEIR VALUE PER TON AT A LOW ESTIMATE.

SUBSTANCES. Manufactured Products.	Dry Matter.	Nitrogen. 18 cts.	Potash. 6 cts.	Phosphoric Acid. 10 cts.	Manure, Value per Ton.
	Lbs.	Lbs.	Lbs.	Lbs.	
1. Linseed Cake	880	45.0	14.7	19.6	$21.88
2. Linseed Meal (extracted)	903	59.8	17.0	25.6	28.68
3. Wheat Bran	865	22.0	14.8	32.3	16.15
4. Rye Bran	875	23.2	19.3	34.2	16.43
5. Millet Meal	860	18.3	2.3	5.5	8.32

GRAINS AND SEEDS.

SUBSTANCES.	Dry Matter.	Nitrogen. 18 cts.	Potash. 6 cts.	Phosphoric Acid. 10 cts.	Manure, Value per Ton.
	Lbs.	Lbs.	Lbs.	Lbs.	
1. Beans	855	41.0	12.0	11.6	$18.52
2. Vetches	864	44.0	6.3	7.9	18.17
3. Flaxseed	905	36.0	12.3	15.4	17.51
4. Peas	857	36.0	9.8	8.8	15.87
5. Oats	870	20.6	4.5	6.2	10.27
6. Wheat	856	18.8	5.4	8.0	9.01
7. Rye	851	17.6	5.4	8.2	8.62
8. Barley	860	17.0	4.9	7.3	8.16

HAY.

Substances.	Dry Matter.	Nitrogen. 18 cts.	Potash. 6 cts.	Phosphoric Acid. 10 cts.	Manure, Value per Ton.
	Lbs.	Lbs.	Lbs.	Lbs.	
1. Green Vetches...............	840	22.7	30.9	9.4	$13.75
2. Green Peas.................	833	22.8	29.6	9.7	13.69
3. White Clover...............	840	23.8	10.6	8.5	11.53
4. Lucern....................	840	23.0	15.2	5.1	11.00
5. Red Clover, in blossom......	840	19.7	19.5	5.6	10.55
6. Green Oats................	855	14.7	24.1	5.1	9.20
7. Timothy..................	856	15.5	17.2	6.8	9.00
8. Meadow Hay...............	857	15.5	16.8	3.8	8.35
9. Red Clover, ripe...........	840	15.0	12.2	3.5	7.56
10. Dead Ripe Hay............	856	12.0	5.0	2.9	5.56

GREEN FODDER.

Substances.	Dry Matter.	Nitrogen. 18 cts.	Potash. 6 cts.	Phosphoric Acid. 10 cts.	Manure, Value per Ton.
	Lbs.	Lbs.	Lbs.	Lbs.	
1. Lucern....................	247	7.0	4.5	1.5	$2.94
2. Hungarian Millet..........	320	5.3	8.6	1.3	2.54
3. Rye, in blossom...........	300	5.3	6.3	2.4	2.51
4. Green Vetches.............	180	4.9	6.6	2.0	2.35
5. Green Peas................	185	5.1	5.6	1.8	2.34
6. Red Clover................	200	5.2	4.6	1.3	2.27
7. Meadow Grass.............	300	4.8	6.0	1.5	2.24
8. Swedish Clover............	185	5.2	3.5	1.0	2.18
9. White Clover..............	190	5.0	2.4	2.0	2.15
10. Oats, coming into bloom.....	180	3.6	7.1	1.7	1.94
11. Timothy..................	300	5.4	6.1	2.3	1.94
12. Oats, in blossom...........	230	3.0	6.5	1.4	1.61

STRAW AND ROOTS.

SUBSTANCES.	Dry Matter.	Nitrogen. 18 cts.	Potash. 6 cts.	Phosphoric Acid. 10 cts.	Manure, Value per Ton.
	Lbs.	Lbs.	Lbs.	Lbs.	
1. Oat Straw...............	830	5.0	10.4	2.5	$3.54
2. Barley Straw..............	850	5.0	9.7	2.0	3.36
3. Wheat Straw	857	4.8	5.8	2.6	2.94
4. Potatoes..................	250	3.4	5.6	1.8	2.55
5. Mangolds................	115	1.9	3.9	0.7	1.29
6. Carrots.................	142	1.6	3.2	1.0	1.16

" The above tables are compiled from Professor Stewart's ' Feeding Animals.' The estimates are made for the elements of nitrogen at eighteen cents, phosphoric acid ten cents, and potash six cents a pound. This estimate is a low one, and holds good for the value when both the liquid and solid excrements are saved, and will help to give an approximate estimate of the values of manure from various fodders."

TABLE OF MANURE VALUES FROM JOSEPH HARRIS'S " TALKS ON MANURES."

IN 1000 POUNDS OF MANURE.		Nitrogen.	Potash.	Phosphoric Acid.
		Lbs.	Lbs.	Lbs.
Cow	Solid...........	2.9	1.0	1.7
	Liquid..........	3.4	4.0	1.6
Horse	Solid...........	4.4	3.5	3.5
	Liquid..........	15.5	15.0	
Sheep	Solid...........	5.5	1.5	3.1
	Liquid..........	19.5	22.6	0.1
Swine	Solid...........	6.0	2.6	4.1
	Liquid..........	4.3	8.3	0.7
Hen...........		16.3	8.5	15.4
Guano...........		100.0	23.0	150.0

These figures are given for ordinary care of animals, where the horse gets the best all the year, and the hog rich food a part of the year. If the cow was fed as well as she ought for a full yield of milk and butter, her manure would be as rich as that of the horse.

RELATIVE VALUES OF URINE.

SOURCE.	Water, per cent.	Solid Organic Matter, per cent.	Solid Inorganic Matter, per cent.
Man..........................	96.9	2.34	0.76
Cow (not in milk)...................	93.0	5.00	2.00
Sheep........................	96.0	2.80	1.20
Horse........................	94.0	2.70	3.30
Pig..........................	92.6	5.60	1.80

METHODS OF SAVING MANURE.

That there must be a radical reform in methods of treating and saving manure ought to be apparent to every man that owns an acre of ground, or is interested in the material prosperity of his country.

SOURCES OF MANURES.*

"Manure includes every substance, whether of animal, vegetable, or mineral origin, which, when applied to the soil, has the effect of increasing its fertility.

"In practical agriculture manures are divided into two classes—natural and artificial; the former being originally derived from the soil itself, in the different forms of forage, roots, plants, corn, and purchased food—all of which being consumed by cattle, yield that much-prized substance familiarly known as farmyard manure. Artificial, or, as they are sometimes termed, special or light manures, are, on the contrary, all derived from sources extraneous to the usual products of the farm—that is, they are neither directly the product of vegetable growth nor indirectly the residuum of the consumption of vegetable substances by animals. Thus guano is primarily derived from the ocean, in the fish consumed by the sea fowl, whose excrements, having accumulated on islands and rocks, furnish an almost inexhaustible supply of a manure so powerful and concentrated as to baffle all artificial attempts at imitation.

"Seaweed is another gift of the great deep, and is cast upon our shores in immense quantities by the storms and tides. The earth presents us with another class of manures, not the result of vegetable growth, but the product of great geological events: take, for instance, the limestone rocks, chalk beds, marl beds, and gypseous deposits; the coprolitic and other collections of phosphate of lime; the nitrates of soda and potash, which appear on its surface in efflorescent incrustations in some

---

* John Haxton, Morton's Encyclopædia.

districts of India and upper Peru ; and the sulphur from which that powerful acid, oil of vitriol, is obtained, which so greatly facilitates and economizes the effect of bones and coprolites. The commercial industries are continually adding to our supplies of manure, in the refuse substances of various manufactured articles ; thus the refuse substances of gas-works, consisting of ammoniacal water and the lime used in purifying the gas from sulphurous acid, are now largely employed as fertilizers— the former in a liquid state or the more portable form of sulphate of ammonia, and the latter, after exposure to the air, as sulphate and carbonate of lime. Our salt mines also furnish us with muriatic acid and sulphate of soda, both of which are obtained from salt by various processes in the chemical arts. The manufacture of prussiate of potash yields large quantities of animal carbon derived from the hoofs and horns employed in the process. Bone charcoal is also another refuse product of commerce, and is obtained in the form of a grayish gritty powder from sugar refiners, who employ large quantities of charred bones in clarifying the liquor of dissolved raw sugar before converting it into the whiter and purer sorts.

" Besides these sources of manure *there is one of far higher importance, in a national, sanitary, and economical point of view, than all others, not even excepting guano ; we mean the sewers of all the towns.*

" This source of fertilizing wealth has been strangely overlooked hitherto, a fact which is remarkable when contrasted with the saving and economy displayed in every department of the mechanical arts. Not a rag or shred of clothes is permitted to be lost, but is turned to some use in the making of paper ; not a scrap of rusty, malleable, or cast-iron but is carefully collected, and the one welded together into bars by the ponderous strokes of the steam-hammer, while the other is put into the furnace, whence it issues ready to be formed into any shape which the founder may desire. The gathering and collecting of the odds and ends which constitute the refuse of the useful arts are so important and profitable that they form a large trade in the country ; yet, notwithstanding the examples of success set before our capitalists and speculators by these humble departments of industry, it is only lately that the subject of applying the valuable contents of our city sewers to the purposes of agriculture has attracted anything more than cursory attention. Now, however, there appears something like a systematic attempt to turn to a useful and important purpose that which has so long run to worse than waste, and which, if economized, would not only increase the food of the country, but also render our towns more cleanly and healthful.

" In addition to the natural and artificial sources already specified, there is another class of manures to which the term artificial may be exclusively applied. They consist individually of different substances, mixed in various proportions, according to the special purposes to which they are to be applied, and according to the theoretical opinions of those who compound them.

FARMYARD MANURE.

"According to Dr. Thomson's experiments,

100 lbs. of grass consumed by a cow daily give...................71 lbs. of dung.
80 lbs. of grass and 4¼ lbs. of barley, water *ad libitum*...........78 "   "   "
85 lbs. of grass, 5¾ lbs. of malt, with water as before.............82 "   "   "
25¼ lbs. of hay, 10¼ lbs. crushed malt.........................77 "   "   "

Average..................................77 lbs. of dung.

"From these figures it appears that one hundred pounds of grass, consumed indoors by a cow, produce seventy-one pounds of solid and liquid manure. But a cow also produces from twenty to twenty-five pounds of milk from one hundred pounds of grass; so that were the grass consumed by an ox instead of a cow, we would infer, from the fact of his only increasing a few pounds of live weight daily, that he would void a greater weight of dung than a cow. The quantity and composition of dung, however, are greatly dependent upon the amount of water drank along with the food; but all things being alike, it seems logical, as well as a correct physical deduction, to consider that in the case of a cow and an ox of equal size and capacity, consuming the same amount of food, the one giving a full supply of milk, and the other increasing at a maximum weight, the latter will yield the greatest quantity of manure.

"In stall-feeding the amount of manure will stand thus (for medium-sized cows):

	Tons.	Cwts.	Qts.	Lbs.
Solid dung for 210 days, 55 lbs. daily	5	3	0	24
Solid dung for 155 days, 41¼ lbs. daily..........	2	17	1	20
Litter for 365 days, 14 lbs. daily...............	2	5	2	14
Urine absorbed by litter, 22¼ lbs. daily	3	13	1	8¼
Total solid dung.....................	13	19	2	10¼
Urine which flows into tank	7	18	0	5½
Total manure and litter...............	21	17	2	16

"Estimating the gallon of urine to weigh ten pounds, the whole quantity collected in the tank will amount to seventeen hundred and seventy gallons yearly. According to Sprengel's analysis of cow's urine, this quantity would contain three hundred and thirteen pounds of ammonia, besides other substances of a valuable nature also.

"Under the ordinary system of managing dairy cows the foregoing statements will not harmonize with general experience.

" We scarcely require any chemistry to teach us that the quality of dung voided by any description of fattening stock or milch cows is the difference between the food consumed and that portion of it retained in their bodies, as flesh, fat, etc., or withdrawn in the milk, perspiration, respiration; or, in other words, the dung is the food, minus the flesh, fat, milk, and insensible waste through the lungs and skin. The dung is, therefore, inferior to the food in a fertilizing point of view, just in proportion to the substances extracted from the latter by animals.

MANAGEMENT OF MANURE.

" This may be said without exaggeration to be the most important department of farm practice, and unfortunately one in which there is greater need of improvement than any other. Notwithstanding the fact that the proper management of the manure heap has been explained and enforced by the teachings of agricultural chemistry year after year, the practical application of the lesson remains in a great measure to be made.

" Farmyard manure, as heretofore, continues to be carted out from rain-soaked straw-yards to the distant fields, and there deposited in large, ill-formed heaps, exposed to rain, wind, and sun for weeks and months.

" Many farmers whose practice otherwise is unassailable are yet strangely blinded to the great loss sustained by exposed manure heaps. On the great majority of farms, even in the best-farmed districts, there is a fearful waste of food-producing material.

" Badly constructed homesteads have, no doubt, greatly contributed to this state of things, and it is very seldom that any provision is made, in the construction of new ones, for the preservation of liquid manure or for protecting the straw-yard from being deluged by rain poured into it from the surrounding roofs.

" A loss of manure is equivalent to a diminution of produce, and this again, by lowering the profits of farming, necessarily depreciates the value of land. All manure should be made under cover, either in stalls, boxes, or sheds; if in the former, it must be removed daily, so that a covered shed will be necessary for its protection; if in the second, it may be allowed to accumulate for two or three months; and by the latter mode it may remain until required for laying on the land, provided height of the roof will admit of its being accumulated. How is it that we invariably find box-feeding or house-feeding of some kind or other always accompanied by bulky crops of corn, roots, and clover? Just because the manure so made is richer and more abundant than on those farms where the horse-pond receives the drainage of the courts and byres.

" We need only point to what has been already said in regard to the quantity of urine voided by animals to prove that if there be no tank to receive the drainings of stall-fed animals, the loss sustained will amount to one third the weight of the whole dung, or twice that of the liquid part. Few who have not studied this subject are aware of the enormous quantity of fertilizing materials that accompanies the little black stream that oozes from a straw-yard where there is no tank to drain off the surplus liquid.

MANURE HEAPS.

"There being few steadings where the accommodation is sufficient to hold all the manure until wanted for application to the land, it is necessary and particularly convenient to cart it out to the more distant fields, and to make it up in large heaps.

"Wherever this is necessary, the cart should be driven upon the heap before being emptied.

"By so doing, manure is consolidated, air is excluded, and fermentation prevented.

"In finishing the heap, the ends should be raised nearly on a level with the centre, which is easily done by a little attention on the part of the carter. These portions unavoidably left low at both ends for the cart to get on and off the heap can be raised on a level with the rest by backing several cartloads, tilting them up, and throwing up the manure with forks.

"After this the whole heap should be covered with earth from the sides, three or four inches thick, which should be well beaten down with a spade. Road scrapings are even better than common soil, as they are in a very minute state of subdivision, besides always containing a considerable quantity of manure dropped on the roads.

"If these are sufficiently wet to beat into a plaster on the heap, so much the better, as the surface will thereby be more hermetically sealed, both within and without. In addition to all this the whole surface may very profitably be sprinkled with sulphuric acid, so that any ammoniacal gas escaping may be at once arrested by this useful agricultural detective, whose affinity for fugitive alkalies is altogether insatiable. Dissolved bones, having a free acid, may also be employed for fixing ammonia; and if the manure be intended for turnips or mangolds, it is an excellent plan to mix a few hundred weight through the whole heap.

"An excavated site, built on three sides, with a wall four feet high, is the best mode of preserving manure in a field; there would be no risk of loss from evaporation or fermentation, provided the top and open side were covered with earth.

APPLICATION OF MANURE TO THE SOIL.

"The quicker farmyard manure is buried, the better. This is a maxim that holds good everywhere, and under every circumstance: because, when once covered up by three or four inches of earth, it is safe from all risk of being lost, as the soil has both a physical and chemical power of retaining ammonia, while, at the same time, it yields it up readily to the growing plants. The wasteful practice of spreading manure on the surface of the soil, and allowing it to lie bleaching for weeks, and even months, before being plowed in, is still carried on and stoutly defended by hosts of clay-land farmers.

"If the perpetrators of such an enormity be right, science is at fault, analysis is a delusion, and ammonia and all its kindred a family of impostors.

"The practice in Syria of making the dung into cakes and sticking these upon the walls of their houses to dry in the sun, preparatory to their ultimate destination of being burnt as fuel, is not much more wasteful than spreading out farm-yard manure to the winds, rains, and sun for months together.

"A farmer who imports ammonia from the Chincha Islands and dissipates to the winds that furnished by his own farm, is nearly as wasteful as he would be were he to give away his straw for nothing, and to purchase from others what he required for his own use."

ASH HEAP.*

"There is a source of valuable and extremely useful manure on every farm, of which very few farmers avail themselves—the gathering together in one spot of all combustible waste and rubbish, the clippings of hedges, scouring of ditches, grassy accumulation on the sides of roads and fences, combined with a good deal of earth. If these are carted at leisure times into a large circle, or in two rows, to supply the fire kindled in the centre, in a spot frequented by the farm laborers, with a three-pronged fork and shovel attendant, and each passer-by is encouraged to add to the pile whenever he sees the smoke passing away so freely as to indicate rapid combustion, a very large quantity of ashes are collected between March and October. In the latter month the fire may go out; the ashes are then thrown into a long ridge, as high as they will stand, and thatched while dry. This will be found an invaluable store in April, May, and June, capable of supplying from twenty to forty bushels of ashes per acre, according to the care and industry of the collector, to drill with the seeds of the root crop. It is a good practice to dissolve bones with acid in the beginning of February, and when reduced to a pulp to mix them up with the ashes in a large heap, which should be turned over two or three times at intervals, and the bone paste well reduced with the shovel, and thoroughly mixed at each turning; by the month of May a homogeneous compound will be formed that will run freely and evenly through the drill, and form an inviting bed for the seed."

---

* C. Lawrence, Morton's Encyclopædia.

RESTORATION OF SOILS.*

"The two principal means of restoring the fertility of a soil which has been diminished by the continued cropping, are : (1) The mechanical improvement of the soil. (2) The application of manure.

"All plants take away from the soil a certain quantity of mineral matters which are essential to their existence. Some plants require more phosphoric acid than others, which want a greater supply of potash for a healthy growth ; some again require for their perfection much lime, others silica ; but all take up a number of inorganic chemical substances, which the plant can have derived only from the soil on which it was grown. If it is true that these mineral elements are essential to the very constitution of all plants—and there can be no doubt in reference to the function of the inorganic matters of the soil—it follows that sooner or later the most fertile soil must become exhausted to such an extent that it will no longer produce remunerative crops. Experience has long ago proved this, and at the same time pointed out two ways which are pre-eminently calculated to restore the native fertility of a soil deteriorated by long-continued cropping.

"The first includes all those practical operations, such as digging, plowing, rolling, whereby the physical structure of a soil is improved, or its latent fertilizing properties developed by strictly mechanical means.

"The second consists in the application of manures. In all countries where agriculture is practised as an advancing art, the application of manures, together with their preparation and economy, are justly regarded as the most valuable and indispensable means of an improved system of farming. Hence, the great importance which attaches to the subject of manures in general ; to the theory of their action and their rational application ; to the best modes of preserving and increasing the fertilizing value of farm-yard manure, and to the methods which are pointed out from time to time of saving many natural products, which are still in so many instances allowed to run to waste ; or to the means of converting comparatively valueless articles into fertilizers. It is for these and similar reasons that the subject of manures has been treated in this work at great length.

"Whatever acts as a fertilizer, which is brought to the land, may be termed a manure.

"Clay, lime, marl, water, air, and even sand, accordingly come under the denomination of manures, just as well as dung, urine, and guano. It is quite true the beneficial effects resulting from the application of clay, marl, lime, sand, and many other compounds are realized chiefly in the altered physical condition of soils to which the above substances have been applied. In many cases they do not act so

---

* Prof. Voelcker, Morton's Encyclopædia.

much by supplying direct nourishment to the plants as by indirectly facilitating the absorption of the hidden treasures, which, being present in a dormant state in a soil, are thereby rendered available for the use of plants.

" We shall include *all* materials which are added to the soil for the purpose of increasing its productive power under the name of *manure.* A normal manure will be such only as shall furnish to the growing plant all the elements of food which the plant requires for the formation of its roots, stem, leaves, and fruit.

" A rational application of manures to the land is dependent on several circumstances ; and we can entertain the hope of manuring our fields in the most successful and economical manner only when the following four points shall have been determined accurately :

" 1. The wants of the plants intended to be cultivated in reference to the elements of nutrition.

" 2. The wants of plants in reference to the physical condition of the soil.

" 3. The composition of the soil.

" 4. The composition of the manure.

" The organic portion of which the great mass of all cultivated plants is made up is derived principally from the atmosphere ; whereas the inorganic part of plants, remaining behind in the form of ashes when a plant is burnt, can be supplied only by the soil or the manure.

CONSTITUENTS OF MANURES.

" 1. *Nitrogen, in the form of Ammonia or Nitric Acid.*—Nitrogen is one of the most important of all fertilizing substances ; it must be considered as the most valuable, in so far as its commercial price is taken as the test in estimating its value. It is, however, useful to the luxuriant growth of our cultivated plants only in one of the above forms ; for in a free state it is not assimilated by plants to any extent, nor does the nitrogen of organic bodies become available to plants before the nitrogenized matters have undergone a change by fermentation or putrefaction, the result of which change, among other products, is the formation of ammonia or nitric acid. Nitrogen in either of these two forms exercises a most powerful action in manure, particularly when applied to plants at an early stage of their growth ; at a later period of development the application of ammonia or nitric acid appears much less effective, and sometimes even useless. The rapid forcing effects of ammonia, of the ammoniacal liquor of gas-works, of sal-ammoniac, and ammoniacal salts in general, are too well known generally to require reference to the direct numerous practical field experiments which have been made in order to ascertain the efficacy of ammonia as a fertilizer. It will scarcely be necessary to allude to the presence of ammonia in guano, soot, etc., as being one of the causes of the forcing properties which characterize these and other fertilizers. The beneficial effects of ammonia and its salts have been occasionally denied, because the materials containing these fertilizing agents have been improperly used.

"As a general rule, ammonia or its salts should never be used on the farm in a concentrated form. Their caustic properties necessitate their application in a diluted state.

"Every practical man is acquainted with the burning effect of strong liquid manure or the ammoniacal water of gas-works, and therefore never applies the first to his land in dry weather, or the latter, except diluted with much water, or mixed with other substances.

"It has been observed that the nitrogen of matters, such as flesh, bones, hair, and horn-shavings, benefits vegetation only in so far as it becomes changed into ammonia. When these substances putrefy, ammonia is generated in large quantities, and it is principally for these reasons that they act as fertilizers. In a fresh state they are almost entirely useless, but they are rendered the more powerful in their action the further their decomposition has proceeded. Fresh bones, hair-refuse, wool-refuse, unfermented urine, long dung, are much slower in their action than the same materials after having undergone fermentation or putrefaction. In the latter state they contain ammonia ready formed, which the plant can assimilate at once; but in the first case the decomposition of the nitrogenized matter proceeds slowly in the ground, particularly when plowed in deep; and the plants are thus made to wait a long time before they can absorb the ammonia, which is generated during the decomposition of the nitrogenized organic matters. In stiff soils, and in dry seasons, the formation of ammonia proceeds so slowly that the beneficial action of manuring substances is frequently lost in the first year, because if plants have passed the period of their most vigorous growth they derive very little advantage from the ammonia.

"Therefore wool-refuse, bones, and other fertilizers, the action of which depends on the ammonia which is gradually formed on their decomposition, ought never to be applied in spring, when it is intended to benefit the first crop by such application, but at least three or four months, and in many cases even longer, before the crop is sown. On the other hand, manuring substances, such as guano, soot, refuse-water of gas-works, sal-ammoniac, sulphate of ammonia, putrefied liquid manure, which all contain large quantities of ready-formed ammonia, exercise a surprisingly quickening power on grass land, wheat, and all plants at an early stage of their growth.

"The value of ammonia and its salts in manuring substances has been greatly underestimated by Liebig and his followers, who believe with him that there is no necessity for supplying plants with manures containing ammonia, because plenty of it is afforded to them for assimilation by the air. Now, although it cannot be denied that plants absorb the ammonia of the air, and that the air presents to them an almost inexhaustible source, from which they may derive ammonia, it is nevertheless true that this property of absorbing and elaborating the atmospheric ammonia in sufficiently large quantities is shared by comparatively few plants. To most vegetable productions the supply of ammonia from that source proves insufficient; and as we know practically that almost all our cultivated plants are dependent on other sources, from which they can derive nitrogen, and as ammonia and its salts decidedly improve their condition, it would be unreasonable not to attach any value to the presence of these fertilizing materials in the different articles used as manures. In the form of nitric acid, nitrogen becomes also a most valuable manure, and in this state it closely resembles ammonia in its action. The effects of nitrate of soda, for instance, on grass land are strikingly exhibited by the succulent, luxuriant appearance and the deep green color which the grass assumes shortly after its application. Even small quantities of the alkaline nitrates exercise a most surprisingly quick forcing action on grass lands; and it is undoubtedly the case that cattle prefer grass to which top-dressing of nitrate of soda has been applied to grass grown without the intervention of that fertilizer.

"2. *Organic Substances, Humus.*—Organic matters, consisting of carbon, hydrogen, and oxygen only, are present in farm-yard manure and many other fertilizers in large quantities; but their importance as fertilizing agents is not to be compared with that of the nitrogenized organic matters, ammoniacal salts, or nitrates.

"Formerly the value of a manure was estimated according to the proportion of organic matters it contained; the chief fertilizing effects were thus referred to the presence of organic substances, which, on decomposition, furnished humus, the substance which for a long time was regarded as the only material from which plants derived any direct food. The value of the organic or humus-forming matters in manures, accordingly, was overestimated by former physiologists and agriculturists, until the researches of Liebig have placed it in a clear light that the effects produced by the organic portion of manures in comparison with those of their inorganic matters are so trifling that he disregards the organic substances in manures entirely. Although we do not agree with this view of the subject entirely—a view, it may here be observed, lately modified by Liebig himself—we still hold the opinion of those to be correct who regard the *inorganic matters* of manures as the chief fertilizing agents.

"In one important point, however, we must differ from the strict adherents of the mineral theory—namely, in attaching a much greater value to the nitrogenized organic matters than is done by Liebig and his followers. A little consideration will show the comparative insignificance of the humus-forming substances in relation to the nutrition of plants. In the first place, the insufficiency of humus to supply plants with organic food can be demonstrated by an easy calculation; for if we estimate the weight of the organic matters removed in a crop from the soil, and the amount of humus supplied by the manure, we shall find that a small proportion of the first can have been derived from the humus of the manure, even if we estimate the whole of the latter as having passed into the substance of the crop. We know, secondly, by direct experiments, that the great bulk of all plants is derived from the carbonic acid of the atmosphere, which presents plants with an inexhaustible source from which they may draw organic nourishment.

"A practical confirmation of this fact we find, thirdly, in the abundant crops of Indian corn which are raised in Mexico and Peru on soils destitute of all humus, without the application of any organic manure, as well as in the fertility of irrigated meadows, which likewise do not receive any organic manures. It is for these reasons that we do not attach to the non-nitrogenized organic matters the same importance as to the inorganic, which the plants can derive only from the soil or the manures.

"So far as the direct supply of food to plants is concerned, we are thus inclined to consider the importance of the organic matters of manures as insignificant in comparison with that of their inorganic substances. Indirectly, however, organic manures play an important part in relation to the growth of plants, inasmuch as, by their application, the physical condition of soils is materially improved. This function of the humus-forming substances in manures must not be overlooked. They are further useful to vegetation, because they absorb both moisture and ammonia from the atmosphere with great avidity, thus becoming indirectly suppliers of food; and because, on decomposition, they themselves furnish carbonic acid.

"While we ascribe the chief value of the non-nitrogenized organic matters to the alteration in the physical condition of the soil which they effect, and to the indirect food which they furnish to plants, their use as direct suppliers of food cannot be altogether denied, if dependence can be placed on Soubeiran's experiments, made in reference to the absorption of soluble salts of ulmic acid by plants. From these experiments Soubeiran concluded that ulmate of ammonia was taken up by plants; and Mr. Malaguti has confirmed and extended this observation by quantitative analysis.

"3. By far the most valuable *inorganic* constituent of manures is *phosphoric acid*, as it is a substance without which the grain of our cereals cannot come to perfection. Its deficiency in the soil is generally indicated by the poor, thin appearance of the ears of wheat, barley, or oats. Phosphoric acid rarely occurs in soils in sufficient quantities to equal the demands of the crops, and has therefore to be supplied in the form of manures. The beneficial action of bone-dust, superphosphate, coprolites, must be referred chiefly to the phosphoric acid which these fertilizers contain.

"In the same combination in which phosphoric acid is found in bones—that is, in the form of bone-earth or phosphate of lime—it occurs in the solid excrements of all domestic animals ; it consequently constitutes an important ingredient of farm-yard manure, and of all artificial manures which are applied with advantage to the growth of grain and root crops. It is worthy of observation that phosphate of lime, although insoluble by itself in water, is rendered soluble by the addition of a small quantity of ammonia to the water.

" This property of phosphate of lime agrees well with practical experience, which tells us that phosphate of lime, or phosphates in general, exhibit the most energetic effects on vegetation when they are mixed with ammoniacal salts or nitrogenized organic matters, which furnish ammonia on decomposition.

" For the same reason, the most powerful manures will be found those which contain much phosphoric acid and ammoniacal salts, or nitrogenized organic matter.

" 4. *Alkalies, Potash, and Soda.*—Potash and soda, particularly the former, are valuable component parts of farm-yard manure, and of all the better artificial fertilizers.

" Although potash and soda belong to the more widely diffused inorganic substances on the earth, their quantity in most soils is too small to justify us in neglecting the direct supply of salts of potash in some way or other. The solid excrements of horses, cows, sheep, and pigs contain but small quantities of salts of potash, which, being very soluble in water, are chiefly separated with the liquid excrements, or the urine of our domestic animals. *The preservation of their urine thus becomes a duty imperative on all farmers,* because they will otherwise lose all the advantages of the highly fertilizing salts of potash. In its chemical relation potash resembles ammonia closely, and the same is the case with the salts of potash and ammonia. In their effects on vegetation this similarity is observed ; for potash and its salts exercise the same stimulating or forcing action which we have seen is characteristic of ammonia.

" In manures potash occurs partly in combination with chlorine, as chloride of potash, partly in combination with sulphuric and silicic acid, as sulphate and silicate of potash.

" In the urine of carnivorous animals phosphate of potash also is found.

" All cultivated plants, particularly root crops and herbaceous plants, require potash as a necessary article of food, for their ashes contain large quantities of it. The chief reason of the beneficial effects produced by the application of wood ashes, liquid manure, and many natural silicates is, undoubtedly, the greater or smaller quantity of salts of potash which these kinds of manures contain. The principal cause of the fertilizing effects of burnt clay is to be referred also to the soluble potash, which in burnt clay exists in a larger proportion than in the same clay in its natural state. On burning the insoluble alkaline silicates occurring in clay are in a great measure decomposed, and potash is thus rendered soluble. The beneficial effects produced by the application of quicklime on some lands is also due to the liberation of potash in the soil, which previously existed in an insoluble state. Silicate of potash, which is found in farm-yard manure and other fertilizing mixtures, is a very valuable compound, which appears to exercise a beneficial action, particularly on grain crops. Much less effective than potash salts are the salts of soda, of which the more frequently recurring are chloride of sodium and the sulphate and silicate of soda. Generally speaking, the proportion of salts of soda in manures is larger than that of the salts of potash. There are few soils which do not contain naturally so much soda in one form or the other as to satisfy the wants of the crops which are raised upon them. It is for this reason that the value of soda salts as fertilizers is very much less than that of potash salts. It is so inconsiderable, that we need not care to supply the salts of soda by artificial means to the land. The localities where common salt proves most effective are inland places, far removed from the sea; and in such places beneficial effects following its application are intelligible. In the ashes of plants potash occurs almost always in larger quantities than soda, and this affords another proof of the greater value of the former.

" Nitrate of soda, which exercises a most decided and surprisingly quick forcing action on grass land, owes its efficacy principally, we believe, to nitric acid, and not to soda.

" 5. *Lime and Magnesia.*—Almost all manures contain lime and magnesia, which are indispensable for the healthy growth of plants. Farm-yard manure contains lime partly in the state of carbonate, partly as sulphate of lime. The latter compound, or gypsum, is a fertilizer, which frequently constitutes the chief component part of several artificial fertilizers, which have been mentioned; the better sorts of manures do not, or ought not, to contain too large an amount of gypsum.

" Lime and magnesia are among the most widely distributed mineral substances, and can be very economically added to soils in which a deficiency may have been found, in the form of gypsum, marl, quicklime, gaslime, limestone, chalk. As constituents of manures, lime and magnesia are not very important.

" 6. *Silica.*—All ashes of plants contain silica; some, as the ashes of straw, of wheat, barley, a very considerable proportion. Silica, for this reason, is an essential article of food to plants, without which many could not come to perfection. However, it is in but few cases that the farmer need care to apply silica to the soil, because most soils contain a large excess of it already. The only state in which silica can be taken up is in the soluble form, and it is in this soluble state that silica occurs in the solid excrements of animals. These are rich in soluble silica, and therefore particularly well adapted to soils deficient in this element.

" Silica, even in a soluble form, is far less important than any of the substances previously mentioned.

" 7. Sulphuric acid, chlorine, fluorine, oxides of iron and manganese, and sometimes alumina, are also constituents of many manures; but as these compounds are so generally distributed throughout nature, we find few soils which do not contain as much of them as is required to the healthy growth of plants. Their value as constituents of manures can, therefore, with propriety be altogether overlooked.

"These, then, are the constituents which ought all to be present in a universal manure, and which are present in farm-yard manure.

"In order to avoid misunderstanding, we would observe that when speaking of the different values of manures, we refer to their *commercial value.* In one sense all substances which are found in the ashes of plants are valuable, as they are *essential* to the perfection of plants, and in this sense lime or silica is just as valuable as potash or phosphoric acid, because the largest supply of the latter substances would not prevent the plants languishing for want of the former.

"Referring, then, to the commercial value of the fertilizing constituents of manure, it will appear from the above observations that they range in the following order :

"1. Nitrogen or, rather, ammonia and nitric acid.

"2. Phosphoric acid.

"3. Potash.

"4. Lime and magnesia.

"5. Soluble silica.

"6. Humus-forming organic matters.

"7. Sulphuric acid, chlorine, oxide of iron.

"Nitrogen, in the form of ammonia, ammoniacal salts, nitric acid, nitrates, or nitrogenized organic matters, is the most valuable ingredient of manures, because the mineral matters of manure show their full fertilizing effects only when decaying nitrogenized matters or salts of ammonia are present at the same time.

"Next in value follow phosphoric acid and potash, as both belong to the rarest of the mineral matters which serve as food for plants, and as both are required for their healthy growth in larger quantities than any of the other constituents which are usually found in the ashes of plants.

"The high value of nitrogen in manures has been fully recognized by Boussingault and Payen, who determined the quantity of nitrogen in a great many substances used as manures. These nitrogen determinations were used by them as the basis for calculating the principal relative fertilizing effects of different manures. In the second edition of Boussingault's 'Economie Rurale' he enlarges the general utility of the former table by adding to it another column, in which the equivalent weights are determined in relation to the quantity of phosphoric acid which they contain. In the subjoined table farm-yard manure is taken as the standard of comparison, and its equivalent is assumed to be 100.

"Thus, 250.0 pounds of wheat straw are equal in fertilizing effects to 100.0 pounds of common farm-yard manure, as far as the *nitrogen* is concerned ; but with respect to the fertilizing effects of the *phosphoric acid* 266.7 pounds of wheat straw are equal to 100.0 pounds of common farm-yard manure."

EXTRACT FROM TABLE REPRESENTING THE COMPARATIVE VALUE OF DIFFERENT
MANURING SUBSTANCES.

SUBSTANCE.	Water, per cent. Natural State.	Nitrogen, per cent. Natural State.	Phosphoric Acid, per cent. Dry.	Equivalents derived from per cent. of Nitrogen. Natural State.	Equivalents derived from per cent. of Phosphoric Acid. Natural State.	OBSERVATIONS.
Farm-yard Manure......	65.0	0.63	2.25	....	.....	England.
Mixed Manure.........	66.7	0.60	1.45	100.0	100.0	Farm-yard manure.
Wheat Straw..........	19.3	0.24	0.22	250.0	266.7	Alsace.
Oat Straw...........	21.0	0.28	0.21	214.2	300.0	Alsace.
Rye Straw	12.2	0.17	0.15	352.9	369.2	Alsace.
Carrot Leaves.........	70.0	0.85	....	70.6	.....	Green in autumn.
Clover Roots..........	9.7	1.61	....	37.3	.....	Air-dried.
Seaweed..............	39.2	0.86	..	69.8	160.0	Air-dried.
Fir Sawdust....:......	24.0	0.23	....	260.9	2400.0	
Oak Sawdust.........	26.0	0.54	....	111.1	1600.0	
Cow Dung...........	85.9	0.32	....	187.5	480.0	Solid excrements.
Cow Urine............	88.3	0.44	....	136.4	.....	
Excrements of Cow.....	84.3	0.41	0.55	146.3	533.3	Solid and liquid excrements.
Horse Urine..........	79.1	2.61	....	22.9	.....	Concentrated urine.
Horse Excrements......	75.4	0.74	1.12	81.1	178.9	Solid and liquid excrements.
Pig's Excrements.......	93.8	0.37	3.44	162.2	228.6	Solid and liquid excrements.
Sheep's Excrements.....	67.1	0.91	1.32	65.9	111.6	Solid and liquid excrements.
Human Urine.........	93.3	1.45	3.88	41.4	184.6	Berzelius.
Human Excrements.....	91.0	1.33	2.85	45.1	189.6	Solid and liquid excrements.
Unboiled Bones........	8.0	6.22	22.20	9.6	2.3	Containing ten per cent. of fat.
Peruvian Guano........	25.6	5.52	20.00	10.9	3.2	Denham Smith.
African Guano.........	25.0	6.19	17.00	9.7	3.8	Kasten.
Wood Soot............	5.6	1.15	1.00	52.2	51.1	
Oyster Shells..........	17.9	0.32	0.65	187.5	90.6	
Marl................	1.0	0.51	....	117.6	.....	
Seashore Sand.........	0.5	0.13	....	461.5	.....	

NATURAL MANURES.

"The atmospheric air may be regarded as the great storehouse which provides plants with organic food. It presents an inexhaustible source of carbonic acid, which is principally assimilated by the leaves of plants, and elaborated by them into starch, sugar, cellular tissue, etc. The great bulk of all plants, whether entering them by the leaves or the roots, owes its origin to this natural manure.

"Besides carbon, the air supplies plants with ammonia and with moisture. Though small in quantity, the ammonia is a very important constituent of the air, in reference to the nutrition of plants.

"During thunder-storms nitric acid, which unites with the ammonia, is also formed, and as nitrate of ammonia is a very soluble and highly forcing manuring substance, we can explain in some measure the fresh appearance of our fields after a thunder-storm. The moisture contained in the air in an invisible state provides plants with more water than the rain which falls upon the land.

"Rain-water, the purest natural water, is perhaps the most important of all natural manures, as without it vegetable as well as animal life would become impossible.

"Spring waters owe the additional effects which many exhibit, in comparison with pure or distilled water, to the presence of mineral or inorganic matters. Salts of lime, potash and soda, which occur in some waters, render them well adapted for irrigation.

"Some natural waters contain phosphoric acid, which are used with great advantage for irrigating meadows.

"The muddy deposits near the mouths of some rivers may also be called natural manures; the deposits belong to the most valuable fertilizers, and have converted a great part of the very sterile sands of Holland and Belgium into rich garden land.

QUALITY OF FARM-YARD MANURES.

"The quality and quantity of farm-yard manures are affected,

"1. By the quantity of food upon which the animal is fed.

"2. By the quality of the food.

"3. By the amount of water in ration and water drank.

"4. By the age of the animal. Richer in mature animals.

"5. By the purpose for which the animal is used, being increased by fattening and diminished by milk or work.

"6. By the treatment of animals, comfort increasing and hardship diminishing manure.

"7. By the quantity and quality of the litter.

"8. By the length of time the manure is kept, and the method by which it is preserved.

QUANTITY OF EXCREMENT VOIDED BY ANIMALS.

ANIMAL.	Solid Excrements.	Urine.
A cow furnishes annually......................	20,000 lbs.	8,000 lbs.
A horse furnishes annually....................	12,000 "	3,000 "
A pig furnishes annually.....................	1,800 "	1,200 "
A sheep furnishes annually...................	760 "	380 "

LIQUID MANURE.

"Neither the solid nor the liquid excrements, applied separately, constitute a universal manure, or a manure which can be used for the raising of all kinds of crops; liquid manure can never supersede the use of the solid, well-prepared farm-yard manure, if care is not taken to dissolve in it those substances which enter into the composition of the solid excrements of animals. In Flanders and some parts of Holland a most powerful liquid fertilizer is obtained by dissolving and distributing the excrements of animals in the liquid.

"During the fermentation of the liquid the solid matters are for the greater part dissolved, or at all events reduced to a fine mud, which remains easily suspended in the water.

"For the cultivation of flax, beets, and green crops in general such a liquid manure is preferred in Flanders to any other, as it has been found, by long experience, that in the liquid state the excrements of animals are best employed for the growth of these crops. The Flemish farmers accordingly bestow great care upon its preparation, and carefully collect the urine of the stables, which is conducted through drains into separate liquid manure tanks, into which all the drainings of the dung-heap are allowed to flow. In Belgium the urine and solid human excrements are not wasted, as with us. Before its application to the land this liquid manure must first be diluted with much water, as it is so strong that it would burn up and completely destroy the young plants, if the precaution were not taken to dilute the liquid, according to its strength, with three to six times its bulk of water.

"Even so diluted, it is advisable to apply it to the land in wet weather or when the soil is soaked with moisture, because in dry weather the manure is likely to exercise a burning action on vegetation. It appears incredible to continental farmers that our farmers should prefer willingly to pay heavy sums for the imported guano and other artificial manures, while neglecting to reap the benefit from those fertilizers which present themselves at our own doors.

"The urine of animals possesses greater value than the solid dung, and is subject to great loss if not properly treated. The loss of this valuable fertilizer, by evaporation of ammonia, will be greater in hot than in cold weather, in open than in covered places. Hence, the use of covered liquid-manure tanks and the disadvantage of shallow pits exposed to wind and sun. Next to the collection of the liquid excrements of animals, the preservation of its volatile constituents ought to be attended to by every good farmer.

TO PREVENT LOSS OF AMMONIA.

"Sulphuric acid and sulphate of iron, when they can be had at a cheap rate, are by far the most efficient materials for preventing the evaporation of ammonia. On the average, one pound of oil of vitriol will be sufficient for one hundred and fifty pounds of liquid manure. The acid should first be diluted with water before it is poured into the liquid manure tank."

SEWAGE MANURE.*

"*What is sewage?* In it the chemist recognizes rounds of beef and basins of turtle; cargoes of sugar, coffee, and port wine; millions of loaves of bread and thousands of tons of cheese and butter. Therein are not only all the alimentary productions of our own country, but also our enormous alimentary imports, altered in form, but scarcely in utility or value. It is truly a well-known but unworked mine of gold.

"We might call it a stream of liquid guano. It exists in a form of peculiar availability and almost self-portability; its fertilizing powers are enormous. We may estimate its value by the sums expended to compensate for its loss. We pay for guano, oil-cake, and corn many millions, and vast sums are annually abstracted from the agricultural pocket for phosphates and other artificial manures.

"Nationally, this neglect of sewage is a great calamity, but one that, it is to be hoped, may receive a gradual and wholesome correction.

"If it is considered ruinous by the farmer to waste the excrementitious deposits of his animals, with still greater force does the objection apply to the waste of our sewage.

"Experience has taught the writer of this article that there is no material practical difficulty to overcome in its economy and appliance to the soil as a fertilizing agent.

---

* Extracts from article of J. J. Mechi, Morton's Encyclopædia of Agriculture.

"It is not more difficult to convey than the water which intersects our streets, and finds its way into every house. It may, in fact, be considered the venous return of an arterial circulation; and the more abundant its liquefaction, the more valuable it becomes, seeing that water alone contains all the organic elements of our food. It is hardly possible to treat this subject except as a joint question of sewage and irrigation with drainage, artificial or natural.

"We said there was no practical difficulty in economizing this most valuable commodity, excepting the all-important one that public opinion has not yet appreciated its value.

"The force of public opinion must be brought to bear on this great question.

"Teach the farmer that it is liquid guano, brought to his door in its only available form; let him understand that the water of solution is, independently, a means of fructification; point out to him that every valued meadow whose rich crop of hay he covets owes its powers of production principally to the abundant supply of moisture.

"It is a question for our legislators and the country at large.

"When once convinced of its value, recorded registers of supply will be attached to each farm, like our gasometers. Quarterly demands for its use will be cheerfully paid; our towns will be cleansed and our country fertilized. The evidence on this subject is too abundant and distinct to be doubted or denied. It is collated in a document issued by the General Board of Health, Whitehall, London, entitled, 'Minutes of Information Collected on the Practical Application of Sewage Water and Town Manures to Agricultural Production.' The copious instances of cost and return there exemplified induced the writer of this to carry out the system on a farm of one hundred and seventy acres; and an experience of one year has sufficed to convince him of its easy practicability and great pecuniary advantages; he finds it, in fact, the key to profitable farming.

"The evidences are all sufficiently clear that the mere water irrigation of land on this principle of subterranean pipeage is remunerative. How much more so, then, when saturated with the elements of our food!

"The necessity for irrigation is becoming annually more apparent. The extensive removal of woods, fences, and the general clearing and improved cultivation of our country, added to the daily increasing drainage, render our soil and our climate warmer and drier, and consequently less favorable to succulent productions. By the proposed system of irrigation we shall have a warm moisture for our roots and green crops and dryness for our cereals; in fact, a desirable combination of food in abundance for man and for beast.

"Sewage, or liquefied manuring, renders the root and green crops self-supporting, by furnishing a great increase at a diminished cost. It may be compared to growing the ordinary produce of one hundred acres on fifty acres, thus diminishing by fifty per cent. taxes, horse and manual labor, wear and tear of implements, roads, gates, etc. In many instances, as in those of poor grass lands, the writer has no hesitation in saying that the produce would be doubled and greatly improved in feeding quality.

"The facility and promptitude with which a barren soil may be fertilized is surprising. In lands drained naturally or artificially, the writer has seen cabbages and roots luxuriate in a miserable plastic clay brought from the subsoil immediately after its saturation with sewage or liquefied manure. Its effects are alike beneficial to every crop—cereal, bulbous, or leguminous; although, in the case of cereals, a due regard is required as to the necessity for its application and a judicious regulation of the quantity of seed.

"With regard to the form of application, the writer's experience confirms the evidence collated, that the hose and jet present very great advantages in every respect.

"As to the period of growth, or season for application, the writer has applied it at almost every stage: in sunshine and wet; in winter and summer; on fallows in wet weather very strong, in dry weather more amply diluted. During the heats of summer its frequent application to bulbous, leguminous, and green crops is attended with the most profitable results, illustrating, in degree, the rapid vegetation produced by great heat and moisture in tropical climates. With an increasing population, the time is fast approaching when the concentration of capital on land for a greatly increased production will become a necessity. In lieu of two acres producing barely enough for one cow, six sheep, or one bullock, by these means from three to five cattle, or twenty sheep, may be maintained on one acre. In extreme cases enormous results have been produced. The meadows near Edinburgh, some of them once arid and worthless, have, by being flooded with the sewage of that city, risen to an enormous value, and are annually let by public auction at prices varying from £15 to £32 per acre. It is estimated that the quantity of green food cut annually from each acre is from fifty to eighty tons.

"The supply of milk to our great cities would, by similar irrigations, become greatly improved in quantity, quality, and price.

"One of the most important results is the destruction, or driving away, of injurious grubs or insects. Wire-worm, slug, and beetle either perish under the jet, or quickly leave the field. Clovers do not fail, and roots are freed from knobs and fingers and toes."

There is no subject connected with agriculture so generally attracting attention as that of fertilizers, especially the avoidance of waste, so common upon farms, and the utilizing of the sewage of cities. Much attention has been given to the practical investigation of the saving of liquid manure in Great Britain, but the problem is still considered a matter of experiment, because of the great cost of receptacles and apparatus for application to the soil. There is no question of its great value and the advantageous form for promoting rapid plant growth.

I quote from the Encyclopædia Britannica a part of an article that appeared originally in " Minutes of Information," issued by the General Board of Health, detailing the Scotch method :

" The next place visited was the farm of Myremill, near Maybole, in Ayrshire, the property of Mr. Kennedy, who adopted and improved on the method of distribution just described. On this farm, about four hundred imperial acres of which are laid down with pipes, some of the solid as well as the liquid manure has been applied by these means, guano and superphosphate of lime having been thus transmitted in solution, whereby their value is considerably enhanced. This is especially the case with guano, the use of which is thus rendered in great measure independent of the uncertainties of climate, and it is made capable of being applied with equal advantage in dry and wet weather. In some respects the farm labors under peculiar disadvantages, as water for the purpose of diluting the liquid has to be raised from a depth of seventy feet and from a distance of more than four hundred yards from the tanks where it is mixed with the drainage from the byres.

" These tanks are four in number, of the following dimensions respectively : $48 \times 14 \times 12$; $48 \times 14 \times 15$; $72 \times 14 \times 12$; $72 \times 17 \times 12$. They have each a separate communication with the well from which their contents are pumped up, which are used in different degrees of ' ripeness,' a certain amount of fermentation induced by the addition of rapedust being considered desirable. The liquid is diluted, according to circumstances, with three or four times its bulk of water, and delivered at the rate of about four thousand gallons an hour, that being the usual proportion to an acre. The quantity to be applied is determined by a float-gauge in the tank, which warns the engineer—whose business it is to watch it—when to cut off the supply, and this is a signal to the man distributing it in the field to add another length of hose, and to commence manuring a fresh piece of land. The pumps are worked by a twelve-horse-power steam-engine, which performs all the usual work on the farm, thrashing, cutting chaff and turnips, crushing oil-cake, grinding, etc., and pumping.

" The pipes are of iron ; mains, submains, and service pipes, five, three and two inches in diameter respectively, laid eighteen inches or two feet below the surface. At certain points are hydrants, to which gutta-percha hose is attached in lengths of twenty yards, at the end of which is a sharp nozzle, with an orifice ranging from one to one and a half inch, according to the pressure laid on, from which the liquid makes its exit with a jet of from twelve to fifteen yards. All the labor required is that of a man and a boy to adjust the hose and direct the distribution of the manure, and eight or ten acres may thus be watered in a day. There are now seventy acres of Italian rye grass and one hundred and thirty of root crops upon the farm. The quantity they would deliver by a jet from a pump worked by a twelve-horse-power steam-engine would be forty thousand gallons, or one hundred and seventy-eight tons per diem, and the expense per ton about twopence, but a double set of men would reduce the cost. The extreme length of pipe is three quarters of a mile, and with the hose the total extent of delivery is about one million nine hundred thousand yards, or four hundred acres.

"To deliver the same quantity per diem by water-carts to the same extreme distance would be impracticable. One field of rye grass, sown in April, had been cut once, fed off twice with sheep, and was ready (August 20th) to be fed off again.

"In another, after yielding four cuttings within the year, each estimated at nine or ten tons per acre, the value of the aftermath for the keep of sheep was stated at twenty-five shillings an acre. Of the turnips, one lot of swedes, dressed with ten tons of solid farm manure, and about two thousand gallons of the liquid, having six bushels of dissolved bones along with it, was ready for holing ten or twelve days earlier than another lot dressed with double the amount of solid manure without the liquid application, and were fully equal to those in a neighbor's field which had received thirty loads of farm-yard dung, together with three hundredweight guano and sixteen bushels bones per acre; the yield was estimated at forty tons the Scotch acre, and their great luxuriance seemed to me to justify the expectation. From one field of white globe turnips sown later, *and manured solely with liquid manure*, from forty to fifty tons to the Scotch acre were expected. A field of carrots treated in the same manner as the swedes, to which a second application of liquid was given just before thinning, promise from twenty to twenty-five tons the acre. Similarly favorable results have been obtained with cabbages, and that the limit of fertility by these means has not yet been reached was clearly shown in one part of the Italian rye grass which had accidentally received more than its allowance of liquid, and which showed a marked increase of luxuriance over that around it. The exact increase of produce has not been accurately determined, but the number of cattle on the farm has increased very largely, and by means of the Italian rye grass at least *four* times as many beasts as before can be kept now on the same extent of land, *the fertility of the land being at the same time increased*. This plant, of all others, appears to receive its nourishment in this form with most gratitude, and to make most ample returns for it; and great as are the results hitherto obtained, I believe that the maximum of productiveness is not yet reached, and that the present experiment must be carried yet further before we know the full capabilities of this manure. Of one important fact connected with this crop, I am assured that, notwithstanding the rank luxuriance of its growth, animals fed upon it not only are not scoured, but thrive more than on any other kind of grass in cultivation.

"Taking into the irrigation account the whole cost of the engine and the whole of the fuel and wages—although half of these might have been deducted—the following appears to be the capital account and working expenses for fertilizing Myremill farm :

Tanks complete..........................	£300
Steam-engine............................	150
Pumps..................................	80
Iron pipes, laying, and hydrants.........	1,000
Gutta-percha distributing pipes, etc.....	56
	£1,586
Actual interest on £1,586, and wear and tear at 7½ per cent.. £118	19s.
Annual wages......................... 104	0
Fuel................................. 58	10
£281	9s.

"This amount, divided by the number of acres, is equal to the annual sum of fourteen shillings per acre.

"I now come to the practical results of so cheap a mode of fertilizing land.

"Mr. Young informed me that in one of the fields he had himself measured the growth of Italian rye grass, and had found it to be two inches in twenty-four hours; and that within seven months Mr. Kennedy had cut from a field we were passing at the time seventy tons of grass per acre.

"Where the whole is cut, four or five heavy crops are thus taken; but upon some of the land during the past two years twenty sheep to the acre have been penned in hurdles, and moved about the same field from time to time; after each remove the fluid has been applied, and immediately followed by an abundant growth of food. There is not the slightest appearance of exhaustion in the land—its fertility appears to increase. I was informed that before the liquid manure was used the land would not keep more than a bullock or five sheep to the acre, nor will it maintain, if the crops are cut and carried in, five bullocks or twenty sheep to the acre. Some beans, bran and oil-cake are bought for the stock; but, on the other hand, one third or more of the farm is kept in grain, notwithstanding the great number of live stock.

"*Canning Park—Mr. Telfer's Farm, near Ayr.*—This is a small dairy farm of forty acres, near the level of the sea, and about a mile and a half west of the town of Ayr. The subsoil is beach gravel with a slight admixture of clay. Water is too abundant. It lies dead within about twenty inches of the surface, and in winter nearer than that.

"No bedding or litter is used here. The cows lie on cocoanut mats. The ventilation is perfect, and the air sweeter than in the majority of the dwelling-houses of human beings.

"The following appears to be the cost of carrying out the system of Mr. Telfer's farm:

Tank		£30
Engine		60
Iron pipes and hydrants		100
Distributing hose-pipe, etc		20
		£210
Annual interest on £210, and wear and tear at $7\frac{1}{2}$ per cent	£15	15s.
Wages and fuel	11	0
	£26	15s.

" In summer the cows have a quantity of oil-cake as well as grass; and in winter they have turnips or mangel-wurzel, bean or barley meal, and cut hay or grass, the whole mess being steamed together. Miss Bell, the cousin of Mr. Telfer, manages the dairy, and said that last year the hay bought would amount to from £30 to £40, and she should think the grain to not less than £200. In general terms, the other food is produced upon the farm. As to the produce of grass, which is the chief article, the first cutting during the present year was in the latter end of March, about eighteen inches thick. The second was from eighteen inches to two feet thick. The third was from three feet to four feet six inches thick. The fourth nearly the same. The fifth was two feet thick; and the sixth, in process of cutting at the time I was there, we measured at eighteen inches thick. Taking the mean, where two dimensions are given for the same crop, I find the aggregate depth of grass grown and cut off this farm within seven months to be not less than fourteen feet three inches. All this is, however, eaten upon the premises, and the whole marketable produce of the farm is represented by the milk and butter.

" As to the quantity and value of these, Miss Bell stated that the previous week the butter was one hundred and fourteen pounds and one hundred and twenty pounds—together two hundred and thirty-four pounds sold at one shilling per pound. This, she stated, was about the average quantity and price. The amount for butter would therefore be £11 14s. per week, or per annum £608 8s. She informed me, further, that during about eight months in the year the cold milk realizes about the same amount as the butter. In the summer months, during hot weather, the market value of the milk is only about half that of the butter. From these data, the amount of milk sold per annum is £507. The total receipts for the two articles of milk and butter amount to £1115 8s. per annum.

" I only need to add that, previous to the adoption of the present system of farming, these forty acres of land were, barely sufficient to support eight or nine cows, and would have been well let at a rental of thirty shillings an acre."

EXTRACT FROM TABLE SHOWING COST, ETC., OF THE APPLICATION OF SEWAGE WATERS AND LIQUID MANURES.*

Name of Place.	No. of English Acres.	Mode of Application.	Cost of Works and Apparatus.	Annual Inter-est, etc., at 7½ per cent.	Annual Work-ing Expenses.	Total Annual Charge per English Acre.	Observations.
			£ s. d.	£ s. d.	£ s. d.	£ s. d.	
Edinburgh.							
Craigentinny Meadows.....	63	Steam-engine, pumps, and open gutters and panes.	2,000 0 0	150 0 0	117 12 0	4 4 11	Average rental upward of £16 per English acre.
High-level Sea Meadows....	38	Graviation, open gutters and panes.	700 0 0	52 10 0	19 17 6	1 18 11	Worth about £20 per English acre; worthless before.
Nottinghamshire.							
Old Meadows	228	...	2,700 0 0	202 10 0	119 5 0	1 8 2¼	Maximum rental £25 per acre.
The Duke of Portland, Clip-stone Meadows..........	300	Catch-meadow, graviation and open gutters.	96,000 0 0	2,700 0 0	150 0 0	9 10 0	Worth upward of £12; pre-viously 3s. to 5s.
Wiltshire.							
Wiley Meadows..........	150	Beadwork of ridge and furrow, graviation and open gutters.	3,000 0 0	225 0 0	52 10 0	1 17 0	Four heavy crops of grass per annum.
Devonshire.							
Duke of Bedford....	90	Beadwork and catch-meadow, graviation and open gutters.	1,183 0 0	88 14 6	67 10 0	1 14 8¼	Land more than quadrupled in value after only four years' irrigation.
Berkshire.							
Pusey Meadows.	100	Catch-meadow, graviation and open gutters.	445 0 0	33 7 6	37 18 4	0 14 3	Land not previously worth 6s. per acre yielding 6 heavy crops of grass per annum.
Glasgow.							
Mr. Harvey's Farm.........	508	Steam-engine, pumps, underground iron main pipes, and distributing pipes.	1,450 0 9	108 15 0	240 0 0	0 13 9	Ten feet thick of grass cut from an acre in six months.
Ayrshire.							
Myremill Farm...........	508	Steam-engine, pumps, underground iron main pipes, gutta-percha hose.	1,586 0 0	118 19 0	168 10 0	0 11 1	Seventy tons of grass from one acre in six months.
Lancashire.							
Halewood Farm...........	190	Engine, pumps, underground mains, gutta-percha hose and jet pipe.	591 12 0	39 2 5	19 15 2	0 9 9¼	One dressing of liquid equal to 20 or 25 tons of farm-yard manure per acre.

* Encyclopædia Britannica, from "Minutes of Information."

The agricultural editor of the *Encyclopædia Britannica* cautions those who venture upon such experiments not to be rash or too sanguine, and is inclined to doubt the feasibility of such expensive apparatus.

It would seem, however, that the experiment is well worth trying in our dryer climate, as promising great advantages, especially in seasons of long drouth, not only for grass fields, but for maize, sorghum, and other important soiling crops, and in the Southern States, where the pastures are parched by the scorching sun, to raise immense crops of Johnson grass (*Sorghum halapense*) and Millo maize. Satisfactory results have attended the use of the sprinkling-cart on small farms, but a permanent . system of irrigating apparatus ought to return a large dividend when well managed upon good land and with first-rate Jersey stock.

THEORY OF CULTIVATION.*

" The main conditions required in the cultivation of the soil are :

" 1. A thorough pulverization and drainage of the soil.

" 2. A progressive chemical disintegration or liberation of insoluble ingredients.

" 3. A renewal, by means of manure, of those substances which have been removed from the soil by successive crops.

" The art of cultivation consists in aiding nature to accomplish these conditions with greater celerity than, unaided, would be accomplished.

" By means of the plow and harrow the soil is mechanically pulverized, and fresh surfaces exposed to the disintegrating action of the air. Many soils, especially clayey varieties, contain a very large amount of alkalies, which, by the action of carbonic acid, are liberated and become soluble. In such cases it is more economical to depend upon this vast magazine of supply for the necessary alkalies than to import them in the form of manures. But, as the disintegration of the soil and liberation of the ingredients proceed with slowness, it is necessary not only to offer every facility by increasing the surfaces, but also to admit the air and fresh supplies of rain-water, so as to render the treasures available within the prescribed period ; this is effected by drainage. It is to such rich soils only that the Roman methods of culture apply.

" Cato gave good advice ; for plowing is both the first and second operation of good farming, and manuring is the more advantageous the more thoroughly prepared and pulverized the soil is ; for manure, like land, requires disintegration to render its constituents thoroughly available. The plow and the harrow are, therefore, both mechanically and chemically advantageous. They are mechanically useful in fitting the soil for the reception and growth of plants, and chemically by increasing

---

* Lyon Playfair, Morton's Encyclopædia of Agriculture.

the absorptive powers of the soil for aerial food, and also by admitting those atmospheric influences which disintegrate the soil and liberate the mineral food.

"If the subsoil do not contain an excess of iron, and be not sufficiently tenacious to alter the character of the upper soil, trench-plowing is useful, by presenting to atmospheric influences a new and unexhausted magazine of mineral food. The oxygen, carbonic acid, and rain-water acting on this freshly upturned soil render soluble the alkalies and other ingredients formerly present in an insoluble form; but when the subsoil is either too slowly acted upon by the air or too tenacious, it may act injuriously by preventing that very disintegration which it is intended to promote.

"The lower oxide of iron, if it be present, absorbs the oxygen, which ought to find its way to the roots of the plants; or the tenacity of the soil acts mechanically, by preventing that access of air which the iron refuses to allow to pass by its chemical properties. In all such cases subsoiling is preferable to trench-plowing, because the subsoil, being loosened, is progressively acted upon by disintegrating influences, and, in a few years, changes its character sufficiently to enable it to be mixed with the surface soil without danger. This subsoiling cannot, however, be advantageously done without a previous natural or artificial drainage; for unless the soil be sufficiently free from moisture it cannot be acted upon by the atmospheric causes of change. The more accessible the soil is to air and to the free passage of rain-water the quicker will it become fitted for the wants of vegetation.

"The term cultivation properly includes the abnormal growth or increase of particular ingredients in plants, such as the gluten in the cereals and the starch in the potato."

PLOWING.*

"Wherever farming is conducted on an extensive scale, plowing constitutes the principal operation, as being the preliminary process necessary to prepare the soil for the subsequent series of processes by which systematic cultivation is effected. For this purpose oxen, asses, mules, and horses have been variously employed by different nations to draw the plow, ever since the cultivation of the soil became the necessary consequence of a settled state of society. Up to very recent times oxen appear to have been principally employed for this purpose; and their docility, strength, and endurance, combined with the simplicity of the apparatus required to yoke them, were properties which, in the estimation of the unscientific and uninventive tiller of the soil, gave them a superiority in field labor over all other animals of draught.

"The employment of horses in plowing and other agricultural operations,

---

* John Haxton, Morton's Encyclopædia of Agriculture.

and the introduction of the *iron plow*, are, undoubtedly, among the greatest improvements effected in agriculture.

"When land has been well plowed, and cultivated to a proper depth in preparing for green crops, deep plowing for the subsequent grain crop is not only unnecessary, but oftentimes injurious. This is particularly to be observed in the cultivation of wheat, in which experience has taught us that the firmer the soil is in which the roots of the young plants are embedded, the better are they able to withstand the changes and shiftings produced in the immediate surface soil by the effects of alternate frosts and thaws. Thus it is that the peculiar habit of growth of plants must be studied, and a cultivation adopted which accords, as nearly as possible, with the requirements of nature ; and this knowledge is necessarily the result of observation."

AMERICAN CULTIVATION.

The greatest improvements (with the exception of the steam plow) in recent years have been made by American inventors. For those who are interested in the subject of the history of plows I refer them to the Annual Report of the New York State Agricultural Society for 1867. The more recent history of plows and plowing must be studied in the productions of the past few years that are offered for sale by the best dealers.

Among the modern improvements are the plows for turning *flat furrows ;* the better adjustment for .power, especially in the sulky plows ; the use of better material, as in the best steel plows ; the greater pulverizing power, as exhibited in the Sackett plow ; non-liability to choke in stubble ; lightness of draught ; ease of holding ; durability ; cheapness ; excellence of workmanship ; even distribution of wear ; effective service in burial of weeds and stubble ; regularity of turning flat furrows. Whatever force is used for propelling the plow, the wheel plows undoubtedly have the advantage of easier draught, better quality of work, effective work in drouth-baked land. The efficiency of the wheel plow is independent of the skill of the plowman, and when once properly adjusted will cut every furrow of an equal width and depth, and lay them all over uniformly level. The Sackett plow is the best of its kind, as it serves the purpose of both plow and harrow, doing much better work than can be done with both of those implements ; but it can only be used in fine land that is free from stones and rocks.

HARROWS.

Great improvement has been made in harrows. The best implement has an iron frame with steel spring seat, and slanting, reversible steel teeth, which have a cutting edge for pulverization and a round edge for smoothing and cultivating all kinds of crops, and the frame also in sections, to which plow-handles can be attached, and each section used as a cultivator between rows or drills. This implement is the most effective and useful of its class, and absolutely indispensable to every farmer. It is the best pulverizer, the best cultivator, the best for the purpose of scarifying old pastures and meadows that need renovation.

With this harrow wheat may have three harrowings in early spring; oats and barley two or more, or until three inches high; corn can be harrowed every week, until twelve inches high. The round edge is also used for covering clover seed. Among the harrows for grass seed that require covering only one eighth of an inch, is the chain-harrow, an implement which consists of a draught-bar to which are attached pairs of square-linked chains, each seven and a half feet long, connected by cross-links, and kept expanded by two movable stretchers.

ROLLERS.

These are usually hollow cylinders of cast iron, of diverse weights, for one or two horses. They may have a smooth surface or may be formed of a series of corrugated rings or discs having serrated edges and side-way projecting teeth. Some require three horses abreast to work them. They are very effective for breaking clods, consolidating loose soils, checking the ravages of the wire worm, and covering in clover and grass seeds. For grass seeds the smooth roller is best, with the brush-harrow or chain-harrow attached. Another form of roller is made of a series of eccentric fluted discs, which is said to possess many advantages over any other implement of its class.

CULTIVATORS.

An implement is needed for the effective cultivation of maize that shall *finely pulverize the surface to a depth of one inch*, and work smoothly without plunging or destroying the rootlets of the growing plants. Maize needs, especially in a drouth, a mulch of soil like fine flour, of a depth not to exceed two inches, and the culture should always be level and smooth.

DRILLS.

These implements secure straight rows, and thereby assist in clean culture for all kinds of grain and root crops. Drills are of various patterns, some of them constructed for planting several rows at once, and dropping manure at the same time; while others have added an irrigating apparatus for moistening manure and seed in dry ground, and are more effective in promoting germination, even upon damp ground, and also intensifying the effects of the superphosphates.

OTHER IMPLEMENTS.

Among the many useful machines and implements for tillage, harvesting and feeding, a completely equipped dairy farmer needs trench plows; subsoil plow; a manure-spreader; a horse-hoe; carrot-thinner; reaper and binder; mowing-machine; hay-tedder; horse-rake; hay-loader; thresher; fanning-mill; grinding-mill;

root-cutter; hay-cutter; corn-stalk crusher; standard weighers, scales and measures; also improved wagons and farm-carts.

FIELD TILLAGE.

As a general rule, no tillage operations can be performed when the soil is wet. Clay soils especially are liable to great injury in this way. Plowing or harrowing land when wet is destructive to crops.

Rotation of Crops.*

" The arrangement of a certain succession of crops, by which each shall follow in such a rotation as shall best economize the resources of the farm, has long been an object of primary consideration among agriculturists. The fact that certain crops impoverish the soil in a greater degree than others is very much dependent on the use that is made of them.

" If a crop is entirely removed from the farm on which it is produced the land will obviously be deprived of some of those elements which would be restored to it by the consumption of the whole or a portion of the same crop on the ground. The manner in which a crop is cultivated will also influence the condition of the soil. A succession of grain crops, grown in such a manner as would not admit of the soil being kept free from weeds, even though they did not of themselves draw from the soil a greater supply of the elements of fertility, would be more injurious to it than a succession of well-hoed crops, the intervals between the rows of the latter and the comparatively late period of the season at which they are sown admitting of the complete eradication of weeds.

" The inorganic matter abstracted from the soil by any one crop is so small in amount as to render the choice of a particular crop, in reference to that point, of little moment. A deficiency is generally rectified by the manure applied in the ordinary routine of cultivation. Without entering, in this place, on the scientific investigation of this subject, it will be found that, for practical purposes, the principle to be kept in view, in fixing on a rotation of crops, is, WHAT SUCCESSION IS BEST SUITED IN A GIVEN LOCALITY TO DRAW FROM THE SOIL THE LARGEST NET RETURN, WHILE THE CAPABILITIES OF THE LAND ARE, AT THE SAME TIME, MAINTAINED AND INCREASED.

" There are three conditions, namely, climate, nature of soil, and local position, which must first be observed in dealing with this subject. Some plants are best adapted to a dry, some to a moist climate; one is suitable to a stiff clay soil, another to a loam, and a third to a sand. The local demand for a particular crop may render its culture on a particular soil remunerative; while the absence of such a demand may make the same crop on a similar and suitable soil of little value.

---

* James Caird Morton, Encyclopædia of Agriculture.

THE ENGLISH SYSTEM.

"*Norfolk.*—Here the four-course system had its origin, and here it is still practised in the best style. But this county, which was the first to break through the old system of cropping as long as the land would yield grain, is now beginning to amend its own improvements. The ease with which artificial and other manures can now be procured, and the readiness with which they may be applied to the land at any period of the rotation, have taught the enterprising farmers of this county that the matter for their consideration, in fixing on a course of crops, is simply which, with a given outlay, will produce the largest return, and, at the same time, most enrich the land. Instead of the four-course, the following is adopted by some first-rate farmers, namely: 1, clover, trefoil or peas; 2, wheat; 3, oats; 4, turnips; 5, wheat or barley. Every crop is manured for, either by direct application or by sheep-feeding.

"And on a large farm, where this system has supplanted the four-course, the average produce of all the grain crops has increased, in ten years, between thirty and forty per cent.; the extent of land on this farm in wheat having, during that period, annually increased, till it has now become one-third greater than it was then. The four-course is conducted thus: the clover lay, after being mown, is dunged. A rapid growth of aftermath is produced, which is plowed in to enrich the ground for the wheat crop. In spring the young wheat receives a dressing of one hundred weight of nitrate of soda and two hundredweight common salt mixed, and sown by hand in two applications, at an interval of three weeks, beginning in March and ending in April. When the wheat is removed the ground is plowed and sown with rye, which is eaten off in spring, and followed by the turnip crop. Dung, superphosphate, and guano are applied to the turnips, the greater proportion of which are consumed on the ground by sheep which are also cake-fed. The land is thus prepared for barley, which is sown out with red clover, and with trefoil and white clover alternately. No rye grass is sown with the clovers, as it is reckoned injurious to the following wheat crop. Many of the best Norfolk farmers do not hoe their wheat crops in spring, as hoeing has been found to increase the proportion of inferior grain. The wheat fields are rolled in spring with advantage to the crop."

AMERICAN ROTATION.

A system of rotation for Jersey dairy farms in America must depend upon greatly varying conditions of soil, climate, and proximity to market. Soiling or pasturing of stock also necessitates a variation of crops. Where soiling is practised, especially in a case where the land is both underdrained and irrigated with liquid manure, the rotation may often include two grain crops and one root crop in a single season from the same piece of ground. For the ordinary method of dairy practice the following rotation may prove useful : 1, clover ; 2, carrots or mangolds ; 3, rye or barley ; 4, sweet corn ; 5, oats ; 6, barley. Or this : 1, oats ; 2, sweet corn ; 3, rye forage ; 4, Hungarian grass ; 5, clover and grasses ; 6, carrots or parsnips.

ROTATION OF SOILING CROPS.

1. Winter barley, winter wheat, and winter rye sown in September and October for the May and June feeding, may follow on land that has fed a crop of corn fodder.

2. Barley, oats, oats and peas, oats and vetches for early spring sowing, for June and July feeding. Lucern, red clover, large clover, alsike, alsike and timothy, are also included in this list for second growth for June and July feeding.

3. Millet, Hungarian grass, dent corn, flint corn and sweet corn for July feeding.

4. Sweet corn, flint or dent corn grown on the ground that was occupied by barley and rye, for August feeding.

5. Sweet corn grown on the ground that was occupied by the barley, oats and peas, for September feeding.

6. Barley and rye, grown on the ground that furnished the millet, Hungarian grass, and early corn fodder, for October and November feeding.

7. Carrots, parsnips and mangolds to follow clover and lucern every second or third year.

It would be impossible to specify any course of crops which can be recommended as the best under all circumstances. The agriculturist may select without much difficulty the course of crops most suitable to his soil and locality, and those best adapted to his needs, which are elsewhere mentioned under the list of soiling crops in another section of this work. He may therefore, by saving his manures in tight vats and continually enriching his soil, grow any crop suitable to his climate and soil, in such a succession as he pleases, the conditions needful to success being that the land must be kept *dry*, *clean*, and *rich*.

PLANNING FOR ROTATION.

In order to plan for a rotation of soiling crops it is necessary to know how much a full-sized cow requires for the season.

If your soil is of average good quality the daily allowance for one cow will be one square rod of grass, clover, or lucern ; three fourths of a rod of barley, oats, oats and peas, rye or millet ; and about one half a rod of maize or sweet corn. Rich land will require less.

Estimate the amount of ground you will need for the season according to the number of cows and the variety of crops to be cultivated.

For all the annuals there must be a regular planting in periods, of every seventh day, so that there shall be provided a succession of young, tender, juicy herbage, ready to be cut while in bloom, that the waste from toughness may be reduced to the lowest degree. It would be preferable to double the frequency of seed-sowing rather than to lengthen the above-named period.

PUNCTUAL PERFORMANCE OF ALL FARM-WORK.

" A stitch in time saves nine."—Old Proverb.

The farmer who excels the average achievements of agriculturists must needs be free from the fetters of prejudice and a merely routine agricultural education. He must bring to bear upon his calling all the tact and business ability with which he is gifted and experience enables him to develop in himself.

Energetic industry and sound common sense, combined with systematic and thorough methods and extreme punctuality in all operations, are the elements upon which depends the success of the farmer.

It is well for every farmer to have a calendar of operations for the year made to suit his locality and the special work upon his farm.

JANUARY.

Take account of stock and balance the books.

This is generally the coldest month of the year. Those who have access to beds of marl or other natural fertilizers, as well as factory waste, may cart them upon lands which are suitable, during the whole winter, but the earlier the better, so as to get the beneficial action of the frost upon them. Thrashing of grain, composting manures, preparation of bone manures. Tools and implements should be looked over to see that they are in good condition.

Cattle should be made very comfortable in good stables that are both warm and well ventilated.

Harness should be kept well oiled, bright and clean, and not allowed to freeze or crack from getting wet.

Water-meadows must be closely watched, where the English method is adopted; obstructions from dead leaves removed; let the water flow until a scum appears upon the grass, an indication that the soil is surfeited with water.

Prune trees.

Breed cows for winter butter.

Cows should be kept in good condition and full flow of milk for winter butter.

FEBRUARY.

The weather is generally very irregular during this month. Where the climate admits, oats, barley, peas and spring wheat are to be sown.

Parsnips may be sown in our Middle and Southern States the last of this month.

Seeds of all kinds must be procured this month, if they are not grown upon the farm—grass, clover, maize, carrot, mangold, parsnip, rutabaga, rye, barley, oats, peas, millet, vetch, sweet corn—everything needed for forage crops, pastures and meadows. Grass land intended for oats or barley should be plowed as soon as they are dry enough. Oil the wood and metal of tools and machines with petroleum. Guano and superphosphate or other artificial manures should be purchased and stored this month. Finish pruning apple-trees. Allow no cattle to go upon wet grass lands, as they will seriously injure the sod by trampling it.

The water-meadows, if they have been successfully irrigated, will begin to show green in the South and central States.

Look well after the young calves. The calf-cribs should have one occupant only, with plenty of bedding. Keep them warm and dry, and with a constant supply of pure air, always putting a little rennet in the warm mixed milk to prevent any trouble from indigestion. Breed cows for winter dairy.

MARCH.

This is the first month of agricultural spring over the greater portion of America. The soil dries rapidly. Toward the end of the month young wheat and rye will require hoeing, or the slanting-tooth smoothing-harrow may be used, followed by the roller, if the land is sufficiently dry.

Oats may be sown this month as soon as the ground can be made ready, barley a few days after oats, one to two bushels of seed per acre. Pickle four bushels of oats or barley in a gallon of water containing two ounces of sulphate of copper, as a safeguard against bunt and smut.

Grass and clover seeds may be sown near the end of the month, best on ground especially prepared for them, and not with grain crops. After sowing the clover and grass seeds, go over the ground with the roller and brush-harrow attached, or use the chain or web-harrow.

Carrot ground should now be prepared by deep plowing and thorough manuring. Rich ground thoroughly pulverized, and mixed with manure to the depth of twelve inches, will give good returns for the butter dairy. Subsoiling, going twice in the furrow, is a good preparation for carrots.

Spring vetch may begin the first sowing this month, with peas.

Sow mixtures for soiling crops.

Plant apple-trees this month as early as possible. Also attend to grafting and budding.

Plant hedge fences of prim as an accompaniment to barbed wire.

Dig around and clean young hedges.

Where water-meadows have been regularly irrigated through the winter a good crop of grass may be now expected ; the rye fields also afford good pasture. Use dry hay with this green food to prevent violent scouring.

If cows are properly kept they will not be in much danger of colostrum fever or apoplexy, a disease induced by a high condition, plethora, fat, and a constipated condition of the bowels, probably often complicated by a cold from sudden draught of air. These conditions are made worse by neglect of exercise and insufficient or improper stable ventilation.

The month of March is an important month to look after the destruction of all kinds of field and barn vermin, such as *mice*, *rats*, and *stray dogs*.

The rats and mice destroy a vast amount of farm produce. Soak a box of matches in a half pint of water, and mix the water with flour enough to make a stiff dough ; place this where rats or mice or only small creatures can have access to it. They are very fond of this phosphorated poison, and eat it with avidity to their destruction.

Any mongrel cur or thoroughbred hound straying without his owner on any field thereby forfeits his life. He is easily tempted to injure live stock, and may, by causing fright in a herd, be an agency of producing abortions. One dog will destroy a large flock of choicely bred sheep in a few minutes. Dog-skins make the best gloves, and their carcasses and bones the best manure for meadows and orchards.

Begin to set dog-traps in the month of March. Make a pyramidal frame of slats, leaving a space at the top for them to jump in as they ascend the ladder to get at the bait, which may be a large piece of meat that has been perforated with skewers and the holes filled with powdered strychnine. A large number of dogs may be captured in this way, and thus may be secured a great quantity of the most valuable manure at a little expense, and thereby may be prevented the danger of frightened herds, abortion, or the mangling and destruction of thousands of dollars' worth of sheep throughout the country.

The dairy work is beginning to increase. Provision should be made for prompt performance of every kind of work. The cows should be milked regularly by the minute, two or three times daily, as they require, and all dairy operations, as well as the milking and feeding, should be begun and finished according to a fixed schedule and time-table. Neither good butter nor cheese will be made in this month without oatmeal, parsnips, clover hay, and a little green rye, or water-meadow pasture, combined.

APRIL.

The weather is capricious, with showers, hot sun, cold winds, and nipping frosts, especially in the central and northern States and Canada.

Wheat will require harrowing, after which it may have the roller.

Pull all thistles, docks, daisies and dandelions in grain crops and meadows. These can only be rooted up when the ground is moist after a rain.

Barley sowing may be completed this month.   Carrots may be entirely planted this month.   The land, deeply tilled and rolled as hard as possible, is to be sown in rows eighteen inches apart by the drill, which is to be followed by a light roller to complete the operation.   Five pounds of seed are rubbed, soaked in diluted urine and warm water, mixed with two bushels of ashes or sand, with the drill set to sow two bushels.   A few oats added will earlier show the line of the row, so that weeding may begin before the carrots appear above ground.   Mangold-wurzel may be planted this month.   These are dibbled half an inch deep in richly manured soil, two feet apart by one foot in the drill.   A light roller follows.   Kohl-rabi may be sown for transplantation in May.   Successive beds may be sown for transplanting all through the summer to cultivate like turnips.   Use the wheel-hoe cultivator. Lucern may be sown by the end of April, ten pounds per acre, in rows one foot apart, on deeply tilled, rich, calcareous soil.   Spring vetches or peas, alone or with oats, may be sown during April for soiling in July and August.

Turnip land may have its first plowing in April, after which it should be harrowed and kept clean from weeds.

Paring and burning is the most efficient method of breaking up old grass lands. Spread the ashes, plow, harrow and roll.

April is a good month for laying down grass lands, either by sowing grass seed or by planting bits of turf six by six inches on a well-tilled field.   Follow both grass seed and turf with the roller, and give a coat of fine manure broadcast or saturate with liquid manure.   In the early part of April, in moist or showery weather, sow guano and superphosphates upon grain crops.   More easily soluble manures, as nitrate of potash, may be sown later and in dryer weather.   Hedges and trees may still be planted.   Puddle the roots well.   Keep orchards, hedges, and all crops free from weeds by frequent stirring of the surface or mulching the ground.

Cattle will continue to receive parsnips and mangolds, a portion of green rye and Italian rye grass.   Cheese-making is on the increase.   The cows are turned to grass at the latter part of this month in most of the country, but they need hay until accustomed to the change, or they may receive part soiling of rye and rye grass with roots.

Peas require *lime, bone-powder, or marl*, to insure a good crop.

Cattle need one eighth of an ounce of salt every day.

Rennet is now prepared by the following method: One gallon of thin whey is boiled with a handful of salt and a spoonful of saltpetre; the solution is then strained, and when it is cooled to the temperature of 98° Fahr. four large maws or rennets are put into it; the whole is placed in a covered jar, and may be used after standing fourteen days.   To this may be added one fourth part alcohol for keeping.   A very small quantity of this preparation should be mixed with the food of all young calves, as a preventive against indigestion and consequent diarrhœa.   Four ounces of this solution without the alcohol will be sufficient for a cheese of thirty pounds.

Two jars of rennet should be kept to be used alternately. The rennet-skins may be resalted, dried, and used again the following year if desired. Use no so-called "rennet" manufactured from mineral acids; only a pure article.

MAY.

Wheat, if too heavy, may be topped again, as in April. Weeds must be pulled out of grain crops. Plant maize as soon as the ground is warm enough. Carrots may be cultivated and hoed, and singled out by the end of May, as also parsnips.

Mangolds need tilling and cleaning with hand-hoe work. Use the prong-hoe and single out the plants. Seize the best plant by the left hand, and tear the rest away with the right hand very abruptly.

Blanks may be filled by plants thus taken out.

Rye, barley, vetches, clover, and Italian rye grass will have been the soiling supply for Jersey cattle. The land that is cleared of these crops is to be immediately plowed again.

Rye grass and lucern will make a succession of cuttings till autumn.

Irrigate the late-planted trees. Destroy caterpillars and moths upon trees and crops before they scatter from their webs.

Water the newly-set hedges. Plant maize fields.

Sow buckwheat for green manure, one bushel to the acre, in drills one foot apart, and cultivate once.

Ply the cultivator and hoe against all weeds in all crops.

Early cleanliness and thorough cleanliness are indispensable to success.

Rich spots in pastures should be mown gradually, and cattle will eat the hay which they would not eat as rank grass.

Cattle will receive full allowance of green food by the end of the month or before, and will have finished the mangolds.

There is economy in letting pasture be very forward in growth before stocking it.

If soiling is adopted calculate the quantity of the different crops according to the number of stock on the farm.

If cows are pastured it is economy to use the tether. They should always be housed at night and also during the hottest part of the day, with at least one feed in the stable and great abundance of pure clean water.

If cheese is made, pigs are needed to consume the whey and buttermilk.

JUNE.

Mangolds require the wheel-hoe cultivator, and the second cleaning by the six-tined hand-hoe.

Carrots and parsnips must be singled out as early as practicable, at intervals of eight to twelve inches, and kept clean of weeds. Sow millet for a crop of hay from the first to the tenth of the month. The maize will require the smoothing-harrow until six to twelve inches high, two, three or more harrowings. Orchard grass ready for soiling.

Clovers for soiling, vetches and oats also. Clovers for hay may be mown and early grasses for hay. Look sharp to keep all weeds down this month. Allow no weed to flower and seed. Cut clover and grass for hay *as soon as they begin to flower, then they are most nutritious.*

Thumb-and-finger pruning for apple-trees cannot be neglected this month.

Let every superfluous growth be pinched off while it is tender and small.

Keep the pastures well fed, and mow the grass that is too rank to be eaten by stock. Allow one or two fields for a reserve in case of drouth. Never allow thistles, weeds and briers to encumber grass lands. Thistles may be spudded in dry seasons, or pulled with tweezers in wet weather. Spread all droppings of cattle within three days. Mow the early meadows. Be careful to guard against hoven upon change of feed. Give the working horses and oxen a plentiful feed of oats. This is the month of most abundant pasture and good soiling crops. Cattle need salt every day.

Dairy produce is at its height. Change of pasture as often as practicable increases the flow of milk. Give as great a variety of soiling crops as can be grown.

In the hot days pastured cattle should have one or two feeds of soiling crops in the stalls. The quality of the milk will often vary so much for cheese-making as to require a change in the rennet. Cheeses must be very regularly turned. The temperature of the cheese-room should be kept at 60° Fahr. Hang wet cloths near the windows and doors or ventilators, to aid in cooling. If there is a current of air in the cheese-room the cheeses must be well covered to prevent heaving and cracking.

Sow rutabagas from middle of June onward. Sow soiling crops every week.

JULY.

Wheat is in full ear. The bulk of the hay crop is cured before the middle of this month. The last sowing of millet must be made before the middle of the month. All turnips must be sowed before the first of August.

Plantings of maize may be made every two weeks until the middle of July.

Alsike and timothy are now ready for soiling. Green oats should be combined with them in feeding while they last.

Vetches also, combined with oats and peas. The horse-hoe, or wheel hand cultivator, is to be kept moving in all root crops. The thinning process must be finished for carrots, parsnips and mangolds. Turnips hand-hoed. Weeds kept down in all fields. Maize ground cultivated, shallow, fine, and level.

Apples must be thinned on the trees.

Mow all pastures before seeding to prevent smut.

Pastures must be thinned and soiling increased this month. Thistles destroyed. Keep the dog-trap baited. Make a compost of dogs. Cover the heap with fine earth, and pour on daily diluted sulphuric acid until the bones are all dissolved, or use the bone-mill when the bones are cleaned.

This is a busy month in meadow, field, stable and the dairy.

Do not pasture meadows. Keep the stables and the dairy very, *very* clean and sweet. Milk should be kept by controlling the temperature to the right point.

Sow barley for autumn soiling, every week.

Oil wood-work of all tools and machines with petroleum.

Sow sweet corn every week.

AUGUST.

A good time to renovate old pastures and lay down grass lands.

Wheat, rye and barley grains are generally all harvested before the middle of August in the most northern districts.

Fodder corn is the great soiling crop, and vetches, millet, timothy and rye grass, with second cut of clover and lucern. Plow in buckwheat while in full bloom for green manure. Sweet corn is the best green crop of this month. Keep up the full flow of milk and the routine of butter or cheese-making. Sow late barley and rye for soiling.

Grub up and cut bushes and trees in August to destroy them, as they will not sprout.

SEPTEMBER.

Apply lime, marl, and natural manures of all sorts and clay to sands. Burn all rubbish. . Cut seed clover. Sow the winter vetches, wheat and rye, before the last week in September. Sow winter barley. Sweet corn, vetches, and barley are good soiling feed.

Harvest and shock the maize ; cure corn stover. The pastures have a fresh growth if there are late rains. The full flow of milk must be kept up in the butter dairy.

The stock of cheese is large.

Keep the dog-trap in operation. Dogs make most excellent manure.

OCTOBER.

Gather apples. Feed them to cows lightly at first, beginning with two quarts a day ; increase gradually to a half bushel. Run them through a root-cutter. Never allow the chance of choking animals at any time with apples, roots, or tubers, then you will not need to be on the watch to save the best cow in a dire emergency.

This is the great month for wheat-sowing, as well as the planting of winter barley and rye, with winter vetches, though rye generally does better sown in

September. A half bushel of seed per acre for each grain. In Southern States grains may be planted later. Root-crops are all ready for harvesting.

Drains may be opened. Ditches cleaned.

Hedges have last clipping.

Rats, mice and moles, as well as the dogs, are to be trapped or poisoned this month.

October is a busy month with the irrigator of water-meadows. The rowen hay is cured.

Sweet corn and barley are the best green crops of this month.

Take good care of the stock that they do not suffer from the chilly weather and cold rains. Get all the cattle up early to the stables, so as to be prepared for winter in the northern districts by the month of November. Oil wood-work and metal again when tools and machines are stored for the winter. Always have a place for each implement, and when not in use the implement should be in its place in the tool-house.

NOVEMBER.

Keep up the full flow of milk with green barley and corn stover.

Finish harvesting turnips, carrots and mangolds. Parsnips may remain in the ground. Plant prim hedges. Continue ditching and draining.

Put the water-meadows and irrigating works in perfect order, or build them, if land is suited for irrigation. Carrots pull easily after a soaking rain.

Use a heavy roller for the land during irrigation. Turn off the water on the mild Indian summer mornings, and there will be a beautiful green growth of grass.

Cheese-factory work has generally suspended, but the butter dairy must be made perpetual. Sow wheat in Gulf States. Breed cows for winter dairy.

DECEMBER.

The month is variable. The Indian summer is generally cut short a week before Christmas, sometimes much earlier, by very cold ice-making weather.

This will necessitate the stopping of draining and all field operations, which should be well finished before winter sets in.

Commence ice-harvesting as soon as a sufficient depth is frozen, which is from eight to twelve inches.

If you have water-meadows they need as much care as during any month of the year. Let the water flow unchanged during the severest frost, and change the water upon mild mornings.

Cattle must be kept warm, dry and clean, the butter dairy in full operation, with corn stover, oatmeal and carrots fed in the stalls.

The drinking-water should always be tempered to about 65°, and especially not neglected in autumn and spring months. Keep the stables well ventilated and comfortable for all Jerseys, old and young. Breed cows for winter butter dairy.

BREEDER'S CALENDAR.

AVERAGE TABLE OF GESTATION FOR JERSEYS.

Date of Service.	Due to Calve.	Date of Service.	Due to Calve.	Date of Service.	Due to Calve.
Jan. 1	Oct. 8.	Feb. 4	Nov. 11.	March 10	Dec. 15.
" 2	" 9.	" 5	" 12.	" 11	" 16.
" 3	" 10.	" 6	" 13.	" 12	" 17.
" 4	" 11.	" 7	" 14.	" 13	" 18.
" 5	" 12.	" 8	" 15.	" 14	" 19.
" 6	" 13.	" 9	" 16.	" 15	" 20.
" 7	" 14.	" 10	" 17.	" 16	" 21.
" 8	" 15.	" 11	" 18.	" 17	" 22.
" 9	" 16.	" 12	" 19.	" 18	" 23.
" 10	" 17.	" 13	" 20.	" 19	" 24.
" 11	" 18.	" 14	" 21.	" 20	" 25.
" 12	" 19.	" 15	" 22.	" 21	" 26.
" 13	" 20.	" 16	" 23.	" 22	" 27.
" 14	" 21.	" 17	" 24.	" 23	" 28.
" 15	" 22.	" 18	" 25.	" 24	" 29.
" 16	" 23.	" 19	" 26.	" 25	" 30.
" 17	" 24.	" 20	" 27.	" 26	" 31.
" 18	" 25.	" 21	" 28.	" 27	Jan. 1.
" 19	" 26.	" 22	" 29.	" 28	" 2.
" 20	" 27.	" 23	" 30.	" 29	" 3.
" 21	" 28.	" 24	Dec. 1.	" 30	" 4.
" 22	" 29.	" 25	" 2.	" 31	" 5.
" 23	" 30.	" 26	" 3.	April 1	" 6.
" 24	" 31.	" 27	" 4.	" 2	" 7.
" 25	Nov. 1.	" 28	" 5.	" 3	" 8.
" 26	" 2.	March 1	" 6.	" 4	" 9.
" 27	" 3.	" 2	" 7.	" 5	" 10.
" 28	" 4.	" 3	" 8.	" 6	" 11.
" 29	" 5.	" 4	" 9.	" 7	" 12.
" 30	" 6.	" 5	" 10.	" 8	" 13.
" 31	" 7.	" 6	" 11.	" 9	" 14.
Feb. 1	" 8.	" 7	" 12.	" 10	" 15.
" 2	" 9.	" 8	" 13.	" 11	" 16.
" 3	" 10.	" 9	" 14.	" 12	" 17.

DATE OF SERVICE.	DUE TO CALVE.	DATE OF SERVICE.	DUE TO CALVE.	DATE OF SERVICE.	DUE TO CALVE.
April 13	Jan. 18.	May 22	Feb. 26.	June 30	April 6.
" 14	" 19.	" 23	" 27.	July 1	" 7.
" 15	" 20.	" 24	" 28.	" 2	" 8.
" 16	" 21.	" 25	March 1.	" 3	" 9.
" 17	" 22.	" 26	" 2.	" 4	" 10.
" 18	" 23.	" 27	" 3.	" 5	" 11.
" 19	" 24.	" 28	" 4.	" 6	" 12.
" 20	" 25.	" 29	" 5.	" 7	" 13.
" 21	" 26.	" 30	" 6.	" 8	" 14.
" 22	" 27.	" 31	" 7.	" 9	" 15.
" 23	" 28.	June 1	" 8.	" 10	" 16.
" 24	" 29.	" 2	" 9.	" 11	" 17.
" 25	" 30.	" 3	" 10.	" 12	" 18.
" 26	" 31.	" 4	" 11.	" 13	" 19.
" 27	Feb. 1.	" 5	" 12.	" 14	" 20.
" 28	" 2.	" 6	" 13.	" 15	" 21.
" 29	" 3.	" 7	" 14.	" 16	" 22.
" 30	" 4.	" 8	" 15.	" 17	" 23.
May 1	" 5.	" 9	" 16.	" 18	" 24.
" 2	" 6.	" 10	" 17.	" 19	" 25.
" 3	" 7.	" 11	" 18.	" 20	" 26.
" 4	" 8.	" 12	" 19.	" 21	" 27.
" 5	" 9.	" 13	" 20.	" 22	" 28.
" 6	" 10.	" 14	" 21.	" 23	" 29.
" 7	" 11.	" 15	" 22.	" 24	" 30.
" 8	" 12.	" 16	" 23.	" 25	May 1.
" 9	" 13.	" 17	" 24.	" 26	" 2.
" 10	" 14.	" 18	" 25.	" 27	" 3.
" 11	" 15.	" 19	" 26.	" 28	" 4.
" 12	" 16.	" 20	" 27.	" 29	" 5.
" 13	" 17.	" 21	" 28.	" 30	" 6.
" 14	" 18.	" 22	Feb. 29.	" 31	" 7.
" 15	" 19.	" 23	" 30.	Aug. 1	" 8.
" 16	" 20.	" 24	" 31.	" 2	" 9.
" 17	" 21.	" 25	April 1.	" 3	" 10.
" 18	" 22.	" 26	" 2.	" 4	" 11.
" 19	" 23.	" 27	" 3.	" 5	" 12.
" 20	" 24.	" 28	" 4.	" 6	" 13.
" 21	" 25.	" 29	" 5.	" 7	" 14.

Date of Service.	Due to Calve.	Date of Service.	Due to Calve.	Date of Service.	Due to Calve.
Aug. 8	May 15.	Sept. 16	June 23.	Oct. 25	Aug. 1.
" 9	" 16.	" 17	" 24.	" 26	" 2.
" 10	" 17.	" 18	" 25.	" 27	" 3.
" 11	" 18.	" 19	" 26.	" 28	" 4.
" 12	" 19.	" 20	" 27.	" 29	" 5.
" 13	" 20.	" 21	" 28.	" 30	" 6.
" 14	" 21.	" 22	" 29.	" 31	" 7.
" 15	" 22.	" 23	" 30.	Nov. 1	" 8.
" 16	" 23.	" 24	July 1.	" 2	" 9.
" 17	" 24.	" 25	" 2.	" 3	" 10.
" 18	" 25.	" 26	" 3.	" 4	" 11.
" 19	" 26.	" 27	" 4.	" 5	" 12.
" 20	" 27.	" 28	" 5.	" 6	" 13.
" 21	" 28.	" 29	" 6.	" 7	" 14.
" 22	" 29.	" 30	" 7.	" 8	" 15.
" 23	" 30.	Oct. 1	" 8.	" 9	" 16.
" 24	" 31.	" 2	" 9.	" 10	" 17.
" 25	June 1.	" 3	" 10.	" 11	" 18.
" 26	" 2.	" 4	" 11.	" 12	" 19.
" 27	" 3.	" 5	" 12.	" 13	" 20.
" 28	" 4.	" 6	" 13.	" 14	" 21.
" 29	" 5.	" 7	" 14.	" 15	" 22.
" 30	" 6.	" 8	" 15.	" 16	" 23.
" 31	" 7.	" 9	" 16.	" 17	" 24.
Sept. 1	" 8.	" 10	" 17.	" 19	" 25.
" 2	" 9.	" 11	" 18.	" 20	" 26.
" 3	" 10.	" 12	" 19.	" 21	" 27.
" 4	" 11.	" 13	" 20.	" 22	" 28.
" 5	" 12.	" 14	" 21.	" 23	" 29.
" 6	" 13.	" 15	" 22.	" 24	" 30.
" 7	" 14.	" 16	" 23.	" 25	" 31.
" 8	" 15.	" 17	" 24.	" 26	Sept. 1.
" 9	" 16.	" 18	" 25.	" 27	" 2.
" 10	" 17.	" 19	" 26.	" 28	" 3.
" 11	" 18.	" 20	" 27.	" 29	" 4.
" 12	" 19.	" 21	" 28.	" 30	" 5.
" 13	" 20.	" 22	" 29.	Dec. 1	" 6.
" 14	" 21.	" 23	" 31.	" 2	" 7.
" 15	" 22.	" 24	" 31.	" 3	" 8.

DATE OF SERVICE.	DUE TO CALVE.	DATE OF SERVICE.	DUE TO CALVE.	DATE OF SERVICE.	DUE TO CALVE.
Dec. 4	Sept. 9.	Dec. 14	Sept. 19.	Dec. 23	Sept. 28.
" 5	" 10.	" 15	" 20.	" 24	" 29.
" 6	" 11.	" 16	" 21.	" 25	" 30.
" 7	" 12.	" 17	" 22.	" 26	Oct. 1.
" 8	" 13.	" 18	" 23.	" 27	" 2.
" 9	" 14.	" 19	" 24.	" 28	" 3.
" 10	" 15.	" 20	" 25.	" 29	" 4.
" 11	" 16.	" 21	" 26.	" 30	" 5.
" 12	" 17.	" 22	" 27.	" 31	" 6.
" 13	" 18.				

AVERAGE PERIOD OF GESTATION IN RACES OF ANIMALS.

Elephant	2 years.
Camel	1 year.
Buffalo	1 "
Mare	340 days.
Cow	281 "
Reindeer	240 "
Sheep	144 "
Goat	144 "
Sow	120 "
Dog	63 "
Cat	56 "
Rabbit	28 "
Swan sits	42 "
Goose "	30 "
Duck "	30 "
Pea Hen "	28 "
Turkey "	28 "
Guinea Fowl "	28 "
Hen "	21 "
Canary "	14 "
Pigeon "	14 "

The longest recorded period of gestation in the cow is 313 days.

The shortest period in which the calf survived was a Jersey born at the seventh month.

MEASUREMENTS OF HAY, CORN, ICE AND ROOTS.

One cubic foot of bale hay weighs 9 pounds.

One cubic foot of pressed hay weighs 25 pounds.

Five hundred and twelve cubic feet of hay weigh one ton in mow.

Two cubic feet of sound corn in ear will make one bushel of shelled corn.

One cubic foot of ice weighs $57\frac{1}{2}$ pounds, and sustains a weight of more than 1500 pounds in its natural position.

In building ice-houses allow one ton of ice to thirty-four cubic feet of space. An acre of ice one foot in thickness will yield about 1300 tons.

To find the number of bushels of carrots or mangolds in a bin multiply the length, breadth, and thickness together, and this product by 8, and point off one figure in the product for decimal.

One cubic foot of water measures 8 gallons and weighs 12 pounds.

One quart of milk weighs 2.15 pounds.

NUMBER OF PLANTS FOR AN ACRE.

1 foot by 1 foot	43,560
$1\frac{1}{2}$ feet by $1\frac{1}{2}$ feet	19,630
2 feet by 1 foot	21,780
2 feet by 2 feet	10,890
$2\frac{1}{2}$ feet by $2\frac{1}{2}$ feet	6,960
3 feet by $\frac{1}{2}$ foot	29,040
3 feet by 1 foot	14,520
3 feet by 2 feet	7,260
3 feet by 3 feet	4,840
$3\frac{1}{2}$ feet by $3\frac{1}{2}$ feet	3,555
4 feet by $\frac{1}{2}$ foot	21,780
4 feet by 1 foot	10,890
4 feet by 2 feet	5,445
4 feet by 3 feet	3,630
4 feet by 4 feet	2,722
30 feet by 30 feet	48
33 feet by 33 feet	40
40 feet by 40 feet	27

IRRIGATION.

In a country like our own, containing every variety of soil and climate, and subject to the most variable degrees of rainfall in its different parts—a country of abundant sunshine, a land of streams and great lakes, and yet subject to the severest drouths, extending over large areas—a land, too, where there is a greater waste of fertilizing material than in any other part of the civilized world—in such a country the need of irrigation is great and continually growing.

Drouths occur annually, of more or less severity, in almost every section, while severer drouths occur periodically every third year, and still greater drouths about every decade. Again, large portions of our western domain are under a state of perpetual drouth, but only require that the mountain streams and the mighty rivers of the great valleys be made to flow over them to induce the highest state of fertility.

It is a grievous thing to hear a wail of complaint from every quarter of the land during a time of drouth, while every portion of the country is intersected with brooks, rivulets and mighty floods of water running unheeded and unused to the unfilled sea.

An incident will illustrate how a small stream may sometimes be utilized. On my native homestead, in Connecticut, there flows a small trout-brook, and on one occasion, during a time of the severest drouth ever known in that part of the country, we had a field of corn bordered by the brook. At the time when the maize should have been making its most rapid growth, instead of waving its broad fresh leaves in every breeze, it began to wilt and lose its color, rolling up its leaves, and not receiving moisture enough from the air to allow them to unroll at night. In such a case the maize crop becomes worthless in a few days, unless it can be saved by irrigation. There was an abundant flow of water in the brook, and it was but the work of a few minutes with a shovel and pieces of boards to construct a suitable dam. In a few hours the whole field was saturated to the depth of about six or more inches, when the dam was taken up and the stream allowed to go its way. The result was a crop of one hundred bushels of ears to the acre when other fields unirrigated had a yield so small as to be scarcely worth harvesting.

WATER.*

" This most important of all liquids occurs in nature in all the three states of aggregation which substances are capable of assuming.

" In its solid state, as ice, and in its liquid form, it covers at least three fourths of the entire surface of the earth. It constitutes about three fourths of the weight of living plants and animals, and enters largely into the composition of many mineral matters. In a gaseous form it continually evaporates from the surface of the earth, rises as watery vapors, which, in the colder regions of the atmosphere, become condensed into clouds, and is, without doubt, the most abundant substance we meet with on the face of the earth. It is never found in nature in a state of perfect purity ; but pure water can easily be obtained from almost any kind of natural water, by the simple process of distillation. Distilled or pure water, on evaporation, does not leave the slightest residue, and none of the ordinary chemical tests produce any change in its appearance. Pure or distilled water, from whatsoever natural source

---

* Morton's Encyclopædia, Professor August Voelcker.

it may have been obtained, invariably is a chemical compound of two simple or elementary gases, hydrogen and oxygen. Every nine pounds of water always contain eight pounds of oxygen and one pound of hydrogen; or in one hundred pounds of water there are 88.88 pounds of oxygen and 11.11 pounds of hydrogen.

"Water freezes at 32° Fahr., or at 0° Celcius, and 0° Réaumur, and boils, and becomes converted into watery vapor or steam, at 212° Fahr., or 100° Celcius, or 80° Réaumur. The evaporation of water, however, not only proceeds at an elevated temperature, but takes place, under favorable circumstances, at all degrees of heat; and even in the form of ice, water slowly, it is true, but steadily, evaporates on exposure to a dry atmosphere. The rapidity with which water is changed into vapor depends mainly on the temperature of the surrounding air, its degree of dryness (its hygroscopic condition), its amount of pressure, and the speed with which the air, charged with watery vapors, is replaced by a dry current. Thence the drying effect of a hot sunshine and of a strong and dry wind.

"During the evaporation of water a considerable amount of cold is produced, arising from the circumstance that water, in a gaseous state, contains a much larger amount of *latent* or imperceptible heat, *i.e.*, heat which is not indicated by our thermometers. The heat necessary to change liquid water into vapor is abstracted from surrounding warmer bodies, and consequently we feel the sensation of cold.

"Thus we are liable to catch cold when we sit down in wet clothes, but seldom feel any inconvenience from a shower of rain which may have surprised us, if we take strong bodily exercise, and thereby supply the heat which is removed from our bodies by the evaporation of the moisture from our wet garments.

"The atmosphere always contains water in an invisible form, and is capable of keeping in perfect solution a larger quantity of moisture at a more elevated temperature than at a lower.

"Water is not merely indispensable to animal and vegetable life, but also to the very existence of many purely inorganic compounds.

"The principal varieties of natural waters are: rain water, well-spring water, river water, sea water, and mineral waters.

RAIN WATER.

"Rain water, having undergone a kind of natural distillation, especially when collected in remote country districts, is the purest of all natural waters. On evaporation it scarcely leaves a trace of fixed matters, and is contaminated only with minute traces of impurities, which the rain washes out of the air. The rain water collected in towns is less pure. Besides the usual atmospheric impurities, such rain water contains organic and inorganic matters which the rain washes out of the frequently dense, smoky town atmosphere, or dissolves from the roofs of houses. The organic impurities impart unto it a yellowish color, more observable in water kept some time.

"The same impurities are likewise the cause of the putrid smell which such rain water assumes on keeping.

" The more important of the gaseous impurities collected in rain water are carbonic acid and ammonia, and, especially during thunder-storms, nitric acid. They are washed out of the air by the falling rain, and, as might be expected, the first shower contains a larger amount of carbonic acid and ammonia than the rain which descends after a succession of rainy days.

"The amount of ammonia in the air is ever variable, and for that reason rain water cannot contain always the same quantity of this valuable fertilizing substance. At any rate, the amount of ammonia and nitric acid in rain water is so small that at least twenty gallons are requisite for ascertaining their relative proportions. From the average results of M. Barral's analyses of the rain water collected at Paris, it has been calculated that in the course of a year the following quantities of nitric acid and ammonia are brought down from the air, by the rain falling, on every English acre :

	Lbs.	Nitrogen, lbs.
Ammonia	12.29 =	10.69
Nitric Acid	41.24 =	10.12

"Supposing our annual rainfall to be twenty-eight inches, according to Professor Way's analyses, the following amount of ammonia and nitric acid would be poured down yearly on every English acre :

	Lbs.	Nitrogen, lbs.
Ammonia	28.59 =	23.54
Nitric Acid	68.91 =	17.88

"It thus appears that the rain which falls in a year conveys to the soil a considerable quantity of two of the most beneficial fertilizers.

WELL-SPRING AND RIVER WATERS.

" Water being a solvent for many mineral and organic matters, necessarily must become contaminated with some of the materials of which the strata are composed through which it flows ; and as different strata are composed of a variety of mineral matters, differing greatly in solubility, spring water, according to the nature of the rocks and soils through which it passes, must always contain a smaller or larger quantity of various mineral substances.

"Sometimes spring waters contain so large a quantity of mineral substances in solution that they acquire a saline taste—they are then called mineral waters.

"The purest kinds of spring waters are those which rise in granite districts, or in localities abounding in sands and rocks, which are principally composed of silicious elements. One of the purest natural waters is that of the Laka, in the north of Sweden. It contains only one twenty-sixth of a grain of solid mineral matter in the imperial gallon, and is admirably well adapted for the making of filtering paper.

"On the other hand, water which rises in calcareous districts, or which flows over soils and rocks abounding in lime, is very impure, as it contains invariably a large quantity of mineral matters, more especially lime.

"The drinking-water of Cirencester contains about forty-four grains of solid mineral matters to the imperial gallon, and some other waters a much larger quantity.

"Good drinking-water ought to be perfectly clear, colorless, odorless, tasteless, and uniformly cold at all seasons. The presence of much organic matters renders water disagreeable to the taste, and unwholesome. Inattention to this circumstance has often been productive of serious and fatal disease. Well water is liable to become contaminated with these injurious impurities.

"In sinking a well, the neighborhood of farm-yards, grave-yards, and all places where refuse matters accumulate ought to be avoided, particularly if the soil in the locality is silicious, or of a porous nature, which favors percolation of the surface water. This also shows how desirable it is to prevent the accumulation of the droppings of animals in open yards; for not only will the rains that fall upon them wash out their most valuable constituents, and thus deteriorate the value of the manure, but the well water in the neighborhood is liable to become adulterated with unwholesome impurities.

"In all well-manured and porous soils the organic substances of the manure give rise to the production of nitrates, a class of compounds remarkable for their high fertilizing powers.

"The use of leaden pipes for conducting drinking-water ought to be avoided. Should the water assume a decidedly brownish or black color on the addition of sulphuretted hydrogen water, it may be inferred that it contains in solution a quantity of lead which cannot be introduced in the human organism without causing the most serious consequences.

RIVER WATER.

"Like spring and well waters, river waters contain a variable quantity of solid matters; but, generally speaking, river water in most instances is softer than the well or spring waters in its neighborhood, and for this reason it is better adapted for general purposes than spring or well waters.

"Carbonate and sulphate of lime, or gypsum, constitute the chief portion of the solid matters which are left on evaporation; besides these compounds, ordinary spring, well and river waters usually contain variable quantities of common salt, sulphate of soda, sulphate of potash, carbonate of magnesia, silica, iron, alumina, phosphoric acid, and organic matters.

" River waters generally hold variable quantities of suspended matters, dependent upon locality and state of weather; and thereby are rendered more or less turbid or muddy.

" The particles of suspended matters do not always readily subside, and river water, for that reason, must usually be filtered, or otherwise purified, before it can be employed for domestic purposes. Thus even unwholesome and turbid water, by the use of the water-filter, can be rendered wholesome and clear.

" Simple filtration, however, does not remove to any extent the several constituents contained in natural waters, and cannot, for this reason, be resorted to for the purpose of rendering a hard water soft.

IRRIGATION WATER.

" Prejudicial as are the organic impurities in water to animal life, they materially benefit the growth of plants; consequently a water intended to be used for irrigation will be all the better for containing a good proportion of organic substances. Hence no water is so useful for irrigation as sewage water, or a natural water into which the sewage of towns finds its way; for water of that description invariably contains a considerable amount of putrefying animal and vegetable remains, partly in a state of perfect solution, partly in suspension. But as many natural waters are employed for irrigation with much benefit, although they contain mere traces of organic substances, the beneficial results attending irrigation cannot be due entirely to the organic matters deposited on the soil in the passage of the water over it. The inorganic substances contained in all natural waters certainly must contribute to their general beneficial effects; for several of the mineral constituents of spring and river waters are known to be excellent fertilizers.

" There are few natural waters which do not contain an appreciable quantity of salts of potash and soda, sulphate of lime, and soluble silica; and as all these compounds are calculated to promote the healthy and luxuriant growth of plants, most natural waters must exercise a beneficial action on vegetation, partly on account of their mineral constituents. In many waters known to be well adapted for irrigation we have also detected a small amount of phosphoric acid, or, more correctly speaking, of phosphate of lime or bone-earth; and though the percentage of phosphoric acid in water is but trifling, yet, considering the large quantities which run over irrigated land, an absolute amount of phosphate of lime is conveyed on it, which is equivalent to a good dressing of bones.

" Some natural waters are much richer in alkaline salts than others, and perhaps, partly for that reason, some kinds produce a more marked effect on vegetation than others.

"However, we believe the beneficial effects attending irrigation cannot be referred entirely to the organic and inorganic fertilizing substances which all waters contain. But as we are not inquiring into the full causes of these effects, and are only speaking of the qualities of irrigation water, we shall content ourselves by observing that there is scarcely any natural water, however poor in solid matters, which cannot be employed with advantage for irrigation purposes.

"While experience teaches that all ordinary spring and river waters are capable of being employed with advantage by the irrigator, it also informs us that some kinds of natural waters produce much more striking effects on vegetation in irrigation than others.

"As a general rule, it may be stated that the value of a water, for the purpose of the irrigator, depends first and chiefly on the quality, and second, upon the quantity of the solid matter it contains.

"The streams or springs which flow from or over limestone districts, and especially those which have their origin in the lower strata of limestone rocks, are particularly characterized by fertilizing properties. In the opinion of many practical men it is the lime in these waters which causes the green, fresh, and luxuriant appearance of the herbage; and lime, consequently, is regarded by them as the most valuable ingredient of the water. Lime, no doubt, is a useful fertilizing agent; and water containing a considerable proportion of this substance necessarily must exercise a highly beneficial action when allowed to flow over land naturally deficient in lime.

"Most soils, however, contain proportions of lime amply sufficient to supply all the wants of the growing plant; and almost all natural waters, likewise, contain this substance in considerable quantities.

"In the majority of cases we are not warranted to ascribe to lime alone the whole or even the principal share of the chemical influence which water may exercise when employed for irrigation. Now the springs which rise in the lower strata in limestone rocks we have ascertained invariably contain proportions of alkaline salts and phosphoric acid which are larger than those usually found in natural waters; and as alkaline salts and phosphoric acid belong to the most valuable fertilizing substances, we are inclined to ascribe the superior fertilizing action of these waters, as far as is dependent on chemical substances, not to the presence of lime, but to that of a considerable proportion of alkaline salts and phosphoric acid."

WATER-MEADOWS.*

"Watering meadows, or the system of applying liquid to further the growth of the permanent grasses, is a custom very peculiarly localized, both in England and on the Continent.

---

* Morton's Encyclopædia, Hugh Raynbird.

"While in the latter we see irrigation practised largely in hot climates, as in the south of Spain, in Persia (whence the Persian wheel), in China, where the mechanical contrivances for this purpose are ingenious, though simple; and in Egypt, where the natural and annual irrigation of the Nile leads to productive harvests; yet, with the exception of the rich meadows of Lombardy, and in the mountain slopes of Switzerland, we find that this irrigation is confined principally to the growth of vegetables and the cultivation of grain crops, and does not come under the view of water-meadows.

"In England we find the custom of watering confined to a few of the southern counties; for although successful instances (some of them on a large scale) may be found in other districts, they are only the introductions of large landholders, and come under the head of experiment rather than practice.

"I shall briefly mention the English counties that excel in irrigation. These are: Wiltshire, with its water-meadows on the Avon, and its celebrated Orchiston meads, which are known under the title of the Long Grass Meadows, the crop of hay on which is enormous; Hampshire, with its meadows upon the Avon, the Test and the Itchen, so useful to flock-masters from their vicinity to the Downs and their early produce of herbage for ewes and lambs; Gloucestershire, on the banks of the Severn, Avon and Ledden. Worcestershire has water-meadows on its numerous rivers and small streams, and in many instances the water is brought from a considerable distance by canals, which supply several farms on one estate.

"Dorsetshire possesses six thousand acres of irrigated meadows, including some of the very richest in the vale of Blackmore, watered by the river Stour.

"Devonshire has its hillside or catch meadows, as also many water-meadows, on the alluvial borders of the principal rivers.

"Berkshire has valuable meadows along the river Kennet.

"It is probable that the total quantity of land under this cultivation in England is under one hundred thousand acres, a very small proportion of those tracts that might be improved in this way.

"The most celebrated Wiltshire meadows are on a loose bed of broken flints, with scarcely any earth; the water above feeds the grass. The Hampshire rivers have a hard bottom subsoil of chalk, and the water runs over a gravelly or peaty surface. It seems only necessary that the subsoil should be porous, and the surface soil may be what it will on thoroughly well-drained land; probably it is not of primary importance what the constituents of the soil may be, although the best and healthiest meadows certainly occur where the soil is porous and dry.

QUALITY OF WATER.

"Water which is productive of fish, particularly trout, is generally supposed to be good for water-meadows. Experience seems to declare that, for grass land, the clearer the water the better; that calcareous matter taken up in a form not to render the water turbid is almost the only beneficial admixture.

"When the rivers are turbid from quantities of silt, or of finely divided clay and peat, they injure the grass, especially the former; but streams flowing clear and pure from the hills are of benefit, and especially from hills abounding in lime. But admixture with other soils injures them; as, for instance, in Staffordshire, where the river Dove, flowing from calcareous hills, so enriches its neighboring mead that it is proverbially said: 'In April Dove's flood is worth a king's good;' but when admixed with streams from other sources its benefit ceases.

"On arable land the more thick and turbid the water the better. The Nile water is thick, and the water used in warping land even more so. The Humber, in Lincolnshire, is famous. The basis of the warp soil of this river is fine clay and sand, the latter in the greater proportion, and minutely divided and intimately mixed with the former, with a considerable portion of fine calcareous earth.

"Though not suited for water-meadows, it is probable that warp land, if laid down to grass, would form very fertile natural ones; in fact, it is the mode in which all our rich alluvial meads were originally produced.

"But though an admixture of natural earth with the water is not beneficial, yet that of dissolved animal excrement is. All water is weak liquid manure, and although it might not be economically practicable for every farmer to lay down a water-mead, on the system of Mr. Mechi, of underground iron pipes and steam pumping power, or even advisable, except in particular instances, to imitate the cheaper mode common in Switzerland, and sometimes practised in Cheshire, and also in Devonshire, of turning a rivulet so as to flow through a farm-yard, and thus irrigate meadows situated lower down; yet the example of the Edinburgh meadows shows with what great success liquid manure may be applied to grass, as does the experience of Mr. Dickenson and others with Italian rye grass.

HERBAGE PLANTS SUITABLE FOR IRRIGATION.

"This is a subject which deserves more attention than has yet been applied to it. The nature of the grasses for water-meadows has not been studied, but it has been left to accident to produce and circumstance to alter.

"In Lombardy greater care is taken.

"The Italian rye grass is the principal kind cultivated under this system. It is a native of this district of Italy, and perhaps its larger size originated from its being thus peculiarly cultivated, just as the timothy grass is only an enlarged American variety of the common cat's-tail grass, improved by cultivation and the influence of change of soil and climate. No grass produces earlier or more abundantly under irrigation, and Mr. Rham, in his *Dictionary of the Farm*, mentions having seen an instance of hay made in July from a newly-made water-meadow sown with Italian rye grass in March.

"A Mr. Dickenson, in a letter addressed to the Duke of Richmond in July, 1847, after mentioning that his land was strong clay, thoroughly drained and well pulverized, and sown with Italian rye grass at the rate of four bushels to the acre, mentions that in 1844 he cut a crop ' the first week in March, with about ten inches of grass ; April 13th cut the second time ; May 4th the third time ; May 24th the fourth time ; June 15th the fifth time ; July 22d the sixth time, with ripe seed, and three loads of hay straw to the acre. Immediately after each of these crops the land was watered at once, from a London street water-cart, with two parts of pure urine from the stables and one part of pure water, the produce of each crop increasing with the temperature of the atmosphere, from three fourths of a load per acre, as hay, to three loads per acre.' The land was not watered any more, yet produced four light crops afterward, making ten cuttings in the course of a year.

" In this experiment we may notice the strength of the manure, and also that the crop increased with the temperature of the air. This last is a fact that shows why watering is only useful in hot climates, and accounts for its being almost entirely confined to the southern counties of England, and not practised in Scotland.

QUALITY OF THE GRASS.

" This is a subject not unworthy of notice ; for its increased produce is little gain if, from its laxative and too succulent nature, it produces diarrhœa in young animals feeding on it, though this may be corrected by giving good hay.

" Although the introduction of water-meadows into a district where before unknown is desirable, yet the introducer must not overlook the difficulties or reasons that prevent such having been previously attempted.

" Climate must be considered, nature, and plentifulness of water—and even where plentiful not obtainable, being monopolized by mills ; nature of soil in which to be applied ; and, if all these are suitable, nature of the country ; for one must consider how to take water off as well as how to get it on, and a flat country may have greater difficulties in the latter point than a hilly one to overcome the obstacle of its declivities. We must recollect that all English counties where water-meadows are in vogue are hilly, and all these have a rocky, and at least a fissured subsoil, which collects the water in springs, so as often merely to require easy direction.

" Where land is favorable for water-meadows it cannot better be described than in the words of Mr. Pusey : ' A slight film of water trickling over the surface—for it must not stagnate—rouses the sleeping grass, tinges it with living green amidst snows and frosts, and brings forth a luxuriant crop in early spring, just when it is most wanted, while the other meadows are still bare and brown. It is a cheerful sight to see the wild birds haunting these green spots among the hoar-frost at Christmas; or the lambs, with their mothers, folded on them in March. A water-meadow is the triumph of agricultural art, changing, as it does, the very seasons.'

RIDGE-AND-FURROW IRRIGATION.

"The streams are diverted by means of hatch-work, and the water runs in small gutters or carriers along the tops of the ridges into which the land is shaped, and is made to flow over the entire surface, and falls into another series of gutters, which convey it away, either to the river from whence it came, or to serve the purpose of irrigating meadows that are lower down the stream. The best meadows are those upon a gravelly soil, with a good drainage; the latter is a matter of great importance, although seldom sufficiently attended to in the formation of meadows.

"In extreme cases, where other methods cannot be adopted, the main body of water need not be diverted from its proper channel when it passes from the meadow; but a small drain, with an outlet at a sufficient distance down the stream, generally below a mill, may be opened to convey away the soakage water, the mouth of this drain being, of course, closed when the meadow is watered.

"The process of floating the meadows is intrusted to the care of a man who makes this kind of work his regular employment, and who is usually paid at a certain rate per acre per year for taking charge of the meadows upon one or more farms. He commences in the autumn, by clearing out the gutters, and as soon as the water is turned on he regulates the stops and edges of the gutters, so as to insure an even and regular flow of water over the surface of the meadow. Injury results if the water is allowed to stagnate in any part, or if it remains on too great a length of time; it is therefore turned off and on at short intervals.

"A succession of feed is secured by commencing watering a portion of the meadows upon a farm earlier than the remainder. Some even do not turn the water on till after Christmas, but then the early feed is lost altogether. Upon the best meadows grass is ready to be folded about the middle or end of March, the water being turned off a few days previously. The grass is fed with ewes and lambs, the latter having a run forward; they are not, however, allowed to remain entirely on the meadow, but are removed to a piece of Swedes, or other feed growing upon arable land, at night. When the grass is all fed the watering is again renewed, as before, and is continued until a short time before the grass is in readiness for the scythe, usually at the end of June or beginning of July. The great bulk of the succulent herbage, and the natural dampness of the situation upon which it grows, occasion the process of hay-making to be one that requires much care and attention. The meadow is again watered, and the aftermath fed off with horses and cows, few meadows being safe to feed sheep upon in the autumn.

SEWAGE IRRIGATION.

"The meadows watered with the sewage of Edinburgh afford the best, though in many respects imperfect, examples of the advantage to be obtained by the use of the offscouring of our towns for agricultural purposes.

"The method of application is simple, but it has proved successful, when compared with the elaborate and expensive methods tried in other quarters at present, unfortunately, however, without proportionate results. Although the irrigation is carried on only upon a small scale, and the means used are imperfect—a great quantity of manure running to waste—yet the results appear truly extraordinary. The following sketch will roughly illustrate the method adopted upon a meadow at Lochend farm:

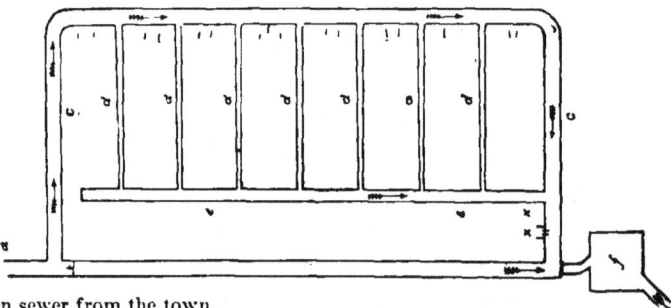

a, The open sewer from the town.

b, Sluice for turning the sewage into the carrier *cc*, which has stops at intervals to turn the liquid into the small carriers *dd*, which are about ten inches by ten inches, having stops to throw the manure water regularly over the grass, to facilitate which small cuts are also made with a spade at the edge of the carriers.

f is a settling-tank into which the liquid runs, having a dam and grating at one end, which prevents the solid portion escaping ; as it accumulates it is removed.

ee is a drain which carries off the waste water, both from the surface and underground drains.

"Of course it will be understood that the wonderful results of this liquid manuring are dependent on other causes than those to which the fertility of the ordinary water-meadow are due. The gross and rapid growth of Lochend meadows is owing simply to excessive manuring; that of English water-meadows may be in a measure attributable to the efficient and rapid supply of food to the grasses by running water; but also, no doubt, in part to an improved temperature—and even if this were not one of the operating causes, feeding by a rapid flow of clear water is very different from the mere drowning of land in thick and sluggish liquid manure.

"At Quarry Hole Farm a meadow of six or seven acres is somewhat differently arranged, with very similar results:

"The liquid, very strongly impregnated with sewage matters, is turned on these meadows for three or four days at a time, at intervals. After mowing, the water is not turned on for six or seven days, or it will rot the roots of the grass.

"The enormous prices given by the Edinburgh cow-keepers for the produce of these meadows is the best evidence of their value.

"The grass is sold in half-acre lots by public auction, and realizes from £30 to £50 per acre : the grass is cut four or five times by the purchaser, and left clear by the 20th of October.

"If such results as these are obtained by the simple means in this case employed to distribute the town sewage, we have reason to anticipate far greater advantages, both in an agricultural and sanitarial point of view, from the more modern means of distribution by the hose and force pump, the practicability of which, for the application of sewage as a fertilizing agent to our fields, is so well and clearly illustrated by Mr. Mechi at Tiptree Farm, and by other spirited agriculturists, and recommended by them as a system far more in character with the improvements of the advancing age than the one just described ; and, we do not doubt, for the application of town manure experience will prove it to be such.

CATCH MEADOWS.

"Catch meadows have the great advantage over ridge-and-furrow meadows of cheapness of formation : the same quantity of water will suffice to irrigate a larger surface, falling as it does from tier to tier of gutters. The hillside affords a more natural surface for the water to flow over than that which is given by the artificial and expensive ridge-and-furrow.

"Philip Pusey gives a very good account of catch meadows in the *Journal of the Royal Agricultural Society :* 'It is to the southwest we must turn, to Somerset and to Devonshire, for patterns of future irrigation. In these two lovely counties, which have the valleys without the Alps of Switzerland, abundant streams roll cheerfully in a rapid descent over stones, or among mossy rocks, and the sheltered sides, shelving rapidly upward, have long since tempted the farmers to lead water along their sloping face in tiers of channels, each of which, receiving the overflow from above as it begins to gather irregularly, receives it in a level trough, to brim over anew, until it reaches the lowest channel, which delivers it back to the river's bed.

" 'The horseman, as he rides along, sees meadows of a few acres rising above his head, bright as emerald, glistening against the sun with their thin film of water, alternating with orchards in which cottages are nestled that seem to cling to the hill, with a canopy of oak copse above, whose russet leaves, a remnant of the last summer, look the ruddier against the narrow space of blue sky that roofs in the glen. These are called catch meadows because each trench thus catches the water from its neighbor above it.'

"Mr. Pusey also quotes examples of catch-work upon level meadows being carried out successfully at a small cost.

" An improved system of irrigation is also described in the *Journal* by John Bickford, of Crediton, Devon, which is also well worthy of perusal by those requiring information on the subject.

" The following monthly directions for the management of water-meadows are given from Boswell Wright, and from original observation.

" But such directions for this and all other agricultural operations depend on the *natural*, not on the nominal season. Those here given are for an average year, and must be altered to suit one more forward or backward, an extraordinarily mild winter, or other peculiarity of time or climate.

JANUARY.

" The land should be floated in frosty weather to protect the grass ; but about once a fortnight air must be given, and the land laid as dry as possible for a few days.

" If the frost has given a complete coat of ice to the meadow, do not float over this, as the attachment of the ice to the surface often draws the soil into heaps and injures the evenness.

FEBRUARY.

" In this month the meadows require much attention. If the water is allowed to flow over the grass several days without intermission a white and very injurious scum is formed ; and if the water is then drawn off, and a severe night-frost attacks the wet grass, it cuts off the herbage. To prevent this scum take the water off by day and lay it on at night, to avoid frost. A less troublesome but inferior plan is to take the water off early in the morning, if a dry day, and let it remain off several days and nights ; for one day's drying is sufficient to enable the grass to resist frost. From the middle of this month water is applied more sparingly than in winter, and more to encourage vegetation than to protect from frost ; and in the last week of the month there probably will be a good bite for ewes and lambs.

MARCH.

" At the beginning of this month old floated meadows will supply abundant food to all kinds of stock. If heavy cattle are turned in the water must be taken off for a week previously, to allow the land to become firm and dry. If the season be cold in the first week give a little hay in the evening, to correct the effects of too moist food. But the grass is best applied for ewes and lambs, and should be hurdled off for them in portions. Peat soils would be damaged if heavier stock than sheep or calves be turned on thus early.

APRIL.

" In this month the use of meadows for ewes and lambs is still greater than in March, and the farmer who possesses a good breed of them will require little else for their keep ; but it must be recollected that they must not be on longer than this month, or the hay crop will be much injured.

MAY.

"Remove ewes and lambs and calves the last day of April : the meadows will be fed bare, and most farmers consider that the barer the ground is left, so much more is the meadow improved, and the quality of the hay superior.

"After clearing water for a week, carefully examine every trench and drain, and so shift the water into other meadows that the land is alternately watered and drained, and the time of the water remaining on the land shortened as the weather gets warmer.

"In five, six, or seven weeks the meadows will be fit to mow for hay.

"This is also a good month for forming new water-meadows, though any time of the year, unless during severe frosts, will answer for the work.

JUNE.

"Mow and make hay. The grass, being of a more succulent nature, requires more careful making, and is more subject to heat if not got up in good order.

"As soon as the grass is off turn in cattle (not sheep) to eat the grass left by the mowers and what grows in the trenches. Then let the water dribble on them as slowly as possible, this being the hottest season of the year; and after two or three days shift this first watering to another meadow. The effect will be very great, and the verdure, compared with unwatered meadows, exceedingly rich ; but recollect not to keep water on too long in warm weather, or a white substance like cream is produced ; and if this is neglected a scum as thick as glue, and nearly as tough as leather, settles on the grass, and quite destroys it.

JULY TO OCTOBER.

"Where the meadows are fed late the beginning of July is the season for making hay on the water-meadows, which, we have before remarked, is an operation requiring much care and attention ; after which the watering is renewed for a short time, and the aftermath fed off with cattle.

NOVEMBER AND DECEMBER.

"Begin to water the meadows ; frequently water can be collected in the higher parts of the farm sufficient to water some of the low meadows, and, by attention to the ditches and water-courses, a free passage can be given from that portion of the farm where injurious to that where it would be highly beneficial and more productive than a coat of manure. It is best to keep the water running over the grass, and not to allow it to stagnate.

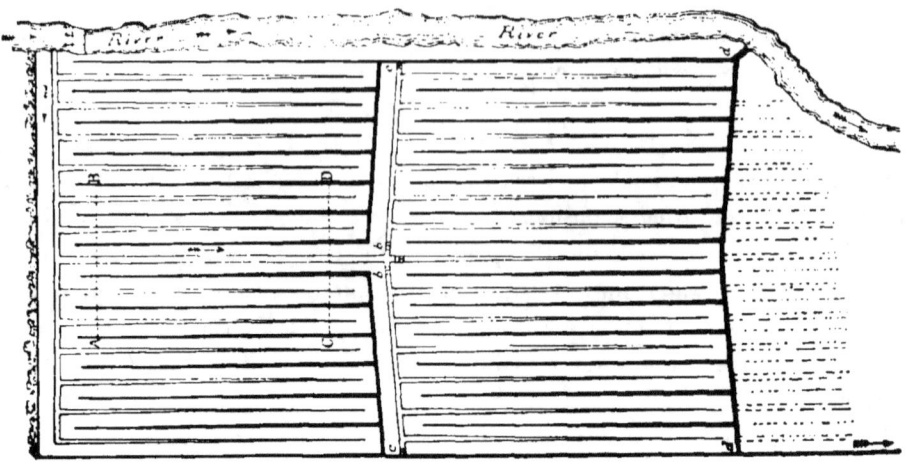

RIDGE-AND-FURROW GROUND PLAN OF A WATER-MEADOW.

Scale of 40 yards to $\frac{4}{10}$ths of an inch.

Light lines, Barriers; black lines, Drains. *a*, Hatch across river; *bb*, Hatches to water lower stem;
cc, Small hatches to draw off the water when the meadow is laid dry. Dotted lines, Barriers from
the main drain of the meadow above, by which the irrigation may be continued lower down the
stream, the water being penned up by the two small hatches *dd*. Small stops of turf are placed
in the small carriers at intervals of about fifteen yards, or as required, to make the water flow
regularly over the surface of the meadow.

MILANESE IRRIGATION—LOMBARDY MEADOWS.

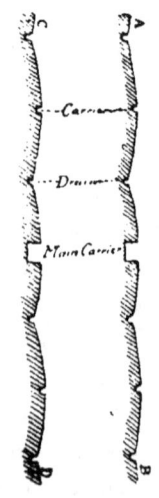

"In Lombardy neither sheep nor cattle are fed upon the
meadows, but the entire produce, whether of permanent grasses or
clovers, is mown and used for soiling cattle in stalls. Manure is
applied to the meadows, and, as they are not trodden by cattle,
their surface is kept smooth with mathematical correctness.

"We believe this example might be followed with advantage,
particularly near the towns, where the produce of grass would be
of so great value. Signor J. Devincenzi, an Italian gentleman,
has favored me with the following details on the Milanese system
of irrigation. He considers that irrigation is neglected in England;
and that many of our canals, now rendered almost useless by the
introduction of railroads, might be employed, at little expense, both
to fertilize tracts now cultivated, and to bring land altogether waste
into profitable cultivation. He also considers that the common idea,
that the Milanese system is unsuited to England, from the difference of climate of
the two countries, is erroneous; for, taking as an instance the work of Professor
Dove, of Berlin, and referring to his maps of the isothermal line of the globe

for each month, we shall find that Milan and England possess the same temperature during the months of January, February, November and December; and during these months the Lombards mow their meadows, called *marcite*, twice or thrice; while in England no such result is obtained.

"No doubt the summer temperature of England is far lower than that of Lombardy, but it may be questioned whether this is not an advantage in the production of grass; and yet in Lombardy they cut eight or nine crops yearly from a meadow.

"Signor Devincenzi's opinion ought to possess some weight, as he has written on the subject, is secretary to the Italian committee on irrigation, appointed by the Milanese Scientific Association, and was reporter to a committee on Milanese agriculture, consisting of first-rate Italian agriculturists, from whose report, so far as relates to irrigation, the following is an abridged extract:*

"'In the province of Milan, as well as the rest of Lombardy, there exist two widely different systems of agriculture, both exceedingly well calculated to suit the varying circumstances of their localities.

"'In Upper Lombardy we find small occupations of arable land, tilled by an industrious population of peasantry; in Lower Lombardy extensive water-meadows, held by wealthy tenant-farmers. If we draw a line from west to east, dividing the province of Milan into two parts, and passing through the capital, we shall very nearly show the correct division of the upper and lower part. In general the cultivation of the land surrounding great cities must be considered by itself, as being quite different and inapplicable to the rest of the country or district; yet that surrounding Milan is but a type of Lombardese valley agriculture, and consists almost entirely of meadows, the tilled land being so small in quantity that it is scarcely worth mentioning.

"'These meadows, though very ancient, are in a most thriving and flourishing condition, and the labor employed in them is merely that of regulating the supply of water and levelling the ground, the grass being that naturally produced by the soil. The meadows lying on the south are irrigated by the sewage water from the city, receive no other manure, and are cut seven, eight, and in many instances nine times a year. Those on the north, partly from copious manuring and partly from spring water, there called *fontalini*, are but little inferior to the former.

"'Winter water-meadows, in the country language, are called *marcite*. They are watered every sixth or eighth day in the summer, and are continually covered by a sheet of flowing water in the winter. By this means vegetation is so encouraged that from November to March two or three abundant crops are cut, so that the cattle fed from and upon them are not deprived of fresh fodder more than thirty or forty days in the year. The rate at which these water-meadows are commonly let in the neighborhood of Milan is from twenty to twenty-five francs a *pertica*, or

---

* Report on Milanese Agriculture.

from £5 to £6 sterling per English acre. The water is applied in summer on meads and all kinds of cultivated plants, as required in winter on the *marcite* only, of which there is sufficient to employ all the water, so that it never runs to waste.

" 'The Lombardese irrigation is worthy of praise, as it has converted what would have been barren sands and unhealthy marshes into fertile meadows, and as combining irrigation, drainage, navigable canals, and motive power for mills and machinery in such a manner that one object does not, or only in a slight degree, interfere injuriously with the other. The Lombardese customs and legislation on irrigation are also deserving of notice and imitation.

" 'Still it must be remembered that Lower Lombardy possesses a peculiar adaptability to irrigating purposes in its immense valley, in the vast reservoirs *above*, the lakes resting on the heights of the mountains, and in a river to carry off the superfluous water.

" 'The farms in the Lower Milanese are generally from two thousand to three thousand *pertiche* in extent (three hundred and thirty-three to five hundred English acres), and they are commonly let on leases of from nine to twelve years. Some of these farms are, from the tenacious nature of the soil, suited for the cultivation of rice. Nearly one tenth is laid down as permanent meadow, and of this very nearly one half is cultivated as winter meadow or *marcite*. The ordinary meadow is manured once every year, the *marcite* often twice; and although water from springs is, from its warmth, the fittest for winter irrigation, *marcite* are nevertheless made with any other kind of water. In the other part of the territory, if the soil be proper for the growth of rice, a nine years' rotation is employed. In the first year wheat is sown with *Trifolium pratense*, which supplies abundant pasturage in the autumn. Manure is applied in the second year; and the *Trifolium repens* and other useful plants spontaneously succeed the *Trifolium pratense* during the third and fourth year, in both which years manure is applied.

" 'In the fifth year the soil is sown partly with flax and partly with maize. The part sown with flax is followed the same year either with millet or with maize *quarantine*, so called from requiring, from its germination, only forty days to grow and ripen in.

" 'In the sixth year maize is cultivated with manure. In the seventh, eighth and ninth years rice is sown, in the last two years with manure. On the fields where rice is not grown the same rotation is practised as for the first six years. Thus in the first rotation the soil is manured six times in nine years, and in the latter four times in six years.

" 'It would be interesting to compare the relation that different kinds of cultivation have to one another, and that which the meadow bears to them all. The Lombard cultivators well understand that the latter system augments, instead of diminishes, the produce of grain. An acre of land produces on an average from twenty to twenty-four bushels of wheat and fifty to seventy bushels of maize.

" 'Among the minor products, though still a valuable one of these meadows, is the mulberry, common both in the permanent and other meadows, which, so far from being injured by irrigation, thrives under it. However, this tree is more largely cultivated in Upper than in Lower Lombardy. The hay of all these meadows is used to feed working cattle and cows. The annual rent of each cow is calculated to average from two hundred and eighty to three hundred francs (£11 to £12). This large sum is not obtained solely near large cities, but is common over the district, the milk being employed entirely in the manufacture of Parmesan cheese and butter, that may be carried to any distance. The general rent of farms in the Lower Milanese is from one hundred and twenty to one hundred and eighty francs per hectare (£2 to £3 per acre). To prove the value of this water, let us examine the estimate that the inhabitants themselves put upon it.

" ' They reduce all measures of flowing water to a common unit, which they call *onica*. The Milanese *onica* is a quantity of water flowing from a hole nearly one hundred and forty-nine millimetres wide and one hundred and ninety-eight high (0.488 by 0.649 of the English foot), and comprising a little less than one third of the English square foot, under a pressure of ninety-nine millimetres (0.324 of an English foot). Now this water *onica* is generally sold at the enormous price of from twenty-five thousand to thirty thousand francs (£1000 to £1200), and often even more. If, however, we should state often double or triple the rent of the farm, we should still be under the mark.

" 'The fertilizing power of water is immense. It changes wild heaths into luxuriant meadows, or, to employ figures, raises the rent of land from a bare thirty or forty francs to three hundred or four hundred.

" 'The Lombardy farmer on these meadows is generally a man wealthy and possessed of considerable capital, which he employs with much profit. As a general rule, forty thousand francs are employed on every one hundred hectares of land (nearly £6 8s. per acre), which would be nearly three years' rent of the land.

" 'In cultivating the soil the farmer employs not only daily laborers, but families of cultivators, who share in the produce ; and cultivation on the large scale does not therefore injuriously affect the moral and economical condition of the lower orders, as it unhappily does in other parts of Europe.' "

SAVING OF WASTE MANURES.

The waste fertilizers of the farm consist of the liquid or drainage from the manure heaps, cattle sheds, stables, and the liquid refuse of the dwelling, the gaseous evaporations from fermenting manure heaps and yards, and all animal and vegetable refuse.

Upon large farms, where steam power is used with a system of iron irrigating pipes, so that all the manure of the farm can be distributed in a liquid form, the method is capable of preventing nearly all waste, where it can be adopted. But upon small farms this method cannot generally be adopted.

Perhaps the most economical of all methods is to remove the manure directly from the stable each day to the field when and where it is to be used. This is certainly advantageous if it can be at once plowed in at a depth of from three to five inches. As far as it is practicable this method may be adopted.

As a general plan for the average farm the following method promises to be economical: Select a shaded place on the north side of a wall or building. It would be well to roof it in. Make the space sufficiently large to hold all the manure made during six or eight months, allowing twelve tons for each animal, so as to form a compost couch, and allowing for a depth of from two to five feet. Make the floor of cement or clay, and inclining so that the liquid may run to the front side. Divide this compost couch into two or three compartments by a partition of concrete.

Make a capacious tank, allowing about thirty hogsheads of space for each animal kept, on any convenient side of the compost couch, and connect the tank by drains with the couch, so that the tank may receive all the liquid manure from the compost heaps. Also connect the tank with the liquid manure gutter of the stables and with the house drainage and water-closets. Also connect the tank with the system of spouts that collect the water from the roofs of stables and barns, so that the rain-water may be conducted at will, either to the compost heap or to the tank.

Fix a pump over the tank, so as to provide for pumping its liquid contents upon any and all parts of the compost couch.

These arrangements may be made at moderate expense, and worked with little trouble by the farmer, so as to give him perfect command over his manure, and to concentrate all the manurial elements of the farm in compost or otherwise, at pleasure.

To prevent waste the manure of the stables and yards is daily removed to the compost couch. One section of the couch forms a place of deposit for all the vegetable refuse which can be gathered together.

The gaseous waste arising from too active fermentation may be prevented in the manure heap by compaction and tramping of animals, and also by a liberal wetting daily from the liquid tank. The liquid in a section of the couch can be retained by closing the pipe or sluice which leads to the tank from that section. By covering a section with ashes, peat, or dry earth to the depth of ten inches, and saturating the surface of this covering with sulphuric acid diluted with twenty parts of water, very little ammonia will escape. Thus the couch and liquid-manure tank, both of which must be preserved water-tight, furnish the means of using compost in any condition desired, and also the tank provides a large amount of liquid manure, to be applied by the sprinkling-cart.

It only remains, for those who desire to be successful cultivators of land, to avail themselves of the means now in their power of remedying a great evil, and of securing and saving a great waste.

IRRIGATION IN CALIFORNIA.*

" The Padres who established the missions here over a century ago are generally credited with having introduced the practice of irrigation, and many of the ditches constructed by them are still in use by the white settlers. In one instance at least, however, in California, a large section of country derives its supply of water for this purpose from a ditch, or *zanja*, which antedates history and presents every appearance of extreme antiquity. In Arizona, too, are found evidences of vast irrigation systems upon lands long since given over to the desert, and which even the traditions of the Indians fail to supply us with anything approaching a history. Several efforts on a large scale have been made to divert the waters of the Salt and Gila rivers upon the plains through which they flow, and, singularly enough, it has been found that the ancient remains of irrigating canals afforded the exact grade and the requisite fall per mile that the best appliances of modern engineering could suggest or construct.

" The common method of irrigation in California may be described as follows : Having settled on a site, which must of necessity be in the vicinity of a running stream, a large triangle is made, having a plumb-line hanging from the apex. With the aid of this primitive appliance a ditch is laid out from the field up-hill at a proper grade, until the stream is reached. When the ditch has been built a rude dam of brush or logs is put across the stream, so as to divert a portion of the water into the mouth of the canal, which is just above the dam. The main ditch is carried across the highest part of the field to be irrigated, so as to have a constant fall from the source of supply to the lower end of the field. Sometimes boxes with a wooden slide are put into the lower bank of the ditch opposite each row of trees, vines or plants. Sometimes one man can attend to twenty or thirty such openings at once, and it is necessary to go up and down through the field and see that the little streams are not lost in squirrel holes, but find their way to the foot of the slope. As may be imagined, this is a very wet and disagreeable job. With grapes, corn, etc., a shallow furrow is plowed each side of the rows, at a distance of a foot or eighteen inches, and through this the water is run as long as needed.

" When alfalfa or grain is to be irrigated the system adopted in the case of large fields is to mark off ' checks ' of five, ten, or twenty acres, with a slight levee, or bank, across the lower end and sides, to keep the water from running off. The ditch at the upper end is then tapped, and the check flooded until the water has reached every part of it.

---

* G. F. W., in *Country Gentleman.*

" The next check is then operated, and so on until all are supplied.

" The method described is only practicable where the water supply is abundant. It is estimated that where open ditches, with no protection to bottom or sides, are used, from half to two thirds of the water is lost by evaporation and seepage. Many ditches have been lined with stone and cemented, so as to prevent loss by seepage, and it will not be long before they will be also protected from loss by evaporation.

" A very successful method of underground irrigation has been devised, far superior in the end, though more costly in the beginning. Pipes, made of cement and sand, are laid throughout the field or orchard at just sufficient depth to escape the plow and cultivator, with taps or plugs at regular intervals. When desired, these plugs are withdrawn, and the water soaks through the ground beneath the surface, and proves far more beneficial than when applied upon the surface. With the growing scarcity of water other sources of supply have been sought, and these are found in surface and artesian wells. With surface wells windmills are used, and the water is often conveyed by iron pipes laid in shallow trenches to all parts of the ranch. From artesian wells the water is either collected in a reservoir, or it is allowed to run continually, being diverted, as occasion requires, in small open ditches, or little V-shaped flumes. By this means one well, if not more than an inch and a half in diameter, can be made to irrigate a very large tract of land.

" As to the result of irrigating, a few actual experiences may be given. An alfalfa field has been cut by the writer eight times in as many successive months, yielding from two to three tons to the acre at each cutting. After the crop was removed one thorough soaking with water was all that was needed to insure an abundant growth at once. Vineyards yield six, eight, or ten tons to the acre, potatoes eight to ten tons. Barley and wheat at the rate of from forty to sixty bushels, and corn at one hundred bushels or more, are harvested; watermelons of one hundred and seventy pounds on ground that would not produce a spear of grass without artificial moisture, and vegetables of all kinds in like proportion. Although the water may be clear as crystal, and apparently free from any organic substance, its use in irrigation seems to render manuring almost unnecessary. The same soil will produce abundant crops year after year, with no apparent diminution, when not a particle of fertilizer of any kind is supplied to it. Irrigation seems to be all that is needed.

" While the use of underground pipes at the East, or where frosts prevail, might not be practicable, still there is hardly a farm where some source of water supply might not be made available, and by a little expenditure in the construction of open ditches a large portion of the loss from drouths might easily be avoided, as well as largely increased yields secured."

FEEDING CATTLE.

ELEMENTS OF NUTRITION.

Feeding is not alone the means of simply sustaining animal life by nourishment, but the basis of a science which has power to transform and reform races of animals. Breeds are made by feed and selection. For a special breed, like Jerseys, that is to be kept up to a certain standard of milk and butter production, a peculiar system of feeding is required. The food must not only contain all the elements of animal bodies, but it must be given in such form and proportions as to develop the milk and butter qualities in the highest degree.

Plants elaborate the elementary principles into complex structures ; cattle appropriate them in a form especially suited to their needs. The organic elements, oxygen, hydrogen, carbon and nitrogen, which form the combustible parts of plants and animals, with potash, soda, magnesia, lime, sulphur, phosphorus, chlorine, silica, iron, and a few other elements which form the incombustible portions, are all to be adapted to the wants of animal sustenance by vegetable growths. These elements are all incorporated into the blood, which always remains the same in composition, and its quantity from six to eight per cent. of the bulk of the body. From the blood all the organs are replenished and built up—the nerves, the muscles, the fat, the bones, the skin, the hair, the horns. The amount and quality of the fat, muscle and bone vary greatly in the dairy and the beef type. The fat ox is said to have about three times as much fat as lean flesh, consisting of forty-nine per cent. water, thirty-three per cent. dry fat, thirteen per cent. of dry nitrogenous matter, muscles separated from fat, hide, etc., and three per cent. of mineral matter ; the lean animal fifty-four per cent. water, twenty-five and one half per cent. of dry fat, seventeen per cent. of dry nitrogenous matter, and three and one half per cent. of mineral matters.

The nutritive chemical compounds are divided into two classes, nitrogenous and non-nitrogenous. The term protein includes the first class, which consists of albumen, gluten, casein, legumin, fibrin, mucedin and gliadin, which resemble each other closely in composition, containing very nearly the same proportions of carbon, hydrogen, nitrogen, oxygen and sulphur. With these are classed certain nitrogenous bodies found in grasses and other plants, called amides, which resemble ammonia. Protein, the material of which flesh and blood largely consists, exists ready-formed in the cereals and leguminous and other plants which cattle eat. All these plants contain but a small proportion of protein, their bulk being made up of cellular fibre, sugar, gum starch and oil, called carbo-hydrates. The first class, nitrogenous nutrients, are called proteids, or albuminoids, and sustain animal life. The second class, non-nitrogenous nutrients, consist of carbon and water. Cellulose is the material which, with lignin, forms the framework of plants, an important part of all fodder, the cellulose being digestible in young and tender plants in proportion of from thirty to seventy per cent., while lignin is not digestible in its crude state. Starch is, next to cellulose, the most abundant carbo-hydrate, and is deposited rapidly near the ripening period; maize contains from sixty to sixty-eight per cent., and wheat from sixty-two to seventy-two per cent. Dextrine is produced from starch by heat. Sugars are of three kinds, cane, grape and fruit sugar. They are all easily digested. Cellulose and starch are supposed to be changed to sugar in the digestive process. Pectin is the jelly of fruits, turnips, beets and carrots, and is believed to aid digestion by gelatinizing the contents of the stomach.

The oils of plants are very important, especially in the rations of dairy cattle. They are estimated to have two and one half times the nutritive value of sugar and starch. Maize contains from four to seven per cent., oats six, the best hay three per cent. of oils. Animals appropriate oil for cream and fat, and also transmute the other carbo-hydrates into fat when needed.

The mineral nutrients are appropriated in the same combination as found in plants.

Table showing the Proportions of Mineral Constituents of some Plants and Grains in One Hundred Pounds of Dry Substance.

HAY.

ONE HUNDRED POUNDS OF SUBSTANCE.	Ash.	Potash.	Soda.	Magnesia.	Lime.	Phosphoric Acid.	Sulphuric Acid.	Silica.	Chlorine.	Sulphur.
	lbs.	lbs.	lbs.	lbs.	lbs.	lbs.	lbs.	lbs.	lbs.	lbs.
Meadow Hay...............	6.66	1.71	0.47	0.33	0.77	0.41	0.34	1.97	0.53	0.17
Dead-ripe Hay.	6.62	0.50	0.19	0.23	0.85	0.29	0.05	4.18	0.38	0.27
Red Clover....	5.65	1.95	0.09	0.69	1.92	0.56	0.17	0.15	0.21	0.21
Swedish Clover............	4.65	1.57	0.07	0.71	1.48	0.47	0.19	0.06	0.13	....
Green Vetches......	7.34	3.09	0.21	0.50	1.93	0.94	0.27	0.13	0.23	0.15
Green Oats................	6.18	2.41	0.20	0.20	0.41	0.51	0.17	2.05	0.25	0.15

GREEN FODDER.

ONE HUNDRED POUNDS OF SUBSTANCE.	Ash.	Potash.	Soda.	Magnesia.	Lime.	Phosphoric Acid.	Sulphuric Acid.	Silica.	Chlorine.	Sulphur.
	lbs.	lbs.	lbs.	lbs.	lbs.	lbs.	lbs.	lbs.	lbs.	lbs.
Meadow Grass in blossom.	2.33	0.60	0.16	0.11	0.27	0.15	0.12	0.69	0.19	0.06
Young Grass..............	2.07	1.16	0.04	0.06	0.22	0.22	0.08	0.21	0.04	0.04
Timothy..................	2.10	0.61	0.06	0.08	0.20	0.23	0.08	0.75	0.11	0.08
Oats beginning to blossom.....	1.70	0.71	0.08	0.06	0.12	0.14	0.06	0.47	0.08	0.03
Barley beginning to blossom...	2.23	0.86	0.04	0.07	0.16	0.23	0.07	0.70	0.12	0.05
Rye Fodder..............	1.63	0.63	0.01	0.05	0.12	0.24	0.02	0.52	....	....
Hungarian Millet....	2.31	0.86	....	0.19	0.25	0.13	0.08	0.67	0.15	....
Red Clover..............	1.34	0.46	0.02	0.16	0.46	0.13	0.04	0.04	0.05	0.05
White Clover....	1.36	0.24	0.11	0.14	0.44	0.20	0.12	0.06	0.04	0.06
Swedish Clover............	1.02	0.35	0.02	0.16	0.32	0.10	0.04	0.01	0.03	....
Lucern..................	1.76	0.45	0.02	0.10	0.85	0.15	0.11	0.04	0.03	0.08
Green Peas..............	1.37	0.56	....	0.11	0.39	0.18	0.05	0.04	0.02	....

ROOT CROPS.

ONE HUNDRED POUNDS OF SUBSTANCE.	Ash.	Potash.	Soda.	Magnesia.	Lime.	Phosphoric Acid.	Sulphuric Acid.	Silica.	Chlorine.	Sulphur.
	lbs.	lbs.	lbs.	lbs.	lbs.	lbs.	lbs.	lbs.	lbs.	lbs.
Potato......................	0.94	0.56	0.01	0.04	0.02	0.18	0.06	0.02	0.03	0.02
Beet.......................	0.80	0.43	0.12	0.04	0.04	0.08	0.03	0.02	0.05	0.01
Turnip.....................	0.75	0.30	0.08	0.03	0.08	0.10	0.11	0.02	0.03	0.04
White Turnip...............	0.61	0.31	0.02	0.01	0.08	0.11	0.04	0.01	0.04	....
Carrot.....................	0.88	0.32	0.19	0.05	0.09	0.11	0.06	0.02	0.03	0.01
Cabbage....................	1.24	0.60	0.05	0.04	0.19	0.20	0.11	0.01	0.03	0.05

STRAW.

One Hundred Pounds of Substance.	Ash.	Potash.	Soda.	Magnesia.	Lime.	Phosphoric Acid.	Sulphuric Acid.	Silica.	Chlorine.	Sulphur.
	lbs.	lbs.	lbs.	lbs.	lbs.	lbs.	lbs.	lbs.	lbs.	lbs.
Wheat.....................	4.26	0.49	0.12	0.11	0.26	0.23	0.12	2.82	....	0.16
Rye.......................	4.07	0.76	0.13	0.13	0.31	0.19	0.08	2.37	....	0.09
Barley....................	4.39	0.92	0.20	0.11	0.33	0.19	0.16	2.36	....	0.13
Oats......................	4.40	0.97	0.23	0.18	0.36	0.18	0.15	2.12	....	0.17
Maize Fodder.............	4.72	1.66	0.05	0.26	0.50	0.38	0.25	1.79	....	0.39
Pea Straw................	4.92	1.07	0.26	0.38	1.86	0.38	0.28	0.28	0.30	0.07

GRAIN, SEEDS, ETC.

One Hundred Pounds of Substance.	Ash.	Potash.	Soda.	Magnesia.	Lime.	Phosphoric Acid.	Sulphuric Acid.	Silica.	Chlorine.	Sulphur.
	lbs.	lbs.	lbs.	lbs.	lbs.	lbs.	lbs.	lbs.	lbs.	lbs.
Wheat.....................	1.77	0.55	0.06	0.22	0.06	0.82	0.04	0.03	....	0.15
Rye.......................	1.73	0.54	0.03	0.19	0.05	0.82	0.04	0.03	....	0.17
Barley....................	2.18	0.48	0.06	0.18	0.05	0.72	0.05	0.59	....	0.14
Oats......................	2.64	0.42	0.10	0.18	0.10	0.55	0.04	1.23	....	0.17
Maize.....................	1.23	0.33	0.02	0.18	0.03	0.55	0.01	0.03	....	0.12
Millet....................	1.23	0.23	0.07	0.23	....	0.66	0.02	....	....	....
Buckwheat.................	0.92	0.21	0.06	0.12	0.03	0.44	0.02	....	0.02	....
Flaxseed..................	3.22	0.04	0.06	0.42	0.27	1.30	0.04	0.04	....	0.17
Peas......................	2.42	0.98	0.09	0.19	0.12	0.88	0.08	0.02	0.06	0.24
Vetches...................	2.07	0.63	0.22	0.18	0.06	0.79	0.09	0.04	0.02	....
Beans.....................	2.96	1.20	0.04	0.20	0.15	1.16	0.15	0.04	0.08	0.23
Wheat Bran................	5.56	1.33	0.03	0.94	0.26	2.88	....	0.06	....	....
Rye Bran..................	7.14	1.93	0.09	1.13	0.25	3.42	....	....	....	....
Linseed Cake..............	5.52	1.29	0.08	0.88	0.47	1.94	0.19	0.36	0.03	....

Table showing average Composition, Nutritive and Money Value of different kinds of Fodder, compiled from Tables of Dr. Wolff for Germany, Dr. Collier for United States, various Analyses of Connecticut Experiment Station, and other American Analyses.

ARTICLES SUITABLE FOR DAIRY CATTLE-FEEDING.

| KINDS OF FODDER. | Water. | Ash. | ORGANIC SUBSTANCES. | | | | DIGESTIBLE NUTRIENTS. | | | Nutritive Ratio. | Value per 100 lbs. |
			Proteids.	Fibre.	Other Carbo-hydrates.	Fat.	Proteids.	Carbo-hydrates, including Fibre.	Fat.		
HAY.	%	%	%	%	%	%	%	%	%	As 1 :	$
Meadow Hay, medium.....	14.3	6.2	9.7	26.3	41.4	2.5	5.4	41.0	1.0	8.0	0.64
Meadow Hay, very good...	15.0	7.0	11.7	21.9	41.6	2.8	7.4	41.7	1.3	6.1	0.75
Meadow Hay, extra.......	16.0	7.7	13.5	19.3	40.4	3.0	9.2	42.8	1.5	5.1	0.85
Red Clover, medium......	16.0	5.3	12.3	26.0	38.2	2.2	7.0	38.1	1.2	5.9	0.70
Red Clover, very good.....	16.5	6.0	13.5	24.0	37.1	2.9	8.5	38.2	1.7	5.0	0.79
Red Clover, extra........	16.5	7.0	15.3	22.2	35.8	3.2	10.7	37.6	2.1	4.0	0.89
Red Clover, aftermath....	16.6	10.5	13.0	25.1	....	....	24.8	41.8	3.7	1.8	....
White Clover, medium....	16.5	6.0	14.5	25.6	33.9	3.5	8.1	35.9	2.0	5.0	0.76
Hay of pure Red Clover....	16.0	5.6	13.4	25.4	36.4	3.2	....	....	....	....	....
Lucern, medium..........	16.0	6.2	14.4	33.0	27.9	2.5	9.4	28.3	1.0	3.3	0.71
Lucern, very good........	16.5	6.8	16.0	26.6	31.6	2.5	12.3	31.4	1.0	2.8	0.86
Swedish Clover, Alsike....	16.0	6.0	15.0	27.0	32.7	3.3	8.6	34.8	1.8	4.6	0.76
Fodder Vetch, medium....	16.7	8.3	14.2	25.5	32.8	2.5	9.4	32.5	1.5	3.9	0.77
Fodder Vetch, very good...	16.7	9.3	19.8	23.4	28.5	2.3	15.1	31.1	1.4	2.3	0.99
Peas, in bloom...........	16.7	7.0	14.3	25.2	34.2	2.6	9.4	33.1	1.6	4.0	0.77
Lupine, very good........	16.7	4.1	23.2	25.2	28.6	2.2	17.2	36.0	0.7	2.2	1.10
Fodder Rye.............	14.3	5.1	0.4	23.1	44.5	2.8	6.6	44.3	1.3	7.2	0.72
Timothy................	14.3	4.5	9.7	22.7	45.8	3.0	5.8	43.4	1.4	8.1	0.70
Early Meadow Grass (*Poa annua*), in blossom......	14.3	2.4	10.1	25.9	47.2	2.9	6.0	42.5	2.1	7.9	0.74
Orchard Grass, in blossom.	14.3	4.6	11.6	28.9	40.7	2.7	6.9	40.3	1.9	6.5	0.74
June Grass (*Poa pratensis*), in blossom..............	14.3	5.1	8.9	32.6	39.1	2.3	5.9	40.0	1.6	7.5	0.68
Sheep Fescue............	2.5	3.6	8.8	25.1	57.1	3.6	8.8	57.1	3.6	6.9	0.85
Red-top, in blossom.......	6.4	6.8	10.3	20.6	53.1	2.6	10.3	53.1	2.6	5.4	0.82

AVERAGE COMPOSITION, ETC., OF FEEDING STUFFS (*continued*).

KINDS OF FODDER.	Organic Substances.						Digestible Nutrients.			Nutritive Ratio.	Value per 100 lbs.
	Water.	Ash.	Proteids.	Fibre.	Other Carbo-hydrates.	Fat.	Proteids.	Carbo-hydrates, including Fibre.	Fat.	As 1 :	$
HAY (*continued*).	%	%	%	%	%	%	%	%	%	As 1:	$
Meadow Foxtail, after blossom.................	8.5	7.4	7.8	23.1	49.6	3.2	7.8	49.6	3.2	6.7	0.62
Meadow Soft Grass, very young...................	9.45	9.0	11.2	16.8	49.3	4.1	11.2	49.3	4.1	4.8	0.85
Fowl Meadow Grass (*Poa serotina*)...............	14.3	4.4	8.8	21.7	49.0	2.9	7.5	49.0	2.9	6.9	0.69
Blue Grass (*Poa compressa*)	14.3	3.6	6.2	17.8	56.4	2.4	5.3	56.4	2.4	10.9	0.66
Blue Grass, early bloom...	5.2	6.2	12.7	19.1	52.7	4.0	10.2	52.7	4.0	4.5	0.83
Foxtail Pigeon Grass, early bloom....	5.0	6.9	8.6	24.4	52.4	2.5	8.5	52.4	2.5	6.4	0.70
Johnson Grass............	14.3	8.4	10.7	20.2	46.0	1.8	9.16	46.0	1.8	5.2	0.71
Bermuda Grass..........	14.3	6.9	....	21.4	44.7	....	10.1	44.7	2.4	4.7	....
Quack Grass.............	14.3	7.8	11.4	16.6	48.2	3.0	9.8	48.2	3.0	5.2	0.76
Gama Grass.............	14.3	5.3	8.6	22.7	48.2	2.0	7.4	48.2	2.0	6.8	0.65
Timothy...............	13.5	3.9	6.2	28.9	45.8	1.7	....	....	....	....	....
Timothy and Red-top......	14.3	5.5	7.6	26.5	44.1	2.0	....	....	....	....	....
Timothy and June Grass...	14.3	4.7	7.0	26.9	45.4	1.7	....	....	....	....	....
Mixed Grasses............	14.3	5.1	7.3	26.7	44.9	1.8	....	....	....	....	....
Containing Clover........	14.3	5.4	10.9	24.1	43.0	2.3	....	....	....	....	....
Low Meadow Hay........	10.0	5.8	7.4	30.8	43.8	2.2	....	....	....	....	....
Salt Marsh Hay....	10.7	7.6	6.1	31.9	41.3	2.4	....	....	....	....	....
Japan Clover............	14.3	3.8	12.9	20.3	44.8	3.7	12.9	44.8	3.7	3.8	....
Italian Rye Grass.........	14.3	7.8	11.2	22.9	40.6	3.2	7.1	41.5	1.4	6.3	0.74
English Rye Grass........	14.3	6.5	10.2	30.2	36.1	2.7	5.1	35.3	0.8	7.3	0.57
Upland Grasses..........	14.3	5.8	9.5	28.7	39.1	2.6	5.3	40.9	1.1	8.2	0.64
Mexican Clover..........	14.3	7.1	....	25.6	45.1	....	5.1	45.2	2.6	9.3	....
Hungarian Grass.	13.4	5.7	10.8	29.4	38.5	2.2	6.1	41.0	0.9	7.1	0.66
Desmodium..............	14.3	6.6	....	21.7	....	2.3	16.2	38.2	7.3	2.5	....
Brown Hay of Clover......	14.0	8.2	16.7	25.4	33.3	2.4	....	....	....	....	....

AVERAGE COMPOSITION, ETC., OF FEEDING STUFFS (*continued*).

KINDS OF FODDER.	Water.	Ash.	ORGANIC SUBSTANCES.				DIGESTIBLE NUTRIENTS.			Nutritive Ratio.	Value per 100 lbs.
			Proteids.	Fibre.	Other Carbo-hydrates.	Fat.	Proteids.	Carbo-hydrates, including Fibre.	Fat.		
HAY (*continued*).	%	%	%	%	%	%	%	%	%	As 1 :	$
Brown Hay of Grasses.....	14.3	6.3	8.6	22.4	45.5	2.9	....	....	....	....	....
Brown Hay of Maize......	79.3	1.5	1.0	7.0	10.1	1.1	....	....	....	....	....
GREEN FODDER.											
Grass, before bloom........	75.0	2.1	3.0	6.0	13.1	0.8	2.0	13.0	0.4	7.0	0.22
Pasture Grass............	80.0	2.0	3.5	4.0	9.7	0.8	2.5	9.9	0.4	4.4	0.21
Rich Pasture Grass........	78.2	2.2	4.5	4.0	10.1	1.0	3.4	10.9	0.6	3.6	0.27
Italian Rye Grass........	73.4	2.8	3.6	7.1	12.1	1.0	2.3	12.6	0.4	5.9	0.23
Timothy Grass............	70.0	2.2	3.4	8.0	16.3	1.1	2.1	16.0	0.5	8.2	0.28
Upland Grasses...........	70.0	2.1	3.4	10.1	13.4	1.0	1.9	14.2	0.5	8.1	0.23
Maize Fodder............	84.0	1.0	1.4	4.7	8.4	0.5	....	....	....	....	....
Green Maize, German.....	83.0	1.0	1.8	4.4	9.3	0.5	1.0	8.4	0.2	8.9	0.13
Fodder Rye..............	76.0	1.6	3.3	7.9	10.4	0.8	1.9	11.0	0.4	6.3	0.20
Fodder Oats.............	81.0	1.4	2.3	6.5	8.3	0.5	1.3	8.9	0.2	7.2	0.15
Hungarian Grass, in blossom..................	75.0	1.8	3.1	8.5	10.9	0.7	1.8	11.8	0.3	7.0	0.20
Pasture Clover, young.....	83.0	1.5	4.6	2.8	7.2	0.9	3.6	7.4	0.6	2.5	0.25
Red Clover, before bloom..	83.0	1.5	3.3	4.5	7.0	0.7	2.3	7.4	0.5	3.8	0.19
Red Clover, full bloom....	80.4	1.3	3.0	5.8	8.9	0.6	1.7	8.7	0.4	5.7	0.17
White Clover, in blossom..	80.5	2.0	3.5	6.0	7.2	0.8	2.2	7.9	0.5	4.2	0.19
Swedish Clover, beginning of bloom	85.0	1.5	3.3	4.5	5.1	0.6	2.1	5.8	0.4	3.2	0.17
Fodder Vetch, beginning of bloom	82.0	1.8	3.5	5.5	6.6	0.6	2.5	6.7	0.3	3.0	0.18
Fodder Peas, in bloom.....	81.5	1.5	3.2	5.6	7.6	0.6	2.2	7.4	0.3	3.7	0.18
Fodder Cabbage..........	84.7	1.6	2.5	2.4	8.1	0.7	1.8	8.2	0.4	5.2	0.17
White Cabbage...........	89.0	1.2	1.5	2.0	5.9	0.4	1.1	6.0	0.2	5.8	0.11
STRAW.											
Winter Wheat Straw......	14.3	4.6	3.0	40.0	36.9	1.2	0.8	35.6	0.4	45.8	0.37
Winter Barley Straw.	14.3	5.5	3.3	43.0	32.5	1.4	0.8	31.4	0.4	40.5	0.33

AVERAGE COMPOSITION, ETC., OF FEEDING STUFFS (*continued*).

| KINDS OF FODDER. | | ORGANIC SUBSTANCES. | | | | DIGESTIBLE NUTRIENTS. | | | | |
	Water.	Ash.	Proteids.	Fibre.	Other Carbo-hydrates.	Fat.	Proteids.	Carbo-hydrates, including Fibre.	Fat.	Nutritive Ratio.	Value per 100 lbs.
STRAW (*continued*).	%	%	%	%	%	%	%	%	%	As 1 :	$
Summer Barley Straw...	14.3	4.1	3.5	40.0	36.7	1.4	1.3	40.6	0.5	32.2	0.44
Barley Straw, with Clover...............	14.3	6.7	6.5	38.0	32.5	2.0	3.3	38.8	0.9	12.4	0.53
Oat Straw............	14.3	4.0	4.0	39.5	36.2	2.0	1.4	40.1	0.6	29.9	0.45
Summer Grain Straws, medium............	14.3	4.1	3.8	39.7	36.4	1.7	1.4	40.4	0.7	31.0	0.45
Summer Grain Straws, very good	14.3	6.7	6.9	36.7	32.9	2.5	2.5	36.9	0.8	15.5	0.47
Fodder Vetch	16.0	4.5	7.5	42.0	29.0	1.0	3.4	31.9	0.5	9.8	0.46
Pea...................	16.0	4.5	6.5	38.0	34.0	1.0	2.9	33.4	0.5	12.0	0.44
Seed Clover..........	16.0	5.6	9.4	42.0	25.0	2.0	4.2	28.5	1.0	7.4	0.49
Corn Stalks.....	15.0	4.2	3.0	40.0	36.7	1.0	1.1	37.0	0.3	34.4	0.39
Wheat...............	14.3	9.2	4.3	36.0	34.6	1.4	1.4	32.8	0.4	24.1	0.37
CHAFF AND HULLS.											
Rye...................	14.3	7.5	3.6	43.5	29.9	1.2	1.1	34.9	0.4	32.6	0.37
Oats..................	14.3	10.1	4.0	34.0	36.2	1.5	1.6	36.6	0.6	23.8	0.39
Barley	14.3	13.0	3.0	30.0	38.2	1.5	1.2	35.0	0.6	30.4	0.38
Vetch................	15.0	8.0	8.5	33.0	33.5	2.0	4.2	34.3	1.2	8.9	0.54
Pea.....	15.0	6.0	8.1	32.0	36.9	2.0	4.0	36.2	1.2	9.8	0.55
Flax.........	11.2	7.2	2.7	45.2	32.6	1.1	0.7	36.8	0.4	53.8	0.38
White Clover	11.5	7.9	18.3	22.4	36.8	3.1	10.7	34.8	1.5	3.6	0.84
ROOTS AND TUBERS.											
Potatoes..............	75.0	0.9	2.1	1.1	20.7	0.2	2.1	21.8	0.2	10.6	0.29
Mangolds............	88.0	0.8	1.1	0.9	9.1	0.1	1.1	10.0	0.1	9.3	0.14
American Mangolds....	92.1	1.04	1.77	0.78	4.23	0.45	....	....	....	....	....
Sugar Beets	81.5	0.7	1.0	1.3	15.4	0.1	1.0	16.7	0.1	17.0	0.19
Rutabagas....	87.0	1.0	1.3	1.1	9.5	0.1	1.3	10.6	0.1	8.3	0.15
Carrots	85.0	0.9	1.4	1.7	10.8	0.2	1.4	12.5	0.2	9.3	0.18
Turnips..............	92.0	0.7	1.1	0.8	5.3	0.1	1.1	6.1	0.1	5.8	0.16

AVERAGE COMPOSITION, ETC., OF FEEDING STUFFS (*continued*).

| KINDS OF FODDER. | Water. | Ash. | Organic Substances. | | | | Digestible Nutrients. | | | Nutritive Ratio. | Value per 100 lbs. |
			Proteids.	Fibre.	Other Carbo-hydrates.	Fat.	Proteids.	Carbo-hydrates, including Fibre.	Fat.		
	%	%	%	%	%	%	%	%	%	As 1 :	$
ROOTS AND TUBERS (*continued*).											
Parsnips..............	88.3	0.7	1.6	1.0	10.2	0.2	1.6	11.2	0.2	7.3	0.18
Sweet Potato.........	69.7	1.1	1.9	1.7	26.3	0.3	0.9	28.0	0.3	31.9	0.30
American Yam........	71.2	0.6	2.1	0.7	25.2	0.2	2.1	25.9	0.2	12.5	0.33
GRAINS AND FRUITS.											
Wheat...............	14.4	1.7	13.0	3.0	66.4	1.5	11.7	64.3	1.2	5.8	1.13
Rye	14.3	1.8	11.0	3.5	67.4	2.0	9.9	65.4	1.6	7.0	1.08
American Winter Rye..	8.7	1.8	12.1	1.4	73.9	2.1	10.8	70.3	1.6	6.8	1.16
Barley..............	14.3	2.2	10.0	7.1	63.9	2.5	8.0	58.9	1.7	7.9	0.95
Oats................	14.3	2.7	12.0	9.3	55.7	6.0	9.0	43.3	4.7	6.1	0.98
Maize....	14.4	1.5	10.0	5.5	62.1	6.5	8.4	60.6	4.8	8.6	1.11
Mammoth Sweet Corn..	6.47	1.92	12.78	1.88	67.95	9.0	....	......	....	...	....
Sweet Corn, average ...	8.59	1.88	12.08	2.04	67.37	8.04	....	......	....	...	....
Stowell's Evergreen Sweet Corn.........	5.98	1.92	11.91	2.66	69.53	8.00	....	......	....	...	....
Millet..............	14.0	3.0	12.7	9.5	57.5	3.3	9.5	45.0	2.6	5.4	0.93
Golden Millet........	13.4	2.8	9.6	11.6	58.6	4.0	7.2	47.0	3.1	7.5	0.87
Rice, hulled..........	14.0	0.5	7.7	2.2	75.2	0.4	6.9	72.7	0.3	10.7	0.96
Peas	14.3	2.4	22.4	6.4	52.5	2.0	20.2	54.4	1.7	2.9	1.44
Vetch..............	14.3	2.7	27.5	6.7	45.8	3.0	24.8	48.2	2.5	2.2	1.63
Cow Peas, American...	20.0	3.1	21.6	4.7	49.3	1.3	19.4	49.6	1.1	2.7	1.33
Flax Seed...........	12.3	3.4	20.5	7.2	19.6	37.0	17.2	18.9	35.2	....	2.47
Apples and Pears	83.1	0.4	0.4	4.3	11.8	....	0.3	12.9	....	43.0	0.13
Roxbury Russet.......	82.2	0.26	0.27	0.95	15.77	0.53	....	......	..	...	....
Pumpkins....	89.1	1.0	0.6	2.7	6.5	0.1	0.4	7.1	0.1	18.4	0.08
Squash, American	88.1	0.7	0.9	1.0	9.1	0.2	0.6	9.0	0.2	15.8	0.11
BY-PRODUCTS.											
Coarse Wheat Bran....	11.4	5.1	12.9	8.1	59.1	3.5	10.0	48.5	3.1	5.6	1.01
Wheat Middlings......	11.8	2.3	11.4	4.8	66.8	2.9	8.9	54.8	2.6	6.9	1.00

AVERAGE COMPOSITION, ETC., OF FEEDING STUFFS (*continued*).

KINDS OF FODDER.	Water.	Ash.	Proteids.	Fibre.	Other Carbo-hydrates.	Fat.	Proteids.	Carbo-hydrates, including Fibre.	Fat.	Nutritive Ratio.	Value per 100 lbs.
			ORGANIC SUBSTANCES.				DIGESTIBLE NUTRIENTS.				
BY-PRODUCTS (*continued*).	%	%	%	%	%	%	%	%	%	As 1:	$
Rye Bran............	12.9	2.9	12.6	2.5	67.0	2.2	10.6	50.0	2.0	5.3	1.00
Buckwheat Bran......	14.0	3.4	17.1	14.7	46.4	4.4	13.5	44.0	3.9	4.1	1.15
Pea Meal Bran........	12.3	4.2	13.1	31.1	37.8	1.5	9.2	45.8	1.2	5.3	0.86
Pea Meal	11.4	3.5	23.7	4.5	54.5	3.5	20.9	55.4	2.8	3.0	1.53
Millet Bran.........	9.5	7.5	6.5	57.6	14.4	4.5	4.5	38.8	2.7	10.1	0.66
Barley Bran	12.0	4.1	14.8	19.4	45.6	4.1	11.5	43.2	3.6	4.5	1.04
Wheat Meal	11.5	3.0	13.9	4.8	63.5	3.3	10.8	54.8	2.9	5.7	1.08
Rice Meal...........	9.9	10.6	10.9	1.1	47.6	9.9	8.6	47.2	8.8	8.0	1.16
Barley Middlings......	12.3	6.2	11.6	14.3	52.9	3.6	9.6	47.0	3.2	6.0	0.93
Oat Bran	9.7	3.7	7.1	19.3	57.9	2.3	5.6	49.8	2.0	9.7	0.77
DAIRY PRODUCTS.											
Cow's Milk..........	87.5	0.7	3.2	....	5.0	3.6	3.2	5.0	3.6	4.4	0.34
Jersey Milk..........	85.2	0.9	3.6	....	4.9	5.2	3.6	4.9	5.2	....	
Skimmed Milk........	90.0	0.8	3.5	....	5.0	0.7	3.5	5.0	0.7	1.9	0.23
Skimmed Milk, by separator.............	91.7	0.7	3.1	....	4.1	0.3	3.1	4.1	0.3	....	....
Buttermilk..........	90.1	0.5	3.0	....	5.4	1.0	3.0	5.4	1.0	2.6	0.22
Whey..............	92.6	0.7	1.0	....	5.1	0.6	1.0	5.1	0.6	6.6	0.11
Cream	62.0	0.6	2.7	....	2.9	31.8	2.7	2.9	31.8	30.5	1.54
Jersey Cream........	36.4	0.2	3.8	....	2.8	56.8	3.8	2.8	56.8	....	....

AVERAGE COMPOSITION, ETC., OF FEEDING STUFFS (*continued*).

Articles of Questionable Utility in Feeding Jersey Cattle.

| KINDS OF FODDER. | Water. | Ash. | ORGANIC SUBSTANCES. | | | | DIGESTIBLE NUTRIENTS. | | | Nutritive Ratio. | Value per 100 lbs. |
			Proteids.	Fibre.	Other Carbo-hydrates.	Fat.	Proteids.	Carbo-hydrates, including Fibre.	Fat.		
	%	%	%	%	%	%	%	%	%	As 1 :	$
Meadow Hay, poor.....	14.3	5.0	7.5	33.5	38.2	1.5	3.4	34.9	0.5	10.6	0.48
Fermented Maize Hay..	83.5	1.1	1.2	5.3	8.9	0.9	0.8	8.6	0.4	12.0	0.13
Cotton Seed	7.7	7.8	22.8	16.0	15.4	30.3	17.1	18.7	27.3	4.6	2.08
Cotton-seed Meal, de-corticated..........	7.2	5.8	41.5	3.1	24.4	18.0	33.2	17.6	16.2	1.8	2.30
Cotton-seed Meal, un-decorticated........	11.3	6.4	23.6	22.0	30.5	6.1	17.5	14.9	5.5	1.7	1.14
Palm Seed...........	7.6	1.8	8.4	6.0	26.8	49.2	8.0	31.2	48.2	18.3	2.75
Palm-nut Cake.......	10.5	4.2	16.9	17.4	41.0	10.0	16.1	55.4	9.5	4.9	1.61
Palm-nut Cake, Ameri-can...............	7.9	4.0	13.5	18.8	41.0	14.8	12.8	56.2	14.0	7.0	1.66
Palm-nut Cake, extract-ed............	10.5	4.0	18.5	20.2	43.5	3.3	17.6	60.4	3.1	3.9	1.44
Linseed Cake	9.1	8.2	32.4	7.3	31.5	11.6	27.6	27.0	10.4	2.0	1.89
Linseed Meal, extract-ed............	9.7	7.3	33.2	8.8	38.7	2.3	27.8	33.9	2.1	1.4	1.61
Sunflower Seed........	8.0	3.0	13.0	28.5	23.9	23.6	10.4	24.6	21.2	7.2	1.59
Sunflower Cake	10.3	8.1	37.3	9.9	26.0	8.4	31.3	24.7	7.6	1.3	1.93
Distillery Slump.......	90.6	0.4	1.8	1.0	5.2	1.0	1.6	5.4	0.8	4.6	0.15
Brewers' Grains.......	75.2	0.3	5.9	3.9	13.2	1.5	4.8	11.3	1.2	3.0	0.36
Malt Sprouts..........	11.6	6.7	25.9	9.3	45.5	1.1	20.8	43.7	0.9	2.2	1.33
Rye Refuse, starch fac-tory	70.0	0.8	6.1	2.7	18.9	1.5	5.2	18.1	1.2	4.1	0.44
Wheat Refuse, starch factory	74.0	0.6	4.4	3.4	15.4	2.2	3.7	15.1	1.8	5.3	0.37
Potato Refuse, starch factory	86.0	0.4	0.8	2.0	11.7	0.1	0.8	13.7	0.1	17.4	0.16

AVERAGE COMPOSITION, ETC., OF FEEDING STUFFS (*continued*).

Articles of Questionable Utility in Feeding Jersey Cattle (continued).

KINDS OF FODDER.	Water.	Ash.	ORGANIC SUBSTANCES.				DIGESTIBLE NUTRIENTS.			Nutritive Ratio.	Value per 100 lbs
			Proteids.	Fibre.	Other Carbo-hydrates.	Fat.	Proteids.	Carbo-hydrates, including Fibre.	Fat.		
	%	%	%	%	%	%	%	%	%	As 1 :	%
Rape Cake............	11.3	7.1	31.6	11.0	29.9	9.6	25.3	23.8	7.7	1.7	1.66
Rape Meal, extracted...	8.5	7.9	33.1	13.4	34.1	3.0	26.5	27.2	2.4	1.3	1.51
Apple Pomace........	77.2	0.5	0.9	3.9	15.7	1.7	....	....			
Ensilage of Maize......	82.0	....	....		....	....	1.0	10.1	0.5	11.4	0.16
Ensilage of Rye.......	76.2						1.9	12.0	0.4	6.8	0.21
Ensilage of Red Clover.	79.2	....	....				3.0	8.1	1.7	4.0	0.28
Ensilage of Sorghum...	77.3	....	....				1.6	11.9	0.3	7.4	0.19
Rye Straw, ripe.......	14.3	4.1	3.0	44.0	33.3	1.3	0.8	36.5	0.4	46.9	0.35
Timothy Ensilage......	70.0	....	....				2.1	16.0	0.5	8.2	0.28
Cow Peas Ensilage.....	76.0	....	....				3.0	9.4	0.2	3.2	0.24
Orchard Grass Ensilage...............	74.0	....	....				2.6	12.4	0.4	5.1	0.24
Maize Ensilage, poor...	74.2	0.8	0.9	4.7	7.0	0.3	....	....			
Carrot Leaves Ensilage...............	82.2	....	....				2.2	7.0	0.5	3.8	0.18
Extra Maize Ensilage...	84.9	1.8	1.9	7.9	13.0	0.9	....	....			
Best Maize Ensilage*...	77.6	1.7	2.0	6.0	12.0	0.54	....	....			

* This was a sample analyzed at the Connecticut Station, made from corn well advanced in ear. It contained acetic acid $\frac{103}{1000}$, "equivalent to one quart of strong vinegar per hundred pounds," and $\frac{335}{1000}$ alcohol, equivalent to one pint of rum per hundred pounds of ensilage.

VALUE OF THE TABLES.

These valuations are only approximate. The standard is average meadow hay figured at sixty-four cents for one hundred pounds. The German estimates by Dr. Wolff on the basis of four and one third cents for one pound of digestible proteids, four and one third cents for one pound of digestible fat, and nine tenths of a cent a pound for the digestible carbo-hydrates.

The feeder will learn to make practical use of the tables by frequent reference to them, and comparing the estimates with the results of his own practice. By the exercise of his own skill he can combine such elements as are best to form rations for his own stock.

REQUISITES TO SUSTAIN LIFE AND HEALTH.

The animal heat must be sustained by the carbo-hydrates and the carbon of the *proteids*. This is believed to be accomplished by oxidation in the cells and capillary vessels of the body, and consumes the starch, gum, sugar and cellulose, and furnishes the carbon given off in breathing.

The fat which is stored up in the body, as well as the great quantity secreted in milk, must be supplied by vegetable oils. The natural wear and waste of muscle and cartilage and the growth of these in young animals must be sufficiently provided for in proteids, as albumen, gluten, casein, legumen, fibrin, mucedin and gliadin. The bones and teeth must be built up and nourished by earthy phosphates, and the processes of digestion, assimilation and excretion aided by saline substances, chlorides, sulphates, and other elements that appear in the various excretions after fulfilling their purpose. Food that furnishes all these essential elements in right proportions and sufficient quantity in the most palatable and digestible forms must be provided by the care and skill which meets all the purposes of the dairyman and the breeder of butter cattle. These objects cannot be accomplished without a great variety of grasses, grains, leguminous plants and roots.

AMERICAN FEEDING STUFFS.

AVERAGE SELECTIONS FROM TABLE OF AMERICAN ANALYSES.

COMPILED BY E. H. JENKINS, PH.D., CONN.

Agricultural Experiment Station Report, 1884.

NAME.	Number of Analyses.	Total Dry Matter.	Protein.	Fat.	Nitrogen, Free Extract.	Fibre.	Ash.
GREEN FODDERS.							
Maize	22	18.86	1.30	.32	10.65	5.37	1.22
Maize Ensilage	31	19.29	1.47	.72	9.88	5.88	1.34
Cow Pea Vine and Pods	2	15.94	3.12	.60	6.91	3.48	1.83
Rye	5	25.30	2.60	.65	5.90	14.30	1.90
HAY.							
CLOVER HAY	12	84.98	11.38	1.98	40.11	26.35	5.15
Timothy	18	87.42	6.36	2.03	44.89	29.93	4.23
Sorghum Leaves	3	27.00	3.10	.....	*15.10	5.20	3.50
Hungarian Grass	8	83.30	6.59	1.81	42.49	27.16	5.24
Oat Straw	3	89.89	3.35	2.07	36.97	42.78	4.72
MAIZE, FIELD-CURED	6	67.95	4.29	1.24	35.96	22.14	4.32
COW PEA VINES	6	88.95	15.68	2.87	42.17	19.82	8.41
ROOTS.							
Beets (red)	2	11.43	1.60	0.18	7.40	1.16	1.08
CARROTS	2	12.68	1.38	0.67	7.28	1.93	1.34
Mangolds	3	7.96	1.70	0.20	4.19	0.82	1.05
Potatoes	2	21.35	1.23	0.13	18.72	0.38	0.89
Sweet Potatoes	3	29.72	0.97	0.31	26.13	1.36	0.93
Turnips	1	11.11	1.34	0.09	8.11	0.86	0.71
Rutabagas	1	12.92	1.15	0.09	9.11	1.16	1.41
FRUITS.							
Apples	1	15.89	0.21	0.28	14.26	0.91	0.23
Squash	2	5.12	0.66	0.28	3.24	0.54	0.40

* Includes fat.

AMERICAN FEEDING STUFFS (*continued*).

Average of American Analyses (continued).

NAME.	Number of Analyses.	Total Dry Matter.	Protein.	Fat.	Nitrogen, Free Extract.	Fibre.	Ash.
GRAINS.							
Peas	1	21.94	4.37	0.55	14.48	1.66	0.88
Barley	9	88.90	12.40	1.80	69.30	2.90	2.50
Cow Pea	5	85.21	20.77	1.43	55.75	4.06	3.20
Soja Bean	3	91.41	36.22	17.92	28.66	4.24	4.37
Maize Kernel (Dent)......	77	89.93	10.36	5.15	70.60	2.29	1.53
Maize Kernel (Flint)	63	88.93	10.67	5.00	70.08	1.71	1.47
Maize Kernel (Sweet).....	24	91.42	11.71	8.31	66.54	2.82	1.93
Maize Kernel (Western) ..	3	80.90	8.30	3.70	66.00	1.75	1.20
Oats	21	89.30	11.30	5.00	61.00	9.00	3.00
Rice	10	87.60	7.40	0.40	79.20	0.20	0.40
Rye..................	6	88.40	10.60	1.70	72.60	1.60	1.90
Wheat (Winter).........	229	89.63	11.82	2.14	72.04	1.77	1.86
Wheat (Spring).........	13	89.63	12.51	2.20	71.19	1.82	1.91
Sorghum Seed..........	9	87.48	8.88	3.65	71.27	1.88	1.80
MEAL AND BRAN.							
BARLEY MEAL...........	3	84.90	11.80	1.70	70.90	0.10	0.50
Hominy...............	2	86.51	8.25	.44	77.12	0.32	0.38
MAIZE MEAL............	20	85.14	9.26	3.82	68.58	2.29	1.54
OAT MEAL.............	6	92.15	14.66	7.06	67.57	0.86	2.00
Cotton-seed Meal........	22	92.17	42.45	13.36	23.49	5.67	7.20
Linseed Meal, old process .	4	91.28	31.23	8.72	37.75	7.34	6.24
Linseed Meal, new process.	6	89.49	33.45	2.83	38.78	8.37	6.06
RYE BRAN..............	3	87.70	15.26	2.19	63.12	3.51	3.62
Wheat Middlings........	9	88.03	12.27	3.23	65.48	4.58	2.47
WHEAT BRAN...........	2	87.98	14.54	3.66	55.16	8.79	5.83
Wheat Shorts	6	88.15	13.14	3.79	51.22	6.12	4.85

RELATIVE VALUE OF FEEDING STUFFS.

	lbs.
100 lbs. Good Hay =	lbs.
Beets	670
Turnips	470
Clover, Green	375
Carrots	371
Mangolds	368
Lucern Hay	89
Clover Hay	88
Corn	62
Oats	59
Barley	58
Rye	53
Wheat	44
Linseed Oil-cake	43

SUMMARY OF FOOD ELEMENTS.*

" 1. The earthy substances contained in food, consisting chiefly of lime and magnesia, present the animal with the materials of which the long skeleton of its body principally consists.

" They may be called, therefore, bone materials.

" 2. The saline substances—chloride of sodium (common salt) and potassium, sulphate and phosphate of potash and soda and some other mineral matters occurring in food—supply the blood, juice of flesh, and various animal juices, with the necessary mineral constituents.

" 3. Albumen, gluten, legumen and nitrogen, containing principles of food, furnish the animal with the materials required for the formation of blood and flesh. They are therefore called flesh-forming substances.

" 4. Fat and oily matters of the food are employed to lay on fat, or to support respiration and animal heat.

" 5. Starch, sugar, gum, and a few other non-nitrogenized substances, consisting of carbon, hydrogen and oxygen, are used to support respiration (hence they are called elements of respiration), as they produce fat when given in excess.

" 6. Starch, sugar, and other elements of respiration alone cannot sustain the animal body.

" 7. Albumen, gluten, or any other albuminous matter alone does not support the life of herbivorous animals.

" 8. Animals fed upon food deficient in earthy phosphates or bone-producing principles grow sickly and remain weak in the bone.

---

* Chemistry of Food, Dr. August Voelcker.

"9. The healthy state of an animal can only be preserved by a mixed food, which contains flesh-forming constituents as well as heat-giving principles, and earthy and saline mineral substances in proportion, determined by experience, and adapted to the different kinds of animals, or to the purposes for which they are kept."

DIGESTION IN CATTLE.

The digestive organs of cattle are very complex. Digestion begins in the mouth and is completed in the large intestine. The mouth is the mill for grinding and salivating food. It contains the tongue, teeth, salivary glands, and also the organs of taste, which latter, aided by the sense of smell, inform the animal in selecting what is good and rejecting what is unsuited to its use. There are five sets of salivary glands—the *parotid* glands, the largest, one on each cheek in front of the ear; the *submaxillary*, under the lower jaw; the *sublingual*, under the tongue; the *molar* glands, parallel to the molar arches; the *lip and palate* glands. These all discharge a thin fluid into the mouth. This fluid, or saliva, contains an element called *ptyalin*, which changes starch to sugar. A large cow is believed to discharge more than a gallon of saliva while chewing the cud one hour. If the cow is in good health and a large milker she must feed largely and secrete an enormous quantity of saliva, and that food is best which is given in a form requiring remastication. Ground feed should be mixed with cut hay to make the whole mass bulky, so as to be raised for cud-chewing.

Bovine animals have a compound stomach with four compartments: the first, or *rumen*, holds about eight bushels in a full-sized ox, and makes up about nine tenths of the bulk of the quadruple stomach. It fills the left side of the belly from the short ribs to the hips, and is lobulate in form, having three compartments.

The second stomach, *reticulum*, is a prolongation forward of the left sac of the rumen, the communicating connection allowing the soft contents to pass freely from the rumen. Its lining membrane has cells like a honeycomb.

These organs are connected with the gullet, and also the third stomach, by a curious structure called the *demi-canal*. This structure forms a common way for the first three stomachs, and has also the power of contracting its walls so as to communicate only with the third stomach.

The third stomach, *omasum*, or manifold, is larger than the second, lies over the reticulum to the right, and above the right fore-sac of the rumen, beneath the short ribs on the right side.

On its convex side a dozen or more leaf-like folds extend nearly across the organ. These are interspersed with shorter folds in alternation, the smallest becoming mere ridges. These present a large amount of apposite surface, and the partitions being endowed with involuntary muscles, for the moving of adjacent surfaces against each other, the third stomach is thus made a triturating apparatus.

The fourth stomach, *rennet*, is of an elongated oval form, tapering backward in the right flank at the lower border of the rumen, to its termination, where it joins the small intestine. Although second in size, it is very small compared with the rumen. This organ corresponds to the one stomach in other animals. This is the organ for secreting and mixing the gastric juice with the softened aliment.

The coarser foods pass as soon as taken into the rumen or to the (reticulum) reticule; finer and softer foods may pass at once to the third and the fourth stomachs.

Liquids with finely divided food may be distributed through the four stomachs, liquids being propelled through the demi-canal into the manifold and rennet by a series of contractions of the reticulum while the animal is drinking. Thus some foods may reach the manifold and rennet and not be returned for rumination. The rumen often holds two hundred pounds of food when an ox is slaughtered. This is one fourth food mixed with three times its weight of saliva and some water. The reticule usually contains liquid. The strong involuntary muscles of these organs give a continuous churning movement to all their contents, rendering all soluble elements into a condition for mixing with the gastric juice of the rennet.

The great bulk of the food in the rumen and a small portion from the reticule are floated back in small quantities to the mouth for mastication. This is done by a muscular compression of the rumen, a contraction of the demi-canal and gullet from below upward, thus forcing a mouthful of liquid mixed with fibrous matter to the tongue and palate, which seize the solid portion in a mass, separating and swallowing the liquid. The solid is then leisurely chewed and remixed with saliva, when it is passed on to the several receptacles, according to its degree of preparation, some of it probably being masticated and remasticated many times before it is fit for the rennet.

The good cow is content to spend a large portion of her time chewing the cud. She has a very capacious rumen, needs from twelve to fifteen gallons of water daily, and perfect quietude. Any worriment or disturbance of the general health interferes with the process of rumination. The manifold presses out the fluid portions of the food, and triturates the residue, still further pulverizing it. In the rennet it is mixed with the gastric juice, which transforms the protein elements into milky peptones, ready for absorption by the lymphatic vessels of the rennet, and the carbo-hydrates already converted into sugar by the saliva are also absorbed by the gastric blood-vessels, while a large portion of the food needs something more than the compound acid and pepsin of the stomach, and must be passed onward to the intestinal canal to complete the process of digestion. At the proper time the pylorus, a sphincter or circular gate, involuntarily opens, allowing the contents of the rennet to flow into the next vessel, a long, thin tube, convoluted and doubled upon itself in many folds and festoons. The first part is the *duodenum*. This organ has its involuntary muscular structure for propulsion, and a mucus lining filled with little follicles which secrete a digestive fluid. Here also through two orifices is received the bile from the liver, and the pancreatic juice from the pancreas, to digest and emulsify the fats and prepare them for the subsequent production of butter. Here also any undissolved starch is, by the combined action of the bile and the *diastase* of the pancreas, rendered into sugar, and any protein elements needing further preparation are served with the proper solvents, and the nutritive fluids are absorbed and pass into the general circulation. The small intestine is about one hundred and twenty feet in length, and leads into the large intestine, which has a length of about thirty feet. Here what is left of assimilative elements in the digested food is absorbed, and the excretory refuse is thrown out upon the surface of the lower or small *colon*, and becomes fecal matter, which is formed into round masses by the propulsive contractions of the tube, and progressively expelled from the rectum.

The following interesting table is taken from Roberts, and gives a general view of the process of digestion :

TABLE OF THE DIGESTIVE JUICES AND THEIR FERMENTS.

Digestive Juices.	Ferments Contained in Them.	Action on Foods.
Saliva	Salivary Diastase or Ptyalin	Changes starch into sugar and dextrin.
Gastric Juice	a. Pepsin	Changes proteids into peptones in acid medium.
	b. Curdling Ferment	Curdles casein of milk.
Pancreatic Juice	a. Trypsin	Changes proteids into peptones in alkaline and neutral media.
	b. Curdling Ferment	Curdles casein of milk.
	c. Pancreatic Diastase	Changes starch into sugar and dextrin.
	d. Emulsive Ferment	Emulsifies and partially saponifies fats.
Bile		Assists in emulsifying fats.
Intestinal Juice	a. Invertin	Changes cane-sugar into invert-sugar.
	b. Curdling Ferment	Curdles casein of milk.

Starch is attacked along the whole line of the alimentary track. Albuminous elements in the rennet and small intestines. The ferments which curdle milk are found in the rennet and pancreas, and possibly in the small intestine. The bile is alkalescent in its reaction, and helps absorption of fatty matters by its emulsifying properties. Healthy bovine digestion depends largely upon appropriate food to suit the complicated series of ferment-actions here illustrated.

The kidneys, the skin and the lungs all assist in excreting the waste materials that have served their purpose. The nitrogen of the food, with the exception of what is appropriated in building up the body or the formation of milk, is believed to be all recovered in the dung and urine. When animals are fed upon rich food the urine will sometimes yield forty per cent. of phosphoric acid, but upon coarse fodder little or none will be found in the urine, while ninety-five per cent. of the soda and potash of the food are excreted in the urine, and also about thirty per cent. of the magnesia, most of the sulphuric acid and chlorine; the silica, with the rest of the ash constituents not utilized in the production of milk or structure of tissues, is excreted in the dung. These facts show the importance of an intelligent selection of all fodders, and also the value and importance of saving all manure, especially the liquid portions.

SOILING CATTLE.

" Turning pasture into tillage makes the man."—English Proverb.

The cutting of green forage plants and feeding to cattle in the stable is commonly termed "soiling," a practice which must become general on all land that is suitable.

THE SEVEN POINTS OF SOILING.

1. It saves land.
2. It saves food.
3. It saves fences.
4. It saves manure.
5. It saves health and condition.
6. It saves the losses of ordinary unproductiveness.
7. It saves the profits of well-employed labor.

THE SAVING OF LAND.

The contrast between feeding luxuriant cultivated crops and ordinary pasturing is as wide as the distinction between civilization and barbarism. Indeed, the one is the result of civilization and progress, while the other is essentially barbaric in its methods.

Soiling utilizes the land for all it can produce of the best crops. Pasturing usually takes what chances to grow, whether good or bad; and, as the animal occupies its own dish and tramples its own food ; scatters its dung and urine upon it; lies upon it and breathes upon it, to the disgust of other animals—it naturally results that a wide space of ground is rendered unprofitable. Besides, the soil that is thus rendered useless ought to be producing crops which in a cold climate must be relied upon for winter support from six to eight months of the year.

It is estimated that under skilful management of pasturing it requires three acres to furnish an equivalent for one acre of forage crops. Under poor pasturing there is a much wider variation, so that fifty acres under the forage system are equal to one hundred and twenty-five acres under the barbaric system of pasturage.

THE SAVING OF FOOD.

The saving of food is the result of having the food under complete control as to growth, selection of kind and quantity, which is or ought to be such that it is wholly eaten, and all the waste caused by the animal from trampling, fouling, lying upon, breathing upon and overgrazing are precluded. Cattle will also eat many weeds, such as daisies and thistles, when cut, in a tender and succulent condition or mixed with other food. The forage system saves everything that is aromatic and edible, and this leads to the extermination of weeds by cutting them before bloom.

The cattle cannot be fed with profit from food used in pasturing, because they are obliged to expend its value in many hours of unprofitable foraging for themselves in the vain effort to utilize dry grass and the branches of trees, so that, whereas the cow should speedily fill the rumen and chew her cud, she must spend sixteen hours to get the amount of a square meal, and then fail in filling her udder.

THE SAVING OF FENCES.

Many a man spends more for fences to keep cattle within the bounds of a poor pasture than the land itself is worth. But on good land the expense is a profitless outlay, as it costs a tax of one dollar per acre as the annual expense of maintaining pasture fences. Where the land admits of it, it were far better to employ one hundred dollars' annual outlay for the fences of a hundred-acre farm in maintaining three times the number of cattle and steadily improving the richness of the farm.

THE SAVING OF MANURE.

The estimates of the value of manure by various authorities warrant the placing it at twenty dollars per cow or upward, provided all the excrement is saved.

The soiling system is the only means of saving this amount, and under this system the manure that is daily dropped in the exercise lot should be gathered in a cart and dumped upon the heap in the manure house. It is wasted if allowed to remain where it is, and would be more than half wasted in the pasture by evaporation or washing into brooks and sloughs.

Those who have had many years' experience in soiling declare that the saving of the manure is more than a full compensation for all the labor in soiling.

THE SAVING OF HEALTH AND CONDITION.

The exercise of the animal, like its food, is all under the control of the owner, and may be so managed as to render the animal healthful and profitable, while the pasturing system may compel the animal to take too much exercise, and that often of a very unprofitable kind.

The health of the animal is benefited by a proper selection of food of abundant quantity. The pasture gives an unequal and often scanty supply of food of an indifferent or bad quality, at a waste of energy in searching for it.

The animal is protected, in the cool stable, from the scorching heat, from the tormenting flies and mosquitoes, and from chilling storms and wet. Surely this protection, combined with good feeding and careful exercise in a convenient yard, cannot but be conducive to the fullest health and comfort.

SAVING BY INCREASED PRODUCTIVENESS.

The dairy cow needs but a very moderate amount of exercise, and both exercise and food must be so controlled as to develop the highest capacity for the production of milk, butter and cheese.

The forage system is adapted to produce such results because it enables the animal to be fed a full ration at all times, so that all her powers are devoted to the one objective point of transforming the greatest possible amount of wholesome fodder into the greatest possible amount of wholesome human food, and this is accomplished by soiling; for actual tests have demonstrated that the productiveness of herds has been more than doubled year by year for long periods. An experiment by Dr. Rhode, of the Eldena Royal Academy of Agriculture of Prussia, conducted through seven years of pasturing and seven years of soiling, gave an average for each cow at pasture for seven years of one thousand five hundred and eighty-three quarts annually, while for soiling the average for each cow was three thousand four hundred and forty-two quarts, giving much more than double productiveness.

And now the saving of land, of food, of fences, of manure, of health, of productiveness, all conduce to the

SAVING OF PROFIT UPON LABOR EMPLOYED.

There is a profit from labor employed because it is expended judiciously in bringing immediate returns and also permanently enriching the farm. Professor Stewart in his excellent work, "Feeding Animals," shows how one man can with proper tools and appliances perform all the hand labor for soiling one hundred head of cattle. The annual expense for fences for pasturing one hundred head of cattle would be not less than three hundred dollars, at a loss of one half their productiveness. The feeder needs a team, mowing-machine, horse-rake, wagon and hay-loader, and a good growth of soiling crops.

SOILING CROPS.

Large crops of the following list of plants are to be recommended for cultivation, to be fed in the order produced, but taking care to combine several of them, when practicable, in one ration.

1. Winter Rye (*Secale cereale*).
2. Winter Barley (*Hordeum*).
3. Red Clover (*Trifolium pratense*).
4. Orchard Grass (*Dactylis glomerata*).
5. Italian Rye Grass (*Lolium Italicum*).
6. Timothy (*Phleum pratense*).
7. Timothy and Large Clover.
8. Timothy and Alsike (*Trifolium hybridum*).
9. Green Oats (*Avena sativa*).
10. Winter Wheat (*Triticum vulgare*).
11. Cow Peas and Oats.
12. Common Millet (*Panicum milliacium*).
13. Hungarian Grass (*Setaria Germanica*).
14. Italian Millet (*Setaria Italica*).
15. Vetch (*Vicia sativa*).
16. Spring Wheat (*Triticum vulgare var.*).
17. Sweet Corn (*Zea mays var.*).
18. Dent Corn (*Zea mays var.*).
19. Flint Corn (*Zea mays var.*).
20. Spring Barley (*Hordeum vulgare var.*).
21. Spring Barley and Rye.
22. Savoy Cabbage (Improved American).
23. Schweinfurt Quintal Cabbage.
24. Sugar Beets.
25. Mangolds.
26. Butman Squash.
27. Pumpkin.
28. Carrots.
29. Rutabagas.
30. Parsnips.

For the Southern and some of the Western States may be added :

31. Lucern.

32. Gama Grass.

33. Alfilaria (*Erodium cicutarium*), an aromatic geranium grown as a forage plant in California.

34. Tall Oat Grass.

35. Texas Millet.

36. JOHNSON GRASS (*Sorghum halapense*), a very important perennial ; use one bushel clean seed per acre.

37. MILLO MAIZE.

38. COW PEAS.

39. Satin Grass.

40. Sweet Potato.

41. Yam.

And many other plants.

SOUTHERN SOILING.*

"With green oats for feed in March, April and May ; corn in May, June and July ; German millet and cow peas in July, August and September, aided by millo maize in times of long dry spells, like the recent one for September, down to and into December, surely one great problem in the possibility of the South becoming a dairying and fine stock section is solved.

"Millo maize is a new plant introduced only two years ago in our vicinity (Mobile), the seed coming from South Carolina, and introduced into this country from Brazil. It evidently belongs to the Sorghum family, closely resembling amber cane. For two years I have watched it on the farms of two of my neighbors, and particularly this year—a year of a most prolonged and disastrous drouth—where the ordinary growth of our pastures, scanty enough at the best, was dry enough to burn.

. "At the home farm of State Senator Smith a field was planted in July. From this field *two* immense crops of rich, juicy food have been taken, and now (November 14th) the *third* crop stands ready for gathering, while from the stubble of some recently cut a luxuriant growth is springing, and this in spite of the fact that within the past fortnight we have had heavy white frosts. Beginning in July, the rainfall till the last of October amounted to almost nothing, yet through all this long drouth this wonderful plant was rank, green, and as dense as a canebrake. It seemed to utterly ignore the dry, hot days that parched and burnt every other living thing, and stood a living oasis in a desert of arid fields.

"It seems essentially a sun plant—those who introduced it here claim this for it —and no better test could have been given than what it underwent this year

---

* George G. Duffee, in "Country Gentleman," December, 1884.

Planted in rich and well-prepared land, in rows eight feet apart, after the ground is thoroughly warm, it grows rapidly to a height of ten feet, and can be cut from two to four times in the season. Mr. Smith's foreman tells me that the mules and cattle eat it with avidity."

Good reports are given by various writers of the excellent results of cultivating Johnson grass (*Sorghum halapense*) in the Southern States. It is a perennial, producing under irrigation four or five enormous crops annually of green fodder. It is regarded as drouth-proof, and well adapted to hot and arid regions.

For green manuring the cow pea is one of the best crops for the Southern States.

GENERAL SOILING CROPS.

WINTER RYE (*Secale cereale*).

This is a hardy, succulent plant, growing from four to six feet high, and flourishing best on sandy or gravelly loam that is moderately rich. If the crop is put in by the first of September in the North, and by the first of November in the Southern States, it will be well rooted before cold weather, and may be fed off or cut with a machine, leaving it from two to three inches high. Rye is ready to cut very early in spring, when it should be mixed with clover hay for feeding. Oatmeal and wheat bran are good accompaniments, as protein is needed to make a good ration. Rye should be sown with a drill at the rate of one bushel to the acre. In the early spring it is greatly benefited by cultivating the ground with a smoothing-harrow. If cut frequently and kept from heading it becomes a perennial, but it is most profitable as a single crop, cut when in blossom and the ground planted again to a late crop. Rye is ready to cut from about May 1st to 15th in latitude 38° to 40°. One square rod is sufficient for one day for each animal, but it is better to mix other food, as the rye alone is not a perfect food. Turn under heavy crops of rye early in spring as green manure.

WINTER BARLEY.

. Winter barley is suited to clay or clay loam. Barley grows best in the cool weather of spring and fall, and helps to give variety to the fodder. It is one of the best of forage plants, also a good grain for butter cows.

RED CLOVER.

This plant is ready for cutting in latitude 38° to 40° from about May 10th to June 1st, and is very succulent. It furnishes green about twenty thousand pounds to the acre, which would feed twenty cows from ten to twelve days. The second and third cuttings will furnish from twelve thousand to fifteen thousand pounds more, sometimes yielding as high as twenty tons in a season. It is one of the most important soiling crops. Its long roots draw fertility from the subsoil and its leaves from the air. Use ten pounds of seed per acre. Also one of the best crops for green manure.

ORCHARD GRASS.

This grass is ready for cutting with clover. It is of great merit, and may be cut three times a season on rich soil. It should be sown thickly, *three bushels of seed to the acre.* It is good alone or mixed with clover, and is well worthy of universal favor as a forage plant.

TIMOTHY.

Timothy will cut as high as ten tons of green fodder before blossoming, and is a very nutritious forage plant. It should be cut when in full bloom or when the bulb is ripe enough to survive.

TIMOTHY AND LARGE RED CLOVER.

The large pea-vine clover makes a good combination with timothy and adds more protein to the ration. This may be cut twice a year; the first cutting may reach as high as sixteen tons to the acre, and will feed thirty cows ten days; the second cutting will feed thirty cows three days.

TIMOTHY AND ALSIKE.

The Swedish hybrid clover is very hardy, and will yield good crops for ten years. It branches much, and the roots penetrate deeply into the soil. It lasts long in bloom, and may be cut for a month. This combination will feed thirty cows per acre twelve days.

Alsike should be thinly seeded, and may be sown with timothy either in the spring or fall. Ten pounds of timothy seed per acre and six of alsike.

GREEN OATS.

Oats require a cool climate and rich, deeply tilled soil for their perfection. They are ready for forage from last of June to middle of July. If cut before heading in June they make a quick second growth. For a soiling crop it must be put in with a drill as early as the ground will admit, with two bushels of seed to the acre. For culture use one or two harrowings with the smoothing-harrow, until they are two or three inches high. Make the first cutting when about a foot high. They are most profitable, however, when cut in the milk, using but one crop.

A better method of cultivation for seed crop, if some of the large varieties of oats are used, is to plant in drills sixteen inches apart, dropping single seeds one foot apart in the drill and tilling with cultivator. Mr. Burpee reports crops of the " Welcome Oat" raised by Mr. Alfred Rose, of Penn Yan, New York, where one ounce of seed produced three thousand seven hundred and eighty-eight ounces of very heavy oats, and two ounces of seed produced ten bushels and three pecks of oats, weighing four hundred and seventy-three and a half pounds.

Roswell Parkhurst, in Montana, raised three hundred and thirty-two pounds, and August Mongin, of Illinois, three hundred and fifteen pounds from two ounces of seed, while six other competitors raised from two hundred to two hundred and seventy-six pounds from two ounces of oats. Specimens were grown six feet four inches high, and as many as seventy-six stalks from one seed. Most of them report hoeing twice and keeping free from weeds. This is doubtless the best way to cultivate wheat, rye and oats where the grain is to be ripened, and it might be well to try the experiment for forage crops. Oat seed needs to be large and heavy to produce good plants. The oat is sure to degenerate in a dry and hot climate, but good seed will produce a good crop the first year. The oat makes most excellent hay if cut green or when in the milk.

PEAS AND OATS.

The pea when combined with the oat makes an excellent milk ration. They grow well together. Plant with drill four bushels per acre of a mixture of forty quarts of oats with two bushels of peas. Cut when the oat is in the milk. If the peas are allowed to get too ripe the butter will not be of so good a flavor. Plant as soon as the ground will admit in spring. Steep seed one night in diluted urine, drain, and roll in mixture of ashes and plaster. This combination has produced as high as fourteen tons to the acre. Peas need lime and bone-powder to insure good crops.

WINTER WHEAT.

Plant in September same as recommended for rye and oats; requires but three pecks of seed per acre if cultivated like maize.

MILLET.

Millet requires a very mellow, rich soil. Clay loam, if thoroughly underdrained and well tilled, will produce the largest crops. Millet grows five feet high, and produces as much as eighteen tons of green fodder, and is a little richer in nutritive value than timothy. Plant with drill sixteen quarts or broadcast twenty-four quarts of seed to the acre, one fourth inch deep, from first of May to July. Cut just before blossoming.

HUNGARIAN GRASS.

This is a millet which grows three feet high and has an abundant foliage and a large quantity of fine seed. It is the most nutritious of green forage grasses. Objection is made to the stiff bristles which surround the seed spikelets, which are said to have caused the death of cattle by penetrating the stomach. Early cutting would avoid this objection. Sow in early June twenty-four quarts of seed per acre for forage or green hay. Cut in early bloom. The richest, by analysis, of green manure crops.

ITALIAN MILLET.

This grows four feet high, has abundant foliage, and yields the greatest quantity of seed; has been reported to produce five times as much grain as wheat. Pure Italian should be yellow when ripe.

VETCH.

The winter vetch may be sown with rye or the spring vetch with oats. Its food value is similar to the pea. It may be cultivated alone, and bears several cuttings of heavy growth. Spring wheat may be cultivated in drills, same as recommended for oats.

SWEET CORN.

This is the best of all forage plants, and pays several per cent. profit with less labor than any other crop.

Good varieties are known as Early Minnesota, Potter's Excelsior, Stowell's Evergreen, Mammoth Late and Egyptian, which form a succession in the order mentioned. Sweet corn is more nutritious than the dent or flint varieties of maize. Plant in drills thirty-two inches apart and one plant every six inches in the rows, so that ears may be formed. It should be combined with clover, oats and peas, in feeding, or wheat bran and middlings can be fed with corn. Cut the corn when the ears are in the milk. The corn should always be run through a feed cutter, reducing it to lengths of from two to four inches.

DENT CORN.

The large varieties of dent produce heavier crops than sweet corn, but the quality is not as good.

There is no plant that produces such a weight of green food, unless it be sorghum cane. Crops have been reported as weighing fifty tons upon an acre. Twenty tons may be easily grown, and with fair tillage thirty tons.

The ground should be plowed about five inches deep and well manured; the entire manure, liquid and solid, is the best for corn. The land may be pulverized with a slanting-tooth harrow, followed by the roller and a second harrowing. The seed ought to be good and about three pecks to the acre; it may be drilled in or planted two and one half inches deep with the hand planter, the drills three feet six inches apart for the largest varieties, and the single plants six inches apart in the row. This gives the largest crop of fodder per acre. No crop pays better returns for good cultivation. After planting use the heavy two-horse roller.

Before the corn breaks the surface go over the field with the smoothing-harrow, using the round side of teeth to break the crust and aerate the ground. As soon as the corn begins to appear above ground repeat the harrowing each week until the plants are one foot in height. Drive the smoothing-harrow over the rows without any trepidation, for it will save much hand labor and greatly increase the crop. After the maize is a foot high cultivate between the rows with a section of the same smoothing-harrow, so as to finely pulverize the surface to the depth of about one inch, or not more than two inches, making a fine powder for a mulch, and always level culture.

FLINT CORN.

In the Northern States and Canada the hardier varieties of flint corn are grown in order to mature a crop of seed before danger from frost. One of the heaviest yields ever reported is by Mr. Davis, of Scituate, Massachusetts. One acre produced two hundred and seventy-five baskets of ears, which weighed eleven thousand and three pounds. The corn was planted in drills three feet apart, four kernels every twenty-two inches. For tillage use the smoothing-harrow until six inches of growth, afterward a fine steel-tooth cultivator and six-prong hoe, giving clean level culture, and making a mulching surface of finely pulverized earth one to two inches in depth.

BARLEY.

Barley is a very important crop, both for green forage and barley meal, in the dairy ration. The Kinver Chevalier may be sowed every two weeks from July to October, the Manshury in March and April. It produces succulent crops in cool weather, and yields a heavy crop of grain in July, as high as ninety bushels per acre on rich clay loam with good tillage.

AMERICAN SAVOY CABBAGE.

Cabbages make an excellent food for dry cows and young stock, but are not desirable in rations for butter, as the flavor can be detected after every precaution. The Savoy variety is very palatable to cattle. The Schweinfurt Quintal yields immense crops, and is the best in quality of the white varieties. Plant eight ounces of seed to the acre, or two ounces for transplanting.

SUGAR BEETS.

The Imperial Sugar Beet is also a good food for dry cows and young stock, but does not give the best flavor to milk and butter. Under high culture thirty tons per acre may be grown, using about five pounds of seed in drills two feet apart and eight inches in the row.

MANGOLDS.

To be used for young stock and dry cows. Mangolds produce the largest of all root-crop yields, as high as seventy-five tons to the acre being reported in England with rich soil and high cultivation, with abundant moisture. Use six pounds of seed to the acre. Keep the surface of the ground fine and mellow and free from weeds. The rows may be two feet apart and plants ten inches in the row.

CARROTS.

The carrot is the dairy root for cows in milk. The Danvers is a stump-rooted variety and easily harvested, but the Long Orange produces larger crops and is said to give a considerable degree of color to butter. In the cultivation of this root adopt those methods which reduce the cost to the lowest degree. The land must be rich, mellow, and free from stones. Saturate with liquid manure, plow twelve inches deep, pulverize with the Acme harrow, smooth with the roller, use the drill planter, making the rows straight and twelve inches apart; plant thickly, using two pounds of seed to the acre. Run the cultivator between the rows as soon as the plants appear, and cross-harrow the field at right angles to the rows with a slanting-tooth harrow once or twice; give clean culture. In harvesting use a swivel plow to turn a deep furrow away from each row. The six-prong hoe may be found useful both in cultivation and in gathering the crop, also wooden forks instead of steel shovels in handling the crop. Some farmers prefer to cut the tops with sharp hoes while in the row, but it is doubtful if that is an advantage, especially with the Long Orange variety. From twenty to thirty tons may be raised upon an acre. Steep the seed in warm water to hasten germination. They are easily harvested by pulling after a heavy rain. In dry weather they may be pulled after irrigating the ground.

PARSNIPS.

The parsnip (Long Smooth) will be found a valuable crop for butter cows, and can be allowed to remain in the ground until early spring. Cultivate and treat as carrots. Steep the seed in tepid water twelve hours before planting.

RUTABAGAS.

The rutabagas are only valuable for young stock and dry cows, because of the flavor they give to milk. Cultivate in drills. One pound of seed to the acre.

MEADOW SOILING CROPS.

It is well to have as great a variety as possible in soiling and also in winter feeding.

MEADOWS.

" The murmur that springs
From the growing of grass."—*Poe.*

For the production of hay crops, fields and meadows containing mixed grasses are desirable.

SEEDING FOR AN EARLY MEADOW.

1. Meadow Foxtail Grass (*Alopecurus pratensis*)........................ 8 lbs.
2. Green Meadow Grass (*Poa pratensis*)............................... 8 "
3. Meadow Oat Grass (*Arrhenatherum avenaceum*)...................... 8 "
4. Red Clover (*Trifolium pratense perenne*)......................... 8 "

                         32 lbs.

OR THIS.

1. Orchard Grass (*Dactylis glomerata*) ....3 bushels or 40 lbs.
2. Red Clover..10 "
3. Lucern... 4 "

                         54 lbs.

SEEDING FOR A LATE MEADOW.

1. Timothy Grass (*Phleum pratense*)...............................10 lbs.
2. Red-top Grass (*Agrostis vulgaris*)............................... 6 "
3. White Bent Grass (*Agrostis alba*)................................ 6 "
4. Alsike Clover (*Trifolium hybridum*).............................. 6 "
5. Meadow Oat Grass (*Arrhenatherum avenaceum*)..................... 6 "

                         34 lbs.

SEED FOR IRRIGATED MEADOWS.

Italian Rye Grass (*Lolium Italicum*) ....10 lbs.
Perennial Rye Grass (*Lolium perenne*)..............................10 "
Timothy Grass (*Phleum pratense*)................................... 4 "
Rough Meadow Grass (*Poa trivialis*)................................ 3 "
Fowl Meadow Grass (*Poa serotina*)................................. 3 "
White Bent Grass (*Agrostis alba*)................................. 2 "
Red-top Grass (*Agrostis vulgaris*)................................ 2 "
Meadow Foxtail Grass (*Alopecurus pratensis*)...................... 2 "
Meadow Fescue Grass (*Festuca elatior*)............................ 2 "
Alsike Clover (*Trifolium hybridum*) .... 3 "

                         40 lbs.

THE CULTIVATION OF MEADOWS.

The importance of thorough tillage as a preparation for laying down grass lands cannot be too strongly presented.

The farmer who keeps himself abreast with the most progressive agriculturists will appreciate the advantages of using only the best machines and implements obtainable in order to keep his land in such a condition as to return him profit and pleasure.

The plowing should be so done as to leave the field in a level state, with no ridges or dead furrows to interfere with the mower and loader. The pulverization is effected by the slanting-tooth harrow, after which the roller prepares the land for the grass seed.

EXTRACT FROM MR. STIRLING'S TABLE SHOWING THE RESULTS OF COVERING SEEDS AT VARYING DEPTHS.

1.	2.	3.	4.	5.	6.
Agrostis stolonifera	13	500,000	0 to ¼	½ to ¾	1
Agrostis vulgaris	12	425,000	....	....	..
Alopecurus pratensis	5	76,000	0 to ½	1 to 1¼	2¼
Arrhenatherum avenaceum	7	21,000	½ to ¾	1½ to 1¾	4
Dactylis glomerata	12	40,000	0 to ¼	¾ to 1	2¼
Festuca elatior	14	20,500	0 to ¼	1 to 1¼	2¾
Lolium Italicum	15	27,000	0 to ¼	1 to 1¼	3¼
Lolium perenne	18–30	15,000	¼ to ½	1½ to 1¾	3½
Milium effusum	25	80,000	¼ to ½	1 to ¼	2¾
Phleum pratense	44	74,000	0 to ¼	¾ to 1	2
Poa nemoralis	15	173,000	0 to ¼	½	1
Poa pratensis	13	243,000	¼	....	..
Medicago lupulina	63	16,000	0 to ¼	¾ to 1	1½
Medicago sativa	60	12,600	....	....	..
Trifolium hybridum	63	45,000	0 to ¼	½ to ¾	1¼
Trifolium pratense	64	16,000	0 to ½	1¼ to 1½	2
Trifolium p. perenne	64	16,000	0 to 1	1¼ to 1½	2
Trifolium repens	65	32,000	0 to ¼	½ to ¾	1½

Column 1. Names of grasses and clovers.

Column 2. Average weight of seeds per bushel.

Column 3. Average number of seeds in one ounce.

Column 4. Depth by inches at which greatest number sprouted.

Column 5. Depth at which only one half sprouted.

Column 6. Least depth at which none germinated.

The seeds were sown in finely sifted loam, which was kept moist throughout the process of germination and under full exposure to light.

By the use of a proper harrow for covering the grass seed a greater proportion will germinate, and consequently a great saving may be made by lessening the amount of seed sown. The table given above showing the experiment of Mr. Stirling would indicate that grass seeds need only a covering of from one eighth to one quarter of an inch in depth. James Smith, of Deanston, the inventor of the modern system of tile-draining on account of these experiments of Stirling's invented a harrow for the covering of small seeds at a shallow and uniform depth. The implement, which may be called the Serrated Disc Web Harrow, combines the operation of roller and harrow. "It consists of an iron chain web, connected together by discs of iron, which, lying obliquely upon their sides when in operation, roll around, thus tearing and abrading the surface of the ground, so as to expose and disturb the surface to depth enough to cover the small seeds strewn upon it.

"Any one who considers how many clover plants, for instance, will suffice to stock an acre, and what a vast number of seeds are contained in the twelve pounds or even twenty pounds which are now sown per acre, will admit the great room there is for the use of some contrivance for avoiding the common waste now permitted. It is only fair to add that the bush harrow forms a good substitute for the more expensive implement. It merely scratches the surface, but it wants to be weighted to make it as effective as the web harrow to compress as well as abrade the surface. The bush harrow is the cheaper, less effective—the web harrow the dearer, but more efficient implement for the purpose of covering small seeds."

OTHER ENGLISH METHODS.*

"When the seeds are to be sown among winter wheat it is expedient to begin by using the horse-hoe (supposing the wheat to have been drilled), as well to loosen the surface and produce a kindly bed for the seeds as to destroy weeds. In the case of broadcasted wheat a turn of the harrows secures the same end. In the case of the more recently sown barley all that is needed is to smooth the surface with the one-horse roller. Over the ground thus prepared the small seeds are distributed by a broadcast sowing-machine, which sows at once a space of fifteen or eighteen feet in width. The covering is then effected by simply rolling with the smooth roller, or by dragging over the surface the chain-harrow, which may either be attached to the sowing-machine or to a separate frame; or by using a roller, with a very light chain-harrow attached to it. On clay soils the chain-web is to be preferred; but on loose soils the roller (Crosskill) imparts a beneficial firmness, and, with its tail-piece of chain-web to fill up the indentations, gives an accuracy of finish which rivals the neatness of a newly raked garden-plot. We have long regarded this covering in of grass seeds as the most important use to which Crosskill's valuable implement is put. The only drawback to it is that it makes a heavy demand on the horse-power of the farm at a pressing season. As it can only be worked in dry weather it is advisable, when the land is in trim, to work it double tides, by means of a relay of horses.

---

* Encyclopædia Britannica.

This mode of procedure is alike applicable to the sowing of mixed clovers and grasses, and to that of the clovers alone, and is the course usually pursued in sowing for one or two years' 'seeds.'

"When it is intended to lay down arable land for several years, or to restore it to permanent pasture or meadow, it is always advisable to sow the seeds without a grain crop.

"This doubtless involves an additional cost at the outset, but it is usually more than repaid by the enhanced value of the pasture thus obtained. To grow the grasses well the soil should be pulverized to the depth of three or four inches only, and be full of manure near the surface. There is no better way of securing these conditions than by first consuming a crop of turnips on the ground by sheep-folding, and then pulverizing the surface by means of the grubber, harrow and roller, *without plowing it.* 'Never sow grass seed in time of drouth. The ground should be moist enough for rapid germination. Sow clover in early spring. Grasses do best sown in early autumn. Choose the morning calm as the best time to make an even seeding.' "

PERMANENT MEADOWS.

It is of the utmost importance that our farmers should give the requisite attention to all their grass lands, and especially to the establishment of permanent meadows.

The time for this work is the month of August. There is greater probability of thorough germination, *a year of time is saved,* and a good stand of grass is made to endure the winter.

This month gives the necessary time also to prepare the land in the best possible manner.

Take a field of oat, rye or wheat stubble or a second growth of clover, having manured the stubble heavily; the clover is equivalent to a liberal supply of barn-yard manure; turn clean furrows not more than five inches deep, and work immediately with the slanting-tooth harrow, making the land fine, mellow and smooth. Go over it again with the poly-section roller, so as to make it firm and obliterate every inequality of surface. A force-feed grass-seeder may be attached to the roller for a second rolling, and a brush or chain-harrow follow, drawn by the same team. Select the seed according to the tables given above or the lists given for pastures. If the plowing has been smooth and of even depth and three or more harrowings given with the cutting edge of the teeth, and the seed covered pretty uniformly to the depth of one eighth of an inch, and the finish given with the fine brush or chain-harrow, you will have provided a mellow seed-bed well firmed, and having a slightly scarified surface ready to receive the first shower of rain that shall promote rapid germination and growth. If the surface has been well fined and levelled by the harrow the rolling may be omitted. In every case the finish must be made by the grass-seed harrow to prevent the crusting, which would be destructive to grass growth.

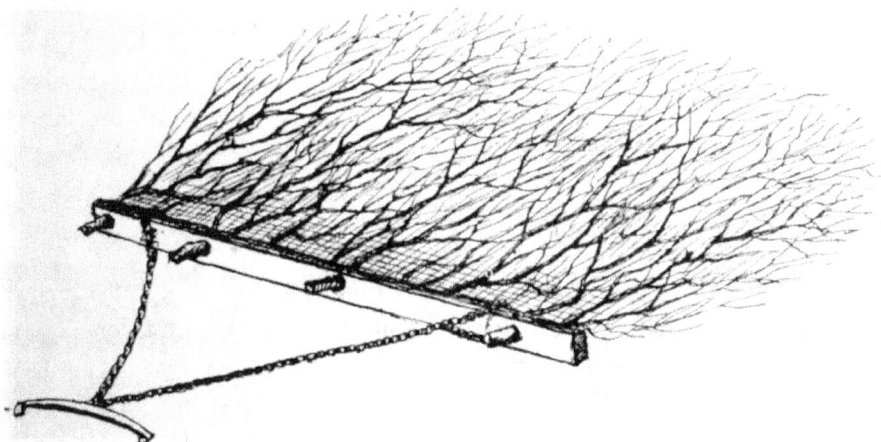

BRUSH HARROW FOR GRASS SEED.

In a month from sowing, if the usual August rains have fallen, the field will be a sheet of vivid green. Do not pasture the ground. The following season will yield one or two crops of hay. Four hundred pounds per acre of bone meal may be harrowed in with the seed with lasting benefit. Subsequent manurings with wood ashes and diluted stable manure from the sprinkling-cart, combined with irrigation, where practicable, will make a permanent meadow yielding a large annual profit.

In the Southern States lucern and cow peas, best planted in narrow drills and kept clean of weeds, make very nutritious hay when cured in shade.

The millets and Johnson Grass are important hay grasses.

In some localities other grasses may be added for permanent meadows. The Italian rye grass (*Lolium Italicum*) is rich in protein, very succulent, and will prove a valuable addition to irrigated meadows.

Barley and rye may be sown together from the middle of July to the first of October, and the barley cut until severe frosts harden the ground. The rye will then furnish an early spring forage. In the culture of meadow grasses the land needs deep and thorough tillage; the seed should be rolled and afterward dragged in by a fine brush harrow, covering the seed lightly. Grasses do not thrive as well when sowed with grain crops, but should have the ground without such shading and choking, while grains yield better crops thinly seeded in drills with light surface cultivation.

In a permanent meadow a mixture of grasses is more profitable, as their roots occupy different areas, and many varieties need company to prevent their dying out. It is believed that timothy, red top and oat grass mutually protect each other.

Another very important point in the preservation of meadows is that they should never be pastured. Deterioration by depasturing and by too late mowing are very speedily ruinous to the best meadows.

For winter protection meadows need a growth of about from four to six inches of aftermath. This autumnal growth corresponds to the depth of root growth and also acts as a winter mulch to prevent killing by the freezing and thawing alternations that destroy grasses having a short top and shallow root. The natural mulch also becomes of great value as a fertilizer in early spring. Meadows preserved by this method and saturated fall and spring with liquid manure will yield large crops perpetually.

BUNT AND SMUT.

"Dombasle's method for treating the bunt fungus in wheat might be applied to the seeds of all grains and grasses to destroy smut of all species with favorable results.

"Thoroughly wet the grain with a solution of sulphate of soda; the wheat or other grain is then mixed with quicklime, which combines with the sulphur to make sulphate of lime (gypsum), which acts as a manure, while the caustic soda destroys the spores of the fungus."

Professor Henslow experimented with sulphate of copper, using two ounces or more to the bushel, which should be used alone. It is not invariably successful in destroying the spores.

Professor Henslow says: "It has always appeared strange to me that practical agriculturists are accustomed to pay so little attention to the raising of pure seed crops. There may be reasons which I do not properly appreciate that would render it inexpedient to cultivate a seed crop; but I should have thought that it was always worth while for every farmer to set aside some portion of ground to be more carefully tended than the rest, for the purpose of securing good and clean seed. Among other reasons for such a practice, he would then be able to weed his crop from every plant infected with bunt or smut before the fungi ripened."

PASTURAGE.

"The waves are a joy to the sea-mew, the meads to the herd."—*Swinburne.*

Where land is cheap, or not specially adapted to soiling, pasturing will be the practice. Some can carry on partial soiling and pasturage with profit.

The essentials of a good pasture are: a soil of more than average richness and sufficiently pervious to rain and flowing water; a persistent growth of sweet, luscious grasses and clovers in great variety; a never-failing supply of pure running water; and fences that will turn not only cattle, but pigs, ducks and turkeys. There should be no quagmires or sloughs, nor streams or ponds where cows can wade deep enough to chill the udder, nor thickets of briers to scratch and wound the teats.

The greatest variety of grasses and aromatic plants edible for cattle is desirable in a pasture. Some of our wild pastures contain more than forty species of grass, besides other plants relished by cows. If the land is arable select the following grasses and clovers for a northern permanent pasture:

For Permanent Pasture. (1.)

1. Meadow Foxtail (*Alopecurus pratensis*)............................	5 lbs.
2. Tall Fescue (*Festuca elatior*).................................	5 "
3. Devon Eaver (*Lolium perenne Devonii*)........................	5 "
4. Green Grass (*Poa pratensis*).................................	2 "
5. Pacy Grass (*Lolium perenne Pacyii*)........................	2 "
6. Red-top (*Agrostis vulgaris*)................................	2 "
7. White Bent Grass (*Agrostis alba*)........................	2 "
8. Blue Grass (*Poa compressa*)................................	2 "
9. Orchard Grass (*Dactylis glomerata*)........................	2 "
10. Fowl Meadow Grass (*Poa serotina*)........................	2 "
11. Rough Meadow Grass (*Poa trivialis*)........................	2 "
12. Meadow Fescue (*Festuca pratensis*)........................	2 "
13. Oat Grass (*Arrhenatherum avenaceum*)........................	2 "
14. Perennial Red Clover (*Trifolium p. perenne*)................	2 "
15. Alsike Clover (*Trifolium hybridum*)........................	2 "
16. White Clover (*Trifolium repens*)........................	1 "
	40 lbs.

Seed for Permanent Cow Pasture. (2)

Perennial Rye Grass (*Devon Eaver*)........................	4 lbs.
Italian Rye Grass..	4 "
Orchard Grass..	4 "
Green Meadow Grass (*Poa pratensis*)........................	4 "
Chicory (*Cichorium Intybus*)............................	4 "
Burnet (*Poterium sanguisorba*)........................	4 "
Alsike Clover..	2 "
Perennial Red Clover....................................	2 "
White Clover..	2 "
Meadow Foxtail Grass....................................	2 "
Timothy Grass...	2 "
Meadow Fescue Grass.....................................	2 "
Red-top Grass...	2 "
Fowl Meadow Grass.......................................	2 "
White Bent Grass..	2 "
Pacy Grass (*Lolium perenne Pacyii*)....................	2 "
Blue Grass (*Poa compressa*)............................	2 "
Oat Grass (*Arrhenatherum avenaceum*)...................	2 "
Lucern (*Medicago sativa*)..............................	2 "
	50 lbs.

The perennial red clover is a variety that should be used in all mixtures.

The green grass sometimes called " Blue Grass" and "June Grass" is in certain localities liable to "smut." Seed should be selected where there is no danger of such disease. *Mow all pastures before seeding of grasses to prevent smut.*

The true blue grass (*Poa compressa*) is especially valuable upon dry soils, as it resists long drouth. Combined with white clover it makes the richest pasturage. Under trees and woods it is best to sow the orchard grass, also drop-seed grass (*Muhlenbergia diffusa*). The latter is a late grass flowering in August and September, and grows to a height of one and a half feet. It is believed to be a good butter grass, and many think it gives a fine flavor to butter. It deserves investigation. It grows only in woods. The sweet-scented vernal grass and vanilla grass, which are sometimes recommended because of their pleasant odors, are probably of little or no value, as cattle do not relish them and they occupy the land as weeds.

SOUTHERN PASTURE.

In the extreme South good pasture grasses are grown with difficulty or not at all. There are some grasses that are very hardy and make terrible pests in cultivated fields, which, however, yield rich pasture. Such are the Johnson grass (*Sorghum halapense*), which is of great value, the Bermuda grass and the crab grass (*Panicum sanguinale*), also the juicy grass (*Paspalum laeve*). The perennial grass (*Paspalum ovatum*) promises to be of great value in the Gulf States, as it is said to thrive on very dry land in the longest drouth.

Texas meadow grass (*Poa arachnifera*), a grass native to the region of the Red River, Louisiana, and elsewhere in the Southwest, is claimed to be more valuable than the green meadow grass (*Poa pratensis*), and is of larger growth. It makes excellent winter pasture, as it has a rapid growth, sometimes making ten inches in as many days in Texas during the winter months.

It is very leafy, makes a dense, permanent sod, and is therefore a fine lawn grass.

It is worthy to be widely introduced and extensively cultivated in all parts of the country where it will prove hardy.

For Southern winter pasture the following list of grasses is recommended :

1. Texas meadow grass (*Poa arachnifera*).
2. Orchard grass (*Dactylis glomerata*).
3. Tall oat grass (*Arrhenatherum avenaceum*).
4. Italian rye grass (*Lolium Italicum*).

The Johnson grass (*Sorghum halapense*) may be pastured or used for a soiling crop. Swine are very fond of its creeping root-stocks.

Lucern requires very rich, warm land. It must be sown in drills eighteen inches apart, using about twenty pounds of seed to the acre, and a dressing of two hundred pounds of bone-powder planted with the seed by the drill. Give thorough cultivation every fall and a rich dressing of cow-manure. Keep down all weeds. It will give four cuttings a year in the Southern States.

PASTURE AND FARM FENCES.

The coming fence is a combination of wire and hedge. Prim is easily grown from cuttings, has a foliage of soft, beautiful green, which remains bright from eight to ten months of the year. A barbed wire fence, of six wires, covered by a prim hedge about eighteen inches wide at the base and five or six inches wide at the top, with a height of five feet, is the ideal fence. The hedge forms a covering or screen from injury by the wire, and the barbs effectually turn all intruders, whether man, beast or fowl. The posts may be set thirty feet apart, and the prim plants nine to twelve inches apart, alternating on either side of the lowest wire, or, if the wires are set on alternate sides of the posts, the plants may be set on a line with the centre of the posts. Five wires may be set, four inches apart at the bottom and widening to sixteen at top. Such a fence would be ornamental to the farm and a very pleasing attraction to the landscape. The wire fence should have an occasional rod of iron set with the posts and soldered to the wires, at least one at each corner of the field, as a protection to cattle from lightning during storms. Other hedge plants worthy of trial are hemlock, spruce, sweetbrier, buckthorn, clethra and althea.

TETHERING.

An economical method of pasturing on small farms is by the use of the tether. It is the practice in the Island of Jersey, and to some extent in America.

This confines the animal to a small area and necessitates a closer and more thorough use of the grass. The removal of the tethering iron or stake a few inches four or five times each day allows the cropping of another space. Water should be supplied every three hours. The tethering iron should have a ring and swivel at the top, and the animal may be secured by a chain of fine links attached to the headstall. The chain should have several link swivels to prevent kinking. A bull may have the chain pass through his muzzle-ring and fasten to a strong leather strap buckled around the base of the horns.

WATER SUPPLY IN PASTURES.

An abundant supply of pure water in every pasture is essential to successful dairying.

One or more troughs or tanks in every field, raised so that animals cannot step or plunge into them, may be filled by pipes conducting from hillside springs of pure water, or from a reservoir filled by a windmill or other power pump. These troughs should be in the open field and most accessible, never in a corner, where they endanger the timid or invite to hooking and goring.

RENOVATION OF PASTURES.

Go over the pastures every year, in August, and root out or cut down every plant, shrub or tree that is unprofitable to you.

The liquid-manure cart will prove of great advantage in seasons of drouth where irrigation is impracticable in any other way. Irrigation and fertilization will do much to keep up a fresh growth in parching weather, provided the ground be well saturated with very dilute manure. Where certain varieties or species of plants and grasses are deficient it is well to go over the pastures in the latter part of August with a steel-toothed harrow, breaking and scarifying the surface sufficiently, when the desired mixture of seeds may be scattered broadcast and rolled or brushed into the scarified ground. If rains do not soon follow, irrigation, by some method, will hasten germination of the seed. All plants having bitter or acrid juices should be cut before seeding in all fields and pastures, so as to exterminate them. Many weeds or stout plants may be destroyed in the early stages of growth by touching the crowns with a wand dipped in a vessel of sulphuric acid. Finely pulverized bones and wood ashes produce sweet grasses, and are the most lasting of manures for pastures.

Clover pastures require a liberal dressing of lime and bone-powder.

RULES FOR PASTURING.

1. Allow no sheep upon new pasture within two years, as they will destroy it.

2. Mow the first growth in early flowering to prevent smut and woodiness.

3. Roll frequently and stock with young cattle only until the second season is over.

4. Never stock pastures in spring until genial weather is fairly established.

5. Never allow the grasses to run to seed or parts of the field to be eaten bare and others to get rank and coarse.

6. Duly spread about all dung, remove all stagnant water, and extirpate all weeds.

7. At midsummer have the pasture grazed or cut so close that there shall be no dead or dry herbage on any part of it.

8. Always adapt the stock, as regards breed, size, condition and numbers, to the actual capabilities of the pasturage.

9. Secure to the stock at all times a full bite of clean, fresh-grown, succulent herbage.

10. In moving stock from field to field take care that it always be to better fare.

11. Have pasture sheds built and furnished with bedding and absorbents, that the manure may be saved while the cattle are sheltered from scorching heat or cold storms.

THE RATION.

At all times the dairy cow, if she be the best type of Jersey, will exhibit a good appetite, the largest digestive power, and great capability for transforming meadow and farm products, in the shape of grass, hay and meal, into milk and cream. Whatsoever is produced in milk, cream, butter and cheese must come from the food which the animal eats. How important, then, that the art of feeding should be thoroughly mastered by all who have charge of dairy cattle.

The cow must be supported by food. It requires two thirds of a full ration to sustain a cow in good condition. This is called the food of support, and is simply appropriated to keep the animal alive. If the animal takes more exercise than is required, or is subjected to very low temperature, or to violent changes of weather and cold storms, or is misused in any way, as by being kicked or beaten by harsh attendants, or worried by dogs, or irritated by being placed with strange cattle, or put under any unusual nervous excitement, there must be a compensation for the loss, for the wear and tear of the system, as far as the law of equivalents can be made to operate, before there is any production of milk whatever.

All the profit must come from the other third of a full ration or from what is used above the two thirds necessary for maintenance. If the cow is not made profitable by right feeding there will be a loss.

The cow must be under the best conditions as to exercise, the maintenance of animal heat and protection from the weather, and then fed so that she will produce the greatest amount of rich milk, and a calf, year by year, to a full age. The cow is the largest producer of food among animals, and consequently the most profitable for economical feeding.

The ration for cows must support animal heat and contribute to maintain all the tissues of the body, and in addition give the largest possible yield of milk of the best quality. If the average temperature is 70° it requires only food enough to raise this temperature to 101°, or to overcome a variation of about 31° between the air and blood-heat. If the stables are kept at an average temperature of 60°, then 10° are added, thus requiring additional food. If the stables are cold and the average winter temperature is 40°, then the temperature must be raised 61° to maintain normal temperature.

Growing cattle require a larger proportion of the elements for maintaining animal heat than milch cows.

In order to save a waste of rations the stable should be made comfortable and of the right temperature in summer and winter, and the animals should be protected from all sources of worriment and annoyance.

The milch cow, the growing animal and the mature bull must each be fed a ration suited to the special requirements of each. The cow must have that proportion of protein, carbo-hydrates and fat suited to the highest productiveness of the best quality of milk and butter and the development of the fœtal calf. The growing heifer must have a ration suited to prepare her to become a perfect cow. The growing bull and the mature bull must each have an appropriate ration, which differs in composition according to age and service.

A PART OF PROFESSOR TANNER'S TABLE OF FODDER VALUES.

WEIGHT REQUIRED TO PRODUCE ONE POUND OF MEAT.

Linseed Cake and Peas, equal parts	4½ lbs.
Linseed Cake	5 or 6 "
Barley	6 "
Rape Cake	6 "
Cotton Cake	6 "
Oats	7 "
Beans	8 "
Peas	8 "
Clover Hay	12 "
Swedes	150 "
Mangolds	150 "
Carrots	160 "

Professor Johnson, of the Connecticut Experiment Station, has translated the feeding standard tables of the German experimenter, Dr. Wolff, which show what has been found to work well on a small scale and may be useful to the Jersey breeder in aiding him to form a better standard suited to his own herd for special purposes. Great variations may be made from these standards, and farther on in the history of individual Jerseys will be given the rations for tests from which the great butter records have been made.

According to Wolff, thirty pounds of young clover hay will keep a cow in fair milk: this contains of dry organic substance twenty-three pounds, of which the digestible substance is: protein, 3.21; carbo-hydrates, 11.28; and fat, 0.63. This varies from the standard in the table by .71 pounds more of protein, .22 pounds less of carbo-hydrates, and .23 pounds more of fat.

The ration must not only contain the correct proportion of nutrient substance, but it must always be combined in such a way as to be most palatable and in the most convenient form for mastication.

Special rations must be fed to cows that are producing a very large amount of butter, and also to service bulls and choice calves. Special feeding that keeps a cow up to the limit of her full capacity has been proven a source of permanent improvement in individual cows, and also a governing factor in the production of better calves, the cow producing her best heifers, according to Stewart, at the period of her highest feeding and greatest productiveness. The best cows of the Jersey breed may be greatly improved in both quantity and quality of their milk by high feeding of well-selected rations.

FEEDING STANDARDS.

PER DAY AND PER THOUSAND POUNDS, LIVE WEIGHT.

AGE. MONTHS.	ANIMALS.	Total Organic Dry Substances.	NUTRITIVE DIGESTIBLE SUBSTANCES.			Total Nutritive Substances.	Nutritive Ratio.
			Protein.	Carbo-hydrates.	Fat.		
		lbs.	lbs.	lbs.	lbs.	lbs.	lbs.
	Oxen moderately worked.....	2.40	1.6	11.3	0.30	13.20	1 : 7.5
	Cows in milk..............	2.40	2.5	12.5	0.40	15.40	1 : 5.4
	GROWING CATTLE.						
	Average Live Weight per Head.						
2 to 3	150 pounds................	22.0	4.0	13.8	2.0	19.8	1 : 4.7
3 to 6	300 "	23.4	3.2	13.5	1.0	17.7	1 : 5.0
6 to 12	500 "	24.0	2.5	13.5	0.6	16.6	1 : 6.0
12 to 18	700 "	24.0	2.0	13.0	0.4	15.4	1 : 7.0
18 to 24	850 "	24.0	1.6	12.0	0.3	13.9	1 : 8.0

PER DAY AND PER HEAD.

AGE. MONTHS.	ANIMALS.	Total Organic Dry Substances.	Protein.	Carbo-hydrates.	Fat.	Total Nutritive Substances.	Nutritive Ratio.
2 to 3	150 pounds................	3.3	0.6	2.1	0.30	3.00	1 : 4.7
3 to 6	300 "	7.0	1.0	4.1	0.30	5.40	1 : 5.0
6 to 12	500 "	12.0	1.3	6.8	0.30	8.40	1 : 6.0
12 to 18	700 "	16.8	1.4	9.1	0.28	10.78	1 : 7.0
18 to 24	850 "	20.4	1.4	19.3	0.26	11.96	1 : 8.0

Professor Horsfall says it requires twenty pounds of good meadow hay, besides the food of support, to produce eighteen quarts (forty pounds) of milk a day. The cow cannot consume this amount of hay above the ration for her maintenance, and the extra food must be sought in more concentrated forms, such as are rich in protein, phosphoric acid and oils, and these be selected with reference to economy. His stables in winter were kept at 60° temperature. His ration for milk consisted of rape cake, five pounds, bran, two pounds, mixed with bean straw, oat straw, and oat shells in equal parts, fed three times a day, all they would eat. These materials were moistened, mixed thoroughly, then steamed and fed warm. Each cow had, in addition, from one to two pounds of bean meal, according to her quantity of milk, and when eaten, green food, consisting of cabbages from October to December, kohl-rabi until February, and mangolds till grass-time. To preserve a good flavor the green food was limited to thirty or thirty-five pounds daily, and after each feed four pounds of meadow hay or twelve pounds daily to each cow, with all the water they would drink twice a day.

This ration was given to produce quantity of milk and prepare cows for the second stage of fattening for the butcher. His cost of feed was twenty-seven cents a day for each cow, and the milk from six cows averaged $46.83 for one hundred and ninety-one days, and the manure was equal to $29.49 per cow for the same length of time.

The English, German and American experiments demonstrate that *two parts of all food are required to keep the cow alive, and one part for production and profit.* They also show that the oil contained in the food is insufficient to supply the needs of the animal, and that the fat must in part be derived from the carbo-hydrates in the food.

RATIONS FOR JERSEY COWS IN MILK WHEN YIELDING FROM TWO TO THREE POUNDS OF BUTTER DAILY.

In the month of May green rye and barley may be cut and mixed with clover hay or extra meadow hay of mixed grasses and clover, twenty-five to fifty pounds of rye, and twelve to sixteen pounds of hay, given in four feeds, at 6 and 9 A.M. and 3 and 6 P.M., allowing the time from 10 A.M. to 3 P.M. for exercise in the open air. Until the cows are accustomed to the green food it is best to graduate the proportions for a few days, giving a ration as follows, for cows of nine hundred pounds live weight:

MAY.

10 lbs. Best Mixed Hay,	
25 lbs. Green Rye and Barley,	Cut and mixed
4 lbs. Wheat Bran,	for two feeds.
2 lbs. Corn Meal.	
10 lbs. Parsnips,	Noon feed.
4 lbs. Barley Meal.	

¼ oz. Salt at each feed.

JUNE.

2 lbs. Best Hay,
10 lbs. Green Rye or Rye Grass,
20 lbs. Green Clover,
25 lbs. Orchard Grass,
25 lbs. Green Barley.

        Mixed for two feeds.

    1 quart of Wheat Bran at each milking.
    ⅛ oz. Salt at each feed.

JULY.

2 lbs. Best Hay,
25 lbs. Green Clover or Clover and Timothy,
25 lbs. Green Oats or Oats and Peas,
20 lbs. Green Wheat or Alsike,
20 lbs. Green Hungarian Grass, in early blossom.

        Two feeds.

    1 quart of Wheat Bran at each milking.
    ⅛ oz. Salt at each feed.

AUGUST.

2 lbs. Best Hay,
20 lbs. Italian Rye Grass,
20 lbs. Italian Millet and Hungarian Grass,
30 lbs. Green Sweet Corn Fodder,
20 lbs. Alsike or Green Wheat.

        Two feeds.

    1 quart of Bran at each milking-time.
    ⅛ oz. Salt at each feed.

SEPTEMBER.

2 lbs. Best Hay,
25 lbs. Sweet Corn (with ears in milk),
10 lbs. Green Barley and Hungarian Grass,
10 lbs. Millet, or ½ bushel Crushed Ripe Apples,
10 lbs. Wheat in early bloom.

        Two feeds.

    1 quart of Bran at milking.
    ⅛ oz. Salt at each feed.

OCTOBER.

2 lbs. Best Hay,
25 lbs. Sweet Corn (with ears in milk),
20 lbs. Green Barley,
10 lbs. Green Wheat.
         } Two feeds.

    1 quart of Rye Bran at milking.
    Given upon 1 peck of Crushed Ripe Apples.
    $\frac{1}{8}$ oz. Salt at each feed.

NOVEMBER.

20 lbs. Green Corn Fodder,
10 lbs. Best Hay,
10 lbs. Green Barley,
2 lbs. Oatmeal,
2 lbs. Barley Meal.
         } Two feeds.

4 lbs. Bran,
10 lbs. Pumpkins.
         } Midday feed.

    $\frac{1}{16}$ oz. Salt at each feed.

DECEMBER.

15 lbs. Best Early Hay of Orchard Grass and
    Clover,
6 lbs. Wheat Bran,
4 lbs. Maize Meal,
2 lbs. Barley Meal,
4 lbs. Oatmeal.
         } Two feeds.

$\frac{1}{4}$ lb. Linseed Meal,
10 lbs. Pumpkins.
         } Noon feed.

    $\frac{1}{16}$ oz. Salt at each feed.

JANUARY.

5 lbs. Green Oat Hay,
15 lbs. Corn Stover,
6 lbs. Rye Bran,
2 lbs. Barley Meal,
4 lbs. Maize Meal,
4 lbs. Oatmeal.
         } Daily ration for two feeds.

$\frac{1}{4}$ lb. Linseed Meal,
10 lbs. Carrots.
         } Noon feed.

    $\frac{1}{16}$ oz. Salt at each feed.

FEBRUARY.

8 lbs. Best Green Clover Hay,
8 lbs. Green Millet Hay,
6 lbs. Rye Bran, or Wheat Shorts,
4 lbs. Maize Meal,
2 lbs. Barley Meal,
4 lbs. Oatmeal.

} Two feeds.

½ lb. Linseed Cake Meal,
10 lbs. Carrots.

} Noon feed.

$\frac{1}{16}$ oz. Salt at each feed.

MARCH.

4 lbs. Green Clover Hay,
15 lbs. Green Corn Stover,
6 lbs. Rye Bran or Wheat Shorts,
4 lbs. Maize Meal,
6 lbs. Oatmeal.

} Two feeds.

10 lbs. Mangolds or Parsnips,
1 lb. Linseed Meal.

} Midday feed.

$\frac{1}{16}$ oz. Salt at each feed.

APRIL.

8 lbs. Green Millet Hay, or 25 lbs. Green Rye,
8 lbs. Cow Pea Hay or Clover Hay,
6 lbs. Rye Bran,
6 lbs. Oatmeal,
2 lbs. Maize Meal.

} Two feeds.

10 lbs. Parsnips,
½ lb. Linseed Meal.

} Midday feed.

$\frac{1}{16}$ oz. Salt at each feed.

A CHEAP WINTER RATION.

10 lbs. Green Corn Stover,
8 lbs. Green Millet Hay,
4 lbs. Oatmeal,
6 lbs. Rye Bran,
3 lbs. Maize Meal,
2 lbs. Linseed Meal.

} Two feeds.

$\frac{1}{16}$ oz. Salt at each feed.

Use the "Crusher" machine for cutting, comminuting and pulverizing all the corn fodder, and the Hay Cutter for all other grasses, then moisten and mix with the ground feed.

RATION FOR WINTER MILK.

3 lbs. Clover Hay.

15 lbs. Corn Stover, well cured.

4 lbs. Oatmeal.

2 lbs. Corn Meal.

8 lbs. Wheat Bran.

2 lbs. Linseed Meal.

$\frac{1}{16}$ oz. Salt at each feed.

OR THIS, FOR WINTER MILK.

		Cost.
18 lbs. Corn Fodder, well cured		4.5 cents.
5 lbs. Best Clover Hay		2.0 "
8 lbs. Wheat Bran		6.0 "
4 lbs. Corn Meal		3.0 "
2 lbs. Linseed Meal		3.0 "
$\frac{1}{16}$ oz. Salt at each feed.		18.5 cents.

STANDARD WINTER RATION FOR A BREEDING HERD.

10 lbs. Corn Fodder, cured green.

5 lbs. Rowen Hay.

1½ qts. Oatmeal.

1 qt. Maize Meal.

1 qt. Wheat Bran.

1 pt. Linseed Meal.

6 qts. Parsnips.

$\frac{1}{16}$ oz. Salt at each feed.

The corn fodder to be cut in four-inch lengths, or crushed in the " Crusher," then well moistened and mixed with the grain, one half at 6 A.M., one half at 6 P.M. (Bundles of corn-stalks may be cut in four-inch lengths with the bucksaw.)

The rowen hay at 12 M.

The parsnips at 3 P.M.

Full watering at 10:30 A.M. and 5:30 P.M., with water at 65° temperature.

WINTER RATION FOR YELLOW BUTTER.

Same as above, provided the corn fodder and hay are cured so as to retain their green color. The parsnips also aid in giving butter color in winter.

Give double the quantity of salt with green succulent crops, and always mix dry hay or oat straw with green clover.

RATION AT ECHO FARM, LITCHFIELD, CONN., AS REPORTED BY CONNECTICUT
AGRICULTURAL EXPERIMENT STATION, 1881.

Daily Ration per Head.

KINDS OF FODDER.		Dry Matter.	DIGESTIBLE.		
			Protein.	Carbo-hydrates.	Fat.
	lbs.	lbs.	lbs.	lbs.	lbs.
Hay..............	23.52	15.72	0.66	8.37	0.14
Provender*..................	4.69	3.88	0.34	2.54	0.13
Bran........	2.50	2.15	0.33	1.16	0.05
Mangolds	7.50	0.64	0.12	0.43	....
Total		22.39	1.45	12.50	0.32
Total per 1000 lbs. live weight		25.16	1.63	14.04	0.36
Standard		24.00	2.50	12.50	0.40

Total digestible matter, 16.03.

Nutritive ratio, 1 : 1.9.

Order of feeding : Morning, Hay, Provender and Bran.

Noon, Hay.

3 P.M., Mangolds.

Night, Hay, Provender and Bran.

RATION RECOMMENDED FOR JERSEY COWS ONE MONTH BEFORE CALVING.

In Winter: 10 lbs. Best Meadow Hay.

10 lbs. Corn Stover.

10 lbs. Rutabagas.

2 lbs. Oatmeal.

½ lb. Oil Cake Meal.

$\frac{7}{8}$ oz. Salt at each feed.

OR THIS.

10 lbs. Rowen Hay.

6 lbs. Oat Straw.

15 lbs. Rutabagas or Cabbage.

2 lbs. Oatmeal.

½ oz. Salt.

* Equal parts ground oats and maize.

DARLINGTON RATION.

At the "Darlington" dairy of grade cows the following ration is fed in order to give a good flavor to the butter, which has a reputation in the market for quality, flavor and uniformity of appearance throughout the year:

Best Clover Hay......	8¼ lbs.
Corn Meal....................................	8¼ "
Wheat Shorts.................................	8¼ "

No cornmeal or shorts are used if in the least degree fermented, but should such fermented meal or bran be sent from any dealer, it is immediately returned, as the feeding of fermented food would destroy the quality of the butter.

DRYING OFF COWS.

If it is desired that a persistent milker shall be thoroughly dry before calving turn her into a box-stall and feed her with a ration of thirty pounds of oat straw. Straw is a good ration to make cows dry off at any time, and therefore not a desirable food except for the sole purpose of drying.

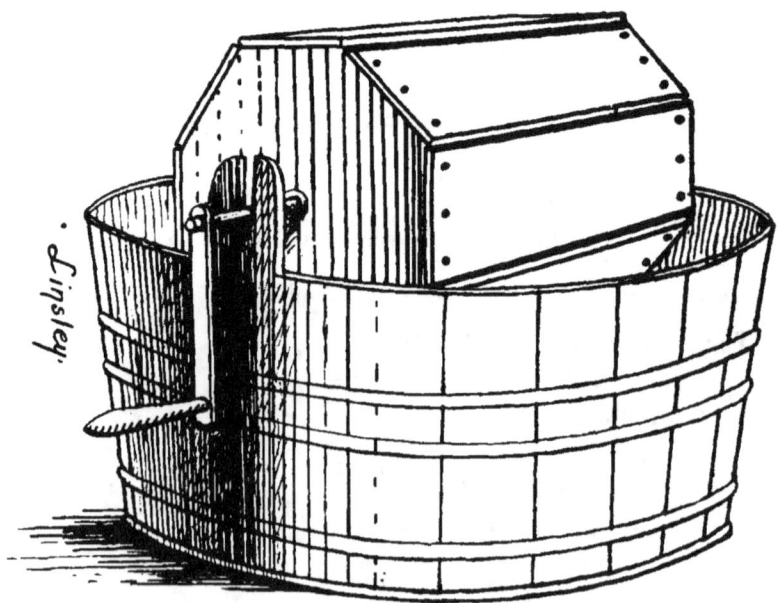

CARROT CLEANER.

SUMMER FEEDING.

In summer give the cow what she will eat of mixed forage plants or good pasture alone. All green crops are better after a few hours of wilting in the sun, especially sweet corn.

No forage crop can be fed when wet by rain or dew without injury, because of the added excess of moisture, which causes indigestion and a diminished yield of milk.

Roots like all the greens and apples must be fed in small quantities at first, gradually increasing to the full proportion in the mixed ration.

RATION FOR THE JERSEY BULL.

SUMMER.

Let him be tethered in good pasture, or give from seventy-five to one hundred pounds of green forage with from one to three pounds of oatmeal, according to size and service. Give salt at each feed.

WINTER.

12 lbs. best Clover or Cow Pea Hay.

3 lbs. Oatmeal.

2 lbs. Linseed Meal; grain mixed with the cut hay.

4 to 6 lbs. Carrots, Mangolds, Rutabagas or Parsnips.

½ oz. Salt.

Feed three times daily, and give water as often.

RATION FOR THE CALF.

The calf is best fed with the utmost regard to punctuality, three times daily, from birth, upon the full milk of its own dam, when practicable. Let the calf suck the dam one day, then remove it to its own softly-bedded stall. The second and third days it may be taught to drink from a pail or bucket by firmly holding the muzzle in the warm milk while two fingers of the right hand are inserted in its mouth. The calf will take from a quart to three pints of the colostrum three times daily, and a gradually increasing quantity of his dam's milk. This should be given as soon as a sufficient quantity is drawn, that it may not lose its normal temperature of 101¼°. After one month, if it is desired to use the cream for butter, the calf can have one third of the dam's milk mixed with two thirds warm skim-milk from the Separator; or, if some other mode of cream-raising is practised, the sweet skim-milk should be warmed to 102° or 103° by careful test. It is better to give six quarts a day in three than the same in two feeds. At the third month the calf can have entire skim-milk, to which should be added a gruel made from flaxseed. It is well to begin the addition of the gruel the second month, to supply the change from whole milk with cream to a partial ration. A pint of flaxseed and a pint of "oilmeal" boiled in ten quarts of water, or flaxseed alone in six times its bulk of water, will make a gruel nearly rich enough to supply the lack of cream in the skim-milk. Mix this, one to two parts, in the skim-milk, and feed at a temperature of 102°. Always use the thermometer, and a good one. From the beginning add a little rennet or lacto-rennetine to the milk before feeding, and a gradually increasing allowance of salt at each feed, beginning with a few grains only.

If the Sucking Feeder is used it may be fastened to the inside of the stall and the milk poured in through the door or a port-hole.

The rennet renders digestion and assimilation easy.

Let the calf be satisfied three times a day until six months old. During the fifth month or earlier teach it to eat a small handful of oats. If the food has a tendency to produce diarrhœa diminish the quantity at once by three fourths, or substitute for a few meals, in the milk, a quart of coarse wheat flour or pure wheat meal instead of the linseed. But always diminish the food at once upon the first symptoms of indigestion. Pea meal may be combined with the flax meal or flaxseed if desired, or when the calf is two months old one pound of oatmeal or wheat middlings may be added to its ration. Twenty pounds of skim-milk may be sufficient in the daily ration until the fourth month. If the milk is diminished the oatmeal must be increased.

OTHER RATIONS FOR CALVES.

WHEY RATION.

There is much valuable nutriment in the whey after cheese-making, but the fat and casein extracted must be supplemented by a sufficient amount of grain food to supply the nitrogenous elements, and thereby render the mixture an equivalent to normal milk. Add a little salt to each feed.

FORMULA FOR YOUNG CALF.

Whey.. 1 gal.
Oil Cake.. ¼ lb.

Mix when hot, and feed at 102°.

FORMULA FOR CALF AT ONE MONTH OLD.

Whey.. 1 gal.
Oil Cake.. ¼ lb.
Oatmeal... ¼ lb.

Mix hot, and feed at 102°.

HAY TEA RATION FOR CALVES ONE MONTH OLD.

Skim-Milk.. 1 gal.

Hay Tea (decoction of early cut hay)............................. 1 gal.

Flaxseed (decoction)... ¼ lb.

Wheat Middlings... ¼ lb.

Mix the milk after boiling the other ingredients well and straining, and feed at 102°.

Gradually increase with age the grain elements. Add a little Salt.

RATION FOR CALF FROM FOURTH TO SEVENTH MONTH.

10 lbs. Skim-Milk.

2 lbs. Oatmeal.

1 lb. Linseed Oil Meal.

Add ¼ teaspoonful of Prepared Rennet; and Whole Oats to eat midway between feeds. Give a little Salt.

OR THIS.

20 lbs. Skim-Milk,

1 lb. Oatmeal,

½ lb. Flaxseed,   } Two Feeds.

20 drops Prepared Rennet.

A half pint Whole Oats, dry, at noon. Add a small quantity of Salt.

This is a ration for a very large Jersey calf; a small calf may thrive on much less than this. During the first year or from six months onward the young heifer should be fed chiefly on hay, so as to expand and develop the digestive organs to a capacious size. Many breeders prefer to keep calves from grass until a year old. Some prefer to keep them upon whole milk for three months, gradually introducing skim-milk until the calf is six months old, returning to whole milk if the calf scours, always reducing the quantity, and giving it at a higher temperature, from 125° to 130° Fahrenheit. Bran has an irritant effect on the bowels of a calf and should not be used. The greatest care in raising calves is necessary, that they may always have just enough, and never too much. Hundreds of valuable calves are killed by overfeeding; especially by persisting in the overfeeding after serious disorder of the bowels threatens to destroy the animal. It is the most important department of feeding. Some breeders rarely or never lose a calf, others have continual disasters from their own mismanagement.

ENSILAGE.

The storing of green forage crops in water-tight vats under enormous pressure is the invention of M. Goffart, of France.

For centuries green crops had been buried in ditches and caverns and subject to great loss by fermentation and decay. The method of Goffart, although a great improvement upon the old, by largely excluding the air and arresting the fermentation at a certain point, still has many serious objections. Its inventor and some of its advocates have undoubtedly claimed too much for the system. For if all their claims are just and tenable, then all fodder should be put into the silo, and every available crop, including apples, squashes, roots, and every grain and grass, would be improved by this process of partial decomposition.

Some claim that it is analogous to the art of canning fruit and vegetables for human consumption and as successful, and that in ensilage they are providing a canned fodder for their cattle.

If this comparison were true the silo and its products would be all and more than any one has claimed for them. But from the chemical analysis of ensilage and the strenuous opposition which many breeders offer against adopting it because of that analysis, and also because of ill effects observed from its use, it is considered to be a very defective fodder for the butter or milk dairy when fed in any considerable quantity.

According to analyses made at the Connecticut Experiment Station, 1882, the best sample ever offered contained acetic acid and alcohol equivalent to " a quart of strong vinegar" and a pint of rum for each hundred pounds of ensilage. In the United States Agricultural Report for 1882 it is stated that " the acidity and alcoholic nature of the ensilage has been of universal remark, and, to a certain extent, of exaggeration." In the sample from C. H. Roberts, of Poughkeepsie, N. Y., the conditions had been such as to make the alcoholic fermentation most prominent, but even under these circumstances alcohol was only recognized in the distillate from the juice by the iodoform test. The juice expressed from the specimen amounted to forty and a half per cent. of the substance taken. The following determinations were made :

Specific gravity, 15° C...1.0335
Total solids... 8.14
Glucose.. .94
Sucrose.. .13
Total acid as acetic... 2.71
Total acid as lactic... 3.08

This sample may be regarded as an extreme of acidity, owing to its having been out of the silo two days before examination. As it requires one tenth of a pound of acetic acid to make one quart of the *strongest vinegar*, one hundred pounds of this ensilage would contain twenty-seven quarts of the very harshest vinegar, beside the three pounds of lactic acid. A specimen from Alexandria, Va., contained acid equivalent to twenty-one quarts of sharp vinegar to the hundred pounds.

It is the vinegar and alcohol, and other products of fermentation, that render ensilage unacceptable as a food for winter soiling. If these products could be avoided or prevented, then winter soiling would be as successful as the summer soiling for dairy cattle. If acetic acid, lactic acid, butyric acid and alcohol in great or small amounts improve the fodder, as some claim, why, then the whole world will have the benefit of it as soon as it can become generally adopted. It is a matter of great consequence to breeders of thoroughbred Jerseys that they adopt nothing that shall hinder the progress of successful breeding and dairying.

Some have boldly risked and lost much in testing the ensilage experiment during recent years, while many of the best breeders and feeders in the Jersey world cannot be induced to try the experiment.

When the silo shall have become as successful in its purpose as the canning of pears or peaches in culinary art all Jersey breeders will adopt ensilage. The feeding is just as important as the breeding. *Prove and hold fast that alone which is best.*

TRAINING HORNS.

One Jersey breeder is very successful in producing cattle of fine form and beautiful colors; another cares little for form, much less for color, but gives his whole attention to filling the churn, while a third is a dabster at training horns.

The horns of a Jersey are ornamental and give a certain character to the animal, varying according to their size, color, shape and texture.

It is best to have Jerseys that breed the true Jersey horn, or at least it is desirable to have something like uniformity in the herd. Some horns crumple, others are tossing, and a few are angular. It is supposed that about one fourth of the Jerseys have horns that either turn inward and downward, or inward and slightly upward, while about three fourths have horns that either flare or assume a nearly vertical direction. It is desirable that they occupy as little space as possible, and also that they be rendered, as far as practicable, harmless from goring. The crumpled form is the best, turning inward and downward.

The process of training should begin at an age before the horn is too hard, usually about one year old.

Bore through the horn half an inch from the tip with a small gimlet. Tie a piece of catgut or a copper wire securely through these holes. With some heifers the weight of the string and its slight pressure may be sufficient; if not, hang a little bag from the centre, having a few birdshot for weight, allowing it to rest upon the forehead of the animal. The weight of the shot must be adjusted to the stiffness of the horn. A young bull might require from half a pound to two pounds, and in rare cases three and four pounds. The process must be watched and the weight adjusted according to the conditions. If one horn is **stiffer** than its mate it may be nicked with a file or rubbed with sand-paper on the inner surface ; at the same time a daily oiling will hasten somewhat the operation. Too much weight or too violent a strain on the wire or catgut will cause a thickening of the horn at the base. The trainer simply needs to exercise judgment and will soon acquire skill in his work. While the training of horns is progressing the animals should be kept in their stanchions, as that secures them from any damage by entanglement or hooking or colliding with fences.

When the horns have the desired turn remove the apparatus and file off and polish the tips of the horns, so as to obliterate the gimlet-holes.

The work will well repay the care and skill expended by a more attractive appearance. Recent improvements in horn-training apparatus obviate the necessity of boring the horn, a button being adjusted to the tip of the horn as a support to the tension and weights.

CARE OF THE BULL.

The bull is "half the herd," and if in himself he is worthy of the place he occupies, must command the most skilful care and attention to preserve his potency and keep him in condition for the transmission of his best qualities to all his progeny. The bull is the breed, and transmits his race characteristics and individual qualities in a greater degree than the female. It is important that he should be kept in a uniformly healthful condition by proper exercise and feeding, and in no case allowed to impair his potency by any excess or too frequent use.

The best bull must possess a vast amount of latent energy and neural force, and consequently should be of a very lively disposition. He therefore needs much exercise in the open air and sunlight and kind treatment, or he may become surly and fierce. He should be nimble in his movements and never lazy, high-spirited and never dull, always ready to respond when properly called upon for service, and unfailing in every effort, provided that the cow is in good health.

He must be kept in a lean and active condition, and yet be well nourished. If he becomes fat impotency will follow. His first service may be at about the age of fifteen months, and may be repeated monthly until he is two years of age. From two to three years old he may give a service bi-monthly, and after three years of age one weekly service is enough to require, if offspring possessing the requisite neural energy is to be secured. The service might be less frequent to the advantage of the progeny, male or female ; and in service the bull should never be allowed to repeat his efforts after one successful copulation, but always removed immediately to his own stall. Nothing is worse than repeated copulations at one interview, or on the same day, for destroying the potency of the male or for tending to produce degenerate offspring. The best progeny must always be procreated when the male is in perfect vigor. One service supplies a superabundance of the sperm cells; a second is less likely to hold, and may be of inferior vital quality.

The question is not how much, but how good service.

BULL EXERCISE.

The bull should have a variety of exercise. The running of an empty tread power at a low rate of speed for one hour or even a half hour each day is of great benefit. He should not be compelled to follow too long. In pleasant weather he may be turned into an open field that is guarded by a barbed wire fence. The bull pays a profound respect to barbed wire. It is a good plan to tether him in a pasture a few hours each day, as advised in chapter on Pasturing, using a strong iron post, which may be driven into the ground to the depth of two feet or more. On the top of this post a sliding ring and swivel admit of the requisite freedom of movement. To this ring is attached a chain, which may be about twenty-five feet long, and composed of steel in fine links, of which a sufficient part are swivelled to prevent kinking while the bull walks. Such a tether gives a circuit of one hundred and fifty feet. The fastenings must be secured by a steel spring or clasp, which may be passed upward through the nose-ring, and secured to the steel ring in his head-stall, that is firmly buckled about the base of the horns.

RINGING THE BULL.

The ring ought to be of the best quality of steel, or of pure copper, and of a size having an outside diameter of two and one half inches.

It is well to apply it when the animal is about one year old, or during the time of his horn-training.

To perform the operation, be provided with the trocar and canule and a ring of the right quality. Secure the bull in the stanchion of his stall, or turn his head and secure him firmly to the post by a strong halter. Grasp the cartilage of the nose and carefully select the point for insertion, which must be as high as possible, so that the ring shall be out of the way of liability to catch upon snags or nails, and also to guard against its tearing out, as may occur in powerful bulls when set too low. Take a good hold upon the nasal cartilage, pass the canule, with the trocar point slightly projecting through the cartilage, let the open ring follow the canule through the incision, clasp it, and insert the screw, turning it down firmly in its place.

For leading, a short steel chain may be attached to the staff hook, always well secured to both hook and ring by a strong lock snap.

The staff needs to be of the finest quality of straight-grained, thoroughly seasoned, well-tested timber.

Care must be observed in the operation of ringing that the trocar and canule are of the best pattern and quality and the trocar always kept sharp and smooth, as a rough or dirty instrument may cause blood-poisoning or other mischief by a ragged incision.

KIND TREATMENT OF THE BULL.

The bull is as susceptible to kind treatment and petting as any other animal, and he resents cruel treatment, oftentimes with the traditional persistency of a bear or the revengefulness of a savage. Instances are recorded where a bull, having been beaten or abused in his stall by a stranger, always entertained a hatred for the person, and knew his step so well that whenever he came within hearing, although he could not see him, he would manifest his displeasure in the most unmistakable manner.

Always treat a bull kindly and manage him with firmness and caution. Never presume upon his friendship, for he sometimes takes a sudden freak of playfulness or a passion for combativeness, and in either case wishes to try the force of his neck, head and horns. Treat him as a pet, and at the same time let him very early be taught to recognize you as his master and to yield implicit obedience to your will and your commands. The bull is inclined to resent a club—at least he cannot be beaten back with a club if he is determined—but he pays respect to the tingling of a tough switch or whip when applied to his muzzle. If early trained he dreads the whip, and pays it as much respect as he does a barbed wire fence that is properly constructed, for that is a barrier he doesn't care to fight against, even under the extremest provocation of the charms of a matable heifer on the opposite side of the fence.

Treat the bull kindly and compel kindness from all his attendants, but allow no man to have charge of him who is a poltroon or a coward.

THE HELMET.

If a valuable bull becomes excitable and hard to manage he may be controlled by the Bull Helmet. This may be made of strong leather and formed to cover the forehead and eyes, and secured firmly around the horns, and by a strong throat-latch made to buckle under the cheeks. The eyes are protected from contact by conical-shaped leather goggles, which are firmly fastened in the helmet. This helmet is a complete blinder, and the bull wearing it is subject to his master's hand, and may be led quietly wherever desired. The helmet ought to be used on all bulls above two years old with absolute safety, and might have saved many a valuable bull from slaughter, as it is the lively fellows that get good stock and transmit neural force to their progeny, and such a bull may remain potent for fifteen years or more if rightly used.

CARE OF BREEDING HEIFERS AND COWS.

The Jersey breeds at an early age.

The heifer should be bred when fifteen months old, as nearly as practicable. Mr. J. W. Vance, Cantrall, Ill., reports heifer dropping calf when eleven months and eighteen days old. It is desirable to have uniformity of size in a herd and to have

Jersey cows of about nine hundred pounds weight, but it is also desirable that they have the characteristic Jersey traits of early development, persistent milking, and cream-producing richness. It seems necessary to breed at the age named to secure and perpetuate these qualities. If bred later they are inclined to lose the Jersey quality and take on fat. If thought desirable, the time of the second calf may be delayed so as to give time for a larger growth of the young cow, but the cow character is to be established as early as the growth and constitution of the animal admit of it. The aptitude for milk production early established, the after-growth and management depend somewhat on the time of the second calf, which it may be well to have timed to the age of three and a quarter or three and a half years, so that the second breeding should be from six to nine months, or even a year from the time of dropping the first calf. The cow should be treated with the utmost gentleness, and the Jersey heifer makes the most attractive pet in the world. Every farm attendant should have the characteristics of a gentleman, and in his mind anything like cruelty or even rudeness to a Jersey cow should be abhorrent, and any person practising such cruelty by a kick or a blow should be banished from the farm at once.

THE PARTURIENT COW.

Two weeks or more before calving the cow should be put in a box-stall where she may have quiet and especial care as to her diet and bedding. The food should be palatable, cooling, and sufficiently laxative to preclude any danger from constipation. The bedding should be sufficient to give her comfort and rest. If the cow is in perfect health she will pass through the ordeal in safety, and afterward, under the right management, give her owner the full benefit of her productive powers in milk and butter.

It may not always be easy to say that a cow is in perfect health, but if she has been properly cared for and had proper feed she will not need any assistance in parturition. Any mechanical interference, unless very skilfully managed, is hazardous, and may destroy the calf and permanently injure the cow. There should always be present a man of experience, who is properly informed in regard to the necessary treatment in case of emergencies, and he should have the good sense to refrain from all unnecessary interference, and yet know how to afford proper aid, either mechanical or medicinal, when such is needful.

The approach of parturition is indicated by the soft and swollen vulva, the fully distended udder, and the day previous a sinking in about the pelvic bones. The cow should now have a quiet stall and no one allowed to come near, except the persons who have charge of her feed and her safety. Her drink must be warm; no cold drinks should be within reach, but pure water 65° or upward, or warm gruels of flaxseed and bran mashes, and cooling, laxative food.

As labor approaches the first stage is ushered in by uneasiness, which gradually increases. The animal must be screened from view by a curtain, and kept from all annoyance. After a few hours the dilatation of the uterine ring is complete, the contractions become violent, the cow gets up and lies down frequently, the belly becomes lank, the cow utters a slow, frequent moan, the expulsive efforts becoming gradually more forcible, and the breathing is quicker.

The bursting of the sac and the discharge of watery fluid indicate that the labor should terminate reasonably within two or three hours. The efforts now proceed in progressive rapidity, till at length a protrusion of the vulva indicates a near termination of labor. The parts gradually dilate with each expulsive pain, the calf presents at the opening vulva in its natural position, the head stretched forward and resting upon and between the knees ; the labor is progressing naturally ; let the cow alone and keep away any intruders, but be ready to attend to the calf, which in a little while is safely expelled and becomes at once a breathing, independent existence. See that the cow is protected from currents of air.

THE CARE OF THE YOUNG CALF.

Examine the navel-cord to see that it does not bleed. If it bleeds tie it with a ligature of soft thread. Place the calf in front of the dam, that she may lick it. When this natural and beneficial process is completed the calf will after a few attempts rise upon its feet and instinctively search for the udder. Allow it for the first day to suck the dam three or more times, and after each sucking milk the udder empty.

Afterward milk the cow regularly three times a day, and teach the calf to drink as directed in the chapter on Feeding Calves. It is very important that the pails, buckets and feeding-troughs of calves should be kept scrupulously clean by daily rinsing in cold water and scalding in hot water, and a subsequent airing and sun-drying. The calves should be kept in separate pens or stalls, as they annoy each other by sucking. Bull calves should not pasture with heifers at any time, as after three months old they are liable to breed. After ten months old bulls annoy each other and should be kept apart.

Difficulties of parturition will be treated of in another chapter on Casualties.

The afterbirth should be conveyed to the manure-house as soon as it is expelled. Never allow the animal to go through the revolting process of trying to hide it by swallowing.

CARE OF THE CALF AT BIRTH.

In cold, wet, or chilly weather many a young calf is lost, or stunted, for want of the requisite attention during the few hours subsequent to its birth. Have several boxes provided of a size large enough to hold the calf, so that it can lie comfortably at rest. When the birth of a calf is expected have a dozen bricks heating in an oven or furnace. As soon as the calf is born place the bricks evenly over the floor of the box, cover them with three or four inches of finely-cut straw, placing a blanket over the straw.

Lay the calf upon the blanket and cover with another. Rub the calf dry with a coarse towel. Such care insures against much loss that would otherwise accrue during cold or inclement weather. Calves that were so feeble as to require feeding with a spoon and constant artificial heat to keep them alive have become wonderfully robust animals.

In one case a birth at the seventh month of a choicely-bred calf required such treatment, and the owner was well repaid for his care in the saving of the life of a valuable Jersey bull.

Calves need as much of sunshine as adult animals. The sun must always be admitted freely to their stalls by very large windows, and their exercise should be sufficient to keep them in good health and prevent the accumulation of fat, a condition unallowable in a Jersey of any age, whether bull, cow or calf.

Never allow any animal to be imprisoned in a dark stall.

Teach every animal to lead by halter, from calfhood.

CASUALTIES.

There is no calling without its casualties.

As in all other human occupations, so in the breeding and management of cattle there are accidents, unforeseen, improbable and strange; accidents from negligence or from carelessness; from ignorance or from improvidence; from mistaken kindness; from overfeeding and from medical malpractice or neglect of correct medical practice.

STRANGE CASUALTIES.

A choice Jersey bull died from swallowing a piece of bale wire in cut feed.

A farmer having a herd of choice Jerseys pastured them in a river meadow. On a day when one of the herd stood quietly chewing her cud upon the river bank a passing steamer blew its whistle, at which sound the cow was so suddenly startled that her one impulse from fear caused the bank to cave in, and she was drowned in a depth of two feet of water.

A farmer turned a heifer (not a Jersey) alone into a pasture, and a few days afterward found her with a broken horn which had bled excessively, and for which no treatment that was used caused any check, the animal at last dying from loss of blood. The injury was probably caused while rubbing herself against a rough rail fence or while attempting to get out from loneliness.

The same farmer left a calf (not a Jersey) tied to a stanchion by a rope around its neck, and turned the dam into the adjoining stanchion to give the calf suck. Returning after a little while, what was his astonishment to find the cow quietly standing in her stall, while her calf, with his rope noose across her back, was hanged by the neck, dead !

How many specific errors can you discover in the management of each of the above cases?

When the cow lay down and the frisky calf jumped over her back he might not have been strangled if he had worn a head-stall instead of a noose ; but who would think of allowing a calf to suck when hampered by any kind of a halter?

A famous prize cow in a noted herd was one day missing. Search was made in every field and building. She was a gentle and sagacious creature, and could open any gate or door, and at last in a little colt barn the searchers, peering down a hatchway into the deep, dark cellar, saw the favorite cow, all in a heap, and grieved to think her dead.

They got her out and found her yet breathing, and with good care she soon began to mend, but her fine, shapely rump was broken, and it took nearly a year to heal. Although she was within three months of calving, she carried her calf to full term, and has since had several choice calves, and at sixteen years of age is producing a large quota of butter, with a prospect of several more valuable calves.

A wealthy gentleman, who might be characterized as more wise than prudent, having a large estate and rich farm which he wished to stock with choice cattle, procured at a large expense two young bulls and several heifers.

He desired that the bulls should be kept in vigor by abundant exercise, and concluded to break them at once to the yoke. To make them speedily familiar with this ancient implement of service he turned them, yoked, into a river pasture, to enjoy close companionship in eating and drinking.

The heifers were also turned into a river pasture. It required but brief time to bring disaster upon disaster, which almost wiped out of existence the new young herd. The bulls waded the stream to drink and play, when one of them floundered and was drowned, and his yoke-fellow was but just able to keep his head above water.

The heifers discovered upon the river's brink some freshly painted boats and paint-pots partly filled with paint, and, having a great relish for linseed oil and the aromatic turpentine with the combination of white lead, were soon eagerly engaged in licking off or lapping up the fresh paint.

Result : all that got access to the paint were speedily attacked with intestinal spasms, colic, general convulsions, and paralysis, which soon put an end to life.

Foresight is better than hindsight. Those who attempt to breed and manage herds of cattle should be able as far as possible to foresee and forefend all such accidents, and also many violent diseases.

LEAD COLIC FROM PAINT.

Several **very** choice animals have died from colic by licking fresh paint in their stables, or from barns and fences. A very choice bull, son of the best Jersey cow ever known, was thus destroyed. He could not have been purchased from his owner for any amount of money, and was of more value to the Jersey interest in America than the cost of all the paint on all the barns of the whole country for a century.

THE DEADLY ARSENITES.

The practice of using Paris green to destroy insects is a reprehensible one, in that it not only endangers the lives of valuable animals, but also human lives. The system cannot be too strongly condemned. It should be abated, and other means as effective, but without hazard to life, substituted therefor.

OVERFEEDING.

Indigestion from overfeeding is very common and fatal among calves, while bulls and cows not infrequently die from the evil effects of the same system of injudicious cramming with ill-assorted rations of rich or indigestible foods.

FATTY DEGENERATION.

Many breeds of cattle have suffered much deterioration from habitual over-feeding, perhaps none more than the Short-horn.

It has frequently been noted that a Short-horn bull kept in full flesh or fat enough for the butcher failed to get any calves, but with gradual decrease of ration and gradual increase of exercise, health was restored and potency returned. Barren cows by being worked in the yoke were freed from surplus fat, and became prolific again. Nothing has been so ruinous as overfeeding for fairs. It destroys the milking properties of a breed and induces a disease, *Fatty Degeneration of all the Muscular Tissues.* Many choice Jerseys have been killed by overfeeding for shows and for tests, and much damage has been done by presenting stock at public sales in a pampered and extremely delicate condition. Such cattle suffer deterioration in quality and may require a year in the new owner's hands to recover ; some are never restored to a normal condition of health. The coat becomes dull, the appetite capricious, the milk falls off in quantity and quality, or fails entirely, and the purchaser suffers great loss.

Skilful feeding is needful for health and success.

HOOVE—METEORISM—TYMPANITIS.

The overfeeding of cattle upon succulent food like green clover or the exces-sively heating maize meal or cottonseed meal, causes very dangerous attacks of indigestion. The filling up with green foods produces a gastric irritation and a rapid fermentation in the rumen, with enormous distention from food gases, a severe and dangerous affection.

GARGET.

Inflammation of the udder may be caused from improper food, as ensilage, cotton-seed meal, or excess of corn meal, or by incomplete milking or too seldom relieving of the udder; from taking cold in a blast of cold air; from dampness and cold; from wading in cold brooks or ponds; by injury to the udder from briers in the pasture; from bites of animals or stings of insects; from attempting to force dry; from acclimation fever or other illness; and, worst of all, from kicks or blows given by brutal attendants. It often renders one or more quarters of the udder useless. Feed wilted grass, warm bran mash, and linseed gruel.

LIGHTNING.

A spark of electricity passing down a tree in a storm is often sufficient to destroy a herd gathered there for shelter from the rain.

So in a barn unprotected by a rod, a bolt may take in its course the bodies of several cattle, or burn the barn and its stored crops, with the cattle; or a rod unskilfully set may conduct the bolt to the cattle instead of the earth.

Any man is culpably negligent who does not see that his buildings and cattle are properly protected from any stray thunderbolt that chances to come within his dominions.

In a city most buildings are well protected by the great amount of metal in them, especially of waste-pipes, water-pipes, and gas-pipes, but in the country barns and stables are very prone to be struck by lightning, with great loss to farmers. Wire fences also conduct the lightning to cattle lying near them. The greatest losses from lightning occur in the tornado region of the United States.

Rules for Adjusting Lightning-Rods.

BY PROFESSOR J. K. MACOMBER, IOWA AGRICULTURAL COLLEGE.

"1. The best material for the rod is iron. Copper is a better conductor, but more costly.

"2. The size of the rod should not be less than three quarters of an inch for solid round iron. A hollow pipe would do equally well if it contained as much metal. *Rods usually sold are too light.*

"3. Insulators of glass or other material are worse than useless. They increase the expense, weaken the support of the rod, and actually do harm by preventing the induced electricity from being drawn from the building by means of the rod.

"4. The rod should be fastened to the building by staples, and be laid up against it as closely as possible. No sharp turns should be made by the rod. All turns should be made by smooth, long curves.

"5. *The rod should be continuous throughout.* If too long to be welded into one rod let it be made in four or five sections, with screws cut so that the parts can be put together strongly. The end should be pointed.

"6. Each prominent chimney or gable should have a rod running to it and a point running up from six to eight feet above the building.

"7. *The rod should be well grounded* and also be metallically connected with all masses of metal within or on the building at their highest points. Eave spouts and metallic roofs should be connected by soldering copper straps thereto. *The point most important and generally neglected is the ground connection.* It is the universal custom for those who put up rods to simply drive a bar of iron into the dry earth a few feet, and shove the rod into it, and call that a sufficient ground connection.

"*A rod put up in such a manner is of no value.* A hole should be dug until permanently moist earth is reached; the rod should run down into this and then bend away from the house.

"Several square feet of metal should be placed in this hole and the rod terminate in this metal. An old copper wash-boiler is a good terminal. One hundred feet of three-quarter-inch iron costs $7.50; painting, $1; couplings, $2; labor for one day in erecting, $2. Total, $12.50.

"A fancy tip, or gilded vane or ball, can be added at a small cost."

Large trees in pastures where cattle remain during storms are sources of danger. A single tree of great size in a pasture should have a rod passing well down beneath its roots.

The barbed wire fence should be made safe by a rod at each corner of the field and at each gateway. These rods should also be set deep enough to reach moist earth. The wires should be wound around or soldered to the rods and all stapled to the fence-posts.

LICE UPON CATTLE.

That farmer cannot be called civilized who would allow cattle to become infested with such a pest as lice while in his own stable. If cattle are kept in clean stables, fed upon suitable rations, and have a good brushing once a day, they will not be so afflicted. But if cattle have in any way been subjected to such a nuisance, it should be remedied as quickly as possible. A decoction of Larkspur (*Delphinium staphisagria*) applied daily until all signs of annoyance disappear is a very effective remedy. The tincture of larkspur can be procured at the pharmacies. If of full strength it can be used by mixing one part to nine parts of hot water, and applying at a temperature of about 130°.

Pyrethrum powder, a teaspoonful to a gallon of hot water, is also effective for the same use.

The cattle should be brushed in a room set apart for that purpose, daily. The Universal Joint Brush is very expeditious by the use of steam power or an ordinary tread power. Such brushing and cleanliness are of great benefit to the cattle.

HEALTH AND ITS CONDITIONS.

In perfect health all bovine animals are sprightly in disposition, good feeders, regularly chew the cud with enjoyment, have a normal pulse and respiration, uniform temperature, soft, mellow skin, and glossy hair, void urine at regular intervals, and also moderately soft fecal excretions.

PULSE.

The bovine pulse is naturally full, soft and rolling to the finger-touch. In disease it may become more frequent or slower; it may have a sharper stroke or a lagging impulse; it may be full and strong, or weak, small and thread-like; hard or soft, oppressed, jerking, intermittent, unequal and thrilling.

The pulse may be felt on the border of the lower jaw; beneath the bony ridge which extends upward from the eye; *over the middle of the first rib or under the tail.*

In adult animals, while lying at rest, the number of beats per minute vary from thirty-eight to forty-two. But after a full feed and in a high temperature the pulse may be excited to sixty or seventy. In young animals it is much more rapid, while in old age it may lessen by five or more beats. Small animals have a faster pulse than larger of the same breed.

The pulse is increased by fear, exertion, nervous exaltation, by pregnancy, in hot, foul air, and by overfeeding. Aside from these conditions, a rapid pulse indicates either fever, debility or some inflammatory action.

The unequal and irregular pulse may indicate a fatty degeneration of the heart and other organs, or dilatation of the heart, or some disease of the heart-valves of the left side of that organ.

The intermittent pulse may indicate merely a disturbance of the heart's action through some disorder of the system; it sometimes accompanies organic disease of the heart.

The jerking pulse indicates disease of the valves at the left side of the heart, and is usually accompanied by a hissing or sighing murmur with the second heart-sound, heard by placing the ear as near to the region of the great vessels as possible.

The action of the heart may be detected by applying the palm of the hand behind the left elbow.

By frequent practice one may learn to detect the slightest variation from a healthy standard, and apply the sanitary and medicinal remedies as needed.

RESPIRATION.

There should be nine or ten full respirations each minute. The nostrils, dilating easily and regularly, admit the air to the larynx, trachea and bronchi, a complex, flexible, and elastic apparatus, retaining a tubular form, and conveying an ample supply of air to the two large, spongy, elastic bodies called lungs, which occupy the right and left portions of the thorax. The lungs are about one-thirtieth of the weight of the body, and are for the absorption of oxygen and expulsion of carbonic acid and other impurities, or the transformation of vitiated or venous blood into bright red arterial blood.

The process of maintaining animal heat is carried on in the minute vessels called capillaries, where the waste of the tissue cells is oxygenated or burned in the processes of repair, and it is in the cells and capillaries of the body that animal heat is produced. The normal bovine temperature does not vary much from $101\frac{1}{2}°$.

ACCLIMATION OF JERSEYS.

If Jerseys are taken into the Southern or Gulf States from the Northern States and Canada, they must necessarily suffer, according to the change of conditions of climate and soil, for some months, during the process of acclimation, a degree of disturbance sometimes rising to quite a high febrile reaction.

It is best to bring animals into the South in the autumn and always locate them, if possible, during the first year upon high lands. The higher and drier the land, the nearer will their condition approximate to that from which they came. Those breeders living upon high and elevated districts are best situated for the importation of Jerseys either from the Island of Jersey or the Northern States and Canada.

It is also advisable to select young animals, from six to twenty months old, as they endure the change better than adult animals or very young calves.

ACCLIMATION FEVER.

Acclimation fever is a term applied to the very marked disturbance of the system caused by a change of climate. Jersey cattle are as easily acclimated, perhaps, as any bovine race.

They thrive with wonderful vigor in Canada and are fast becoming favorites in our Southern States and California. Acclimation fever is most violent at low altitudes in hot weather. Cattle improve when taken from low hot districts to mountain altitudes.

ABORTION (SLINKING).

Abortion is the worst of all the casualties that affect the Jersey breeder, because, if not properly understood and guarded against, he is liable to suffer the greatest loss and disappointment from this dire disaster.

Abortion is the separation and expulsion of the immature ovum from the womb. It may occur at any time between fecundation and the time of normal fulfilment of utero-gestation.

During the first month it is called ovular abortion; from the first to the third month it is called embryonic abortion; from the third to the sixth month it is called fœtal abortion; from the sixth to the ninth month premature birth.

This casualty is mentioned in the oldest literature of the world which we possess. Moses in the book of Genesis makes Jacob allude to the subject in his last interview with Laban. Speaking of the prosperity of the flocks and herds under his care, he mentions that they had been exempt from this scourge during his long sojourn of twenty years, and attributes this security to the favor of Almighty God. Moses in the book of Job, where he utters his complaint in his terrible affliction and makes allusion to the unaccountable prosperity of the wicked, says, "Their bull gendereth and faileth not, their cow calveth and casteth not her calf." From this it would appear that abortion is not a new thing, but was an old-time calamity among the bovine races.

Abortion is very frequent in all breeds of cows, and not more common among Jerseys than among scrubs. It is an evil the more to be dreaded by the Jersey breeder as, besides the loss of a valuable calf, it sometimes occasions also the loss of the dam, or renders her barren by some uterine injury, or subject to repeated abortions.

The causes of abortion are very various, such as: bodily injury by sharp goring; kicks or blows from cruel attendants; fast driving, or running to and from pasture; plunging or jumping down embankments; injury by transportation in carts and rail-cars; from violent efforts at riding with rutting animals; from fright by thunder and lightning, barking dogs and wild animals or any startling sight; from the nervous excitement caused by company of aborting cows; from pasturing with horses; from electric shock by proximity of lightning-stroke; from foul air of non-ventilated stables; from "malaria" or swamp air; from pungent or offensive odors, such as carbolic acid or chlorine gas; from sour, fermented food, as brewers' grains, apple pomace, distillery slump and ensilage; from excess of laxative food; from insufficient or poor quality of food; from excessively rich or stimulating food, as cottonseed meal; from impure water; from mineral waters; from insufficient exercise; from standing on sloping floor; from confinement in dark stables; from lonesomeness; from "acclimation fever"; from many acute diseases;

from poisoning by "smut" of Poa grass, corn smut (*Ustilago maydis*), rye spur (*Secale cornutum*), and other fungous growths ; from other poisonous plants ; from the malpractice of using cathartics, salts and various nostrums ; by contagion from impure vaginal discharges ; from being forced dry ; and from a specific infection which may be communicated from herd to herd by the transportation of animals out of infected herds to healthy herds. This contagion is so virulent that every member of the largest herd may become infected from one animal.

SYMPTOMS OF THREATENED ABORTION.

Sometimes the symptoms are so slight as to be unobserved previous to the culmination of the disaster, especially before the third month ; but in general it is announced by great disturbance of the system, anxious look, depression, sudden diminution of the milk, and by offensive mucous vaginal discharge. It may occur at any period of pregnancy, but especially about the twentieth week, and from that to the thirty-second week. The approach of abortion will be noticeable in the languid gait of the animal, the less active movements of the fetal calf, the diminished appetite, the loss of the cud, the lank, drooping belly, irregular breathing, a yellowish or bloody discharge from the vulva, an irregular or feeble pulse, a springing of the bag and increase of milk.

Always isolate a cow at the first symptoms and give her a separate attendant.

COLOSTRUM APOPLEXY.—MILK FEVER.

These are names applied to one of the most fatal of maladies affecting the cow. The best cows are liable to be destroyed by it within three days from calving. Cows seldom have it with the first calf. Very poor milkers never have it. The danger increases in great milkers as they attain the period of their greatest productiveness, usually from seven to ten years of age.

An excess of fat upon the internal organs and the habit of constipation are conditions which strongly predispose the cow to a fatal attack.

Within twenty-four hours after calving the cow may suddenly fall without any premonitory symptoms having attracted the attention, and, remaining unconscious and unable to swallow, dies in a few hours.

The cases vary much in the severity of the onset, but very few recover spontaneously.

Many show a condition of languor or great depression, cease to chew the cud, and lose all relish for food, and hang the head with a dull expression of countenance ; the muzzle is dry and hot, the horns also hot, the bowels constipated, the urine scanty or suppressed, the pulse fast, and the breathing rapid, with heaving flanks, the milk diminished or checked altogether, the temperature high. If these symptoms are not soon remedied the cow gets rapidly worse. The eyes glisten and become congested, and the white is of a leaden color streaked with the red blood-vessels. The eyes protrude from their sockets, the cow is uneasy, the legs are weak, she continually changes her position, the hind legs become tremulous and can scarcely support her. The discharge (lochia) from the vulva ceases, the pulse becomes slower and the breathing more and more labored; the udder is hard and swollen; the weakness in the hind legs increases, the feet are spread wide apart; she staggers and falls heavily upon the floor, then, struggling to recover herself, is unable to rise. In this condition she tosses and writhes, with lashing of the tail and frequent moaning or bellowing, seems in the greatest distress, the breathing becomes a labored panting, the body is covered with a cold sweat, and the rumen is enormously distended with gas, which more and more increases the difficulty of breathing. The pulse flags, the legs become cold, the cow belches a fetid gas, and life is extinguished.

In other cases the cow may lie stretched upon her side, with the head turned looking backward and resting upon the floor, or the head is thrown upward and backward in a rigid position with the horns pointing over the shoulders, the eyes are glassy and sightless, the pupil widely dilated, the ears limp, the jaw drooping, and the cow scarcely able to swallow and fast losing all sense of touch; the pulse is scarcely perceptible or intermits, the horns, legs and skin become cold, the breath rattles, the belly distends, the udder is swollen and sometimes red, the dung and urine suppressed. The cow dies within two days, often in a few hours. The pathology of this disease is not yet fully understood. It is not yet decided whether it is a disease of debility or of congestion of the brain, or whether it may not be complicated with meningitis. It seems to be the result of the profound disturbance caused by the sudden effort of the system to transform a large part of its tissues into milk. It is remarkable also that the disease occurs only during the colostrum stage, while the milk is yet incomplete, and while the disruption of membrane from the follicles of the udder glands is one of the results of the intense and mighty change, prior to the normal shedding of milk globules and the well-established flow of perfect milk.

Probably in all cases of colostrum apoplexy there is a check of perspiration following labor—in other words, "taking cold."

By a reference to the analysis of colostrum * given below, and comparison of its elements with those of milk analysis in another part of this volume, it does not require very profound reasoning to determine that the organism of the cow, during the week of the *colostrum stage*, endures a great physiological change, which requires but a slight disturbance to become a serious one, bordering on fatally diseased conditions.

---

* First Annual Report N. Y. State Experiment Station, 1882.

According to the amount and richness of the colostrum is the degree of nervous vital expenditure and the danger of a violent irregular circulation of the blood. Consequently the feed and all the sanitary conditions should be regulated for one month before and after parturition with a view to the preservation of a normal action of the whole system.

COLOSTRUM ANALYSIS.*

"Meg, a Jersey cow, calved December 4th. The colostrum was orange yellow, of acid reaction. Specific gravity by weight, 1063. It coagulated into a solid mass by boiling.

Fat	5.22
Casein	7.87
Albumen	7.81
Milk sugar	2.94
Ash	1.23
Loss, etc	.21
Total solids	25.28
Water	74.72
Per cent. nitrogen by combustion, 2.35	100.00 "

GENERAL SUMMARY OF CAUSES OF ACCIDENTS AND DISEASES.

1. Diseases of the bowels and kidneys ordinarily proceed from improper feeding and watering or bad forage.

2. Diseases of the chest from insufficient or improper ventilation, overcrowding, neglect and exposure.

3. Abortion from carelessness of attendants, blows, abuse, or eating smut of grass or maize.

4. Diseases of the skin from want of cleanliness, and sometimes from using barley straw for bedding. Some skin diseases are contagious.

5. Wounds and broken horns usually arise from carelessness of management. If horns are broken in fighting it is often by the introduction of a stranger into the herd. The Jersey horn is fragile, so that special care is needed to guard against an accident that mars the beauty so greatly and gives the animal the appearance of a cripple.

6. Garget and foul foot are often caused by wet yards, muddy pastures, or wading in brooks.

It is best to forefend all such accidents, even including "milk fever," by being

---

* Report N. Y. State Experiment Station, 1882.

well guarded at all times against any form of neglect or carelessness in the routine of management.

TREATMENT OF DISEASES AND SERIOUS CASUALTIES.

" By medicine life may be prolonged,
Yet death will seize the doctor, too."—*Shakespeare.*

An ounce of prevention is worth vastly more than a pound of cure, but when all the precautions of wise forethought and good judgment and well-trained sanitary skill have failed to ward off a much-dreaded malady, then is the opportunity to test the potency of scientific medical skill. The readiness to practice is almost universal, but the skill is much more of a rarity.

Let any man or animal fall a victim to any disease, and every casual visitor of the multitude has a prescription to offer to thrust upon the patient. Everybody likes to doctor except the skilful physician; he holds back until called upon and urged to give advice, which with true modesty and many misgivings he humbly proffers. The world is advancing in civilization, and of all the discoveries of modern times the most beneficent has been that of the great German physician Hahnemann, whose law of cure with its small dose has revolutionized the practice of medicine. But all people have not yet availed themselves of the mild beneficence of scientific medicine. Hahnemann's discovery consisted: First, in finding a universal law for the selection of cures; second, in noting the wonderful susceptibility of the diseased organism to the effects of minute doses; third, a system of preparing medicines so that they may be taken in doses of any degree of division or attenuation desired.

The law for the selection of cures always takes the drug that has the greatest affinity for the organ or organs diseased, and is capable of producing a similar disturbance in the healthy organism. Similars to cure similars. This law will doubtless be demonstrated in the future as an electrical affinity. The small dose avoids aggravation or poisonous effects, but induces a speedy and wonderful reaction. Neither Hahnemann nor any of his followers have been able to ascertain a limit to the curative powers of a drug by any degree of attenuation, and no man is able to say of any drug properly selected that a dose of any degree of limitation ceases to have curative power by reason of its smallness, nor, on the other hand, that any dose has curative power because of its comparative largeness. In other words, the curative power of a drug lies in its *quality* rather than in its quantity. It is not the purpose of this work to supply a text-book on the practice of Veterinary Medicine, but to offer a few suggestions by which breeders will be enabled to combat, with some degree of success and satisfaction, diseases that have hitherto baffled the skill of the old barbaric system that is happily soon to be among the things of the historic past, while a happier day awaits those who shall enjoy the advantages of a higher civilization and a better system of medicine.

Those who would have all the necessary equipments for treating the diseases common to all domestic animals must provide their sanitarium with the best text-book extant, Boericke & Tafel's "Homœopathic Veterinary Practice," and a full list of medicines to meet the requirements of all common diseases. There should be the necessary sanitary apparatus, means for heating water, rubber bags for immersing the udder, rubber sheets or blankets to be used in (milk fever) colostrum apoplexy, sponges of various sizes, air thermometers (accurate) to note the temperature of the stable, hot-water thermometers to test the temperature of external applications and drinks, a fever thermometer to note the animal temperature in each case, an elastic syringe for hot-water injections. Hot-water bags may be useful in various diseases; a rubber probang in case of choking; trocar and canula for dropsy and hoven; and a medicator for placing doses upon the tongue.

All breeders who adopt the medical practice herein set forth will be glad that they live in the nineteenth century rather than the ninth. After a faithful following of the principles and doses as given by the great Hahnemann, both for their animals and themselves, they will never desire to return to the barbaric methods which we have inherited as a legacy from the Dark Ages.

PREVENTIVE TREATMENT FOR ABORTION.

Preventive treatment requires one to guard against all the causes which produce this terrible scourge. Strange cows always have to meet the attacks of the fighting or boss cows. Keep them apart, especially if either the stranger or the boss cow is pregnant. A timid cow is sure to be gored; keep her apart from the others.

Transport cattle by steamers when it can be done. Never allow pregnant cows or heifers to be in the company of non-pregnant cows or in the company of aborted cows. Protect animals from lightning as far as possible by good rods on buildings. Never allow any man or boy to make a cow move faster than a moderate walk. A man that runs the cows for fear of getting his shirt wet in a shower may destroy a thousand-dollar calf and permanently injure a valuable cow. Such a man should be discharged from your service at once for any disobedience of orders.

The stables should always be sweet with a perpetual ventilation and perpetual cleanliness. Offensive odors by all means are to be avoided. Fermented foods are a curse in any dairy. The breeding cow should be kept in good health and given wholesome food at all times. All smut plants must be collected if practicable and burned. A field of green meadow grass or other soil infected with this fungus should be plowed and the ground planted to root crops for two years. To prevent fungous growths all pastures ought to be mowed before the grasses mature their seed. Cathartics or drugs for producing artificial diarrhœa and dysentery and other inflammatory diseases of the digestive organs should not be tolerated in any form; it is mischievous malpractice. All breeders should be cautious about spreading contagion. Those who have abortion in their herds and continue to buy and sell are contributing to spread the disease by every animal that leaves the herd.

The period of incubation after exposure is from three to six months. All cows that abort should be quarantined for from nine to twelve months before being bred again. It is safest to withhold from breeding more than one year rather than less. All strange animals should be quarantined before introducing them to the herd stable at least three months, if there is any uncertainty in regard to their freedom from exposure.

TREATMENT OF THREATENED ABORTION.

Aconite. If the animal has been frightened and the fear remains, or she shows serious after-effects, give a dose or two of the sixth dilution of aconite.

Aletris farinosa. For habitual abortion. Give five drops of the first dilution twice a day during gestation.

Apis mel. Scanty urine with frequent urging to urinate. Constipation. Give sixth or thirtieth dilution.

Arnica. From any mechanical injury, such as goring or any hurt, give ten drops of the first, third or sixth dilution every two hours. Bathe the bruised parts with a lotion of arnica tincture, one part to ten parts of hot water.

Asafœtida. If the cow is very nervously excitable at any time during pregnancy give this remedy daily in the sixth dilution.

Cinchona. Give after abortion to check hemorrhage and to enable the animal to recover from the debility caused by abortion. Ten-drop doses of the first or third dilution.

Helonias. Very important in cases of threatened abortion and for enlarged uterus after abortion or parturition.

Opium. From great disturbance by fright give frequent doses of the thirtieth dilution.

Pulsatilla. If the vulva is swollen and there is an intermittent red flow give the thirtieth dilution every two hours.

Rhus toxicodendron. If the animal is subject to rheumatism or has taken cold in wet weather, subsequent to an injury, give the sixth or thirtieth dilution every four hours.

Ruta. Give to habitual slinkers, in alternation with Sabina.

Sabina. For abortion at the third month. For repeated early abortions, with profuse discharges, use the thirtieth dilution.

Secale cornutum. For thin, scrawny, sickly-looking cows, both before and after abortion, especially if there is violent straining, profuse flow and feeble pulse. Frequent doses of the thirtieth dilution.

Sepia. This may prove useful in restoring where there is a tendency to frequent abortion, especially with disorders of the mucous membranes. Give doses once a day of the thirtieth dilution.

Sulphur. Give a dose of this remedy in the sixth or thirtieth dilution when the system does not respond to any of the above remedies.

Viburnum opulus is also of great value for abortion and hemorrhage.

Viburnum prunifolium. This is a very useful remedy, and may be given as a preventive of abortion. Give once a day ten drops of the first or third dilution at any stage of pregnancy. Many are in the habit of giving enormous doses of this drug, a dram or more of the fluid extract daily, through the whole period of gestation, but this is inexpedient. Do not make your animals drug-sick with even a mild remedy. Try the efficacy of this remedy if you will in ten-drop doses of the tincture or one-drop doses of the fluid extract, and also the first, third, sixth and thirtieth dilutions. This will prove useful in hemorrhages after abortion.

FORMULA FOR TREATMENT OF ABORTION.

If your cow has aborted prepare one gallon of the Hyposulphite of Mercury solution as directed under Germicides, then scrub the floor with the disinfectant, or saturate the surface of the ground with the solution. If the cow does not clean readily give her ten drops, three times daily, of Pulsatilla, third dilution, mixed with four times its bulk of water.

Use the medicator and inject it upon the tongue if the cow does not drink it in water. The placenta will probably come away early enough without mechanical interference. If the medicine does not bring it away within three days it may be carefully removed by the placenta forceps. When she comes in heat give her a vaginal injection of hot water 130° Fahr. and follow with another of Hydrastis Can., ½ ounce; Listerine, ½ ounce; Water, 8 ounces. Mix. Keep the hind feet elevated, and fill the vagina with the injection from an elastic syringe. When her full time of heat is passed give her another vaginal injection.

Then give her three times daily until her next period, Sabina, third dilution, ten drops of a mixture containing four times its bulk of water. If she comes in heat regularly every twenty-one days you need not delay service with this treatment more than three months, but let it be given as soon as the first symptom is observed. The bull used should not be allowed to serve other than " slinkers."

SPECIAL RULES RELATING TO ABORTION.

1. Remove from the herd at once a cow that shows any symptom of impending abortion.

2. Quarantine all aborted cows and all threatening abortion (the two classes in separate buildings) for one year or until producing a healthy calf.

3. Employ special hands to attend each class of cows, and *quarantine these workmen from the rest of the herd.*

4. Use no quack nostrums in treating such cows, but follow carefully the directions given in this work.

5. If the afterbirth does not come away within three days let a skilful veterinary carefully remove it. If a veterinary cannot be had, the herdsman, if he be intelligent, may be able to do it. The arm and hand being well oiled with vaseline or almond oil, introduce the hand and pass it gently forward to the womb. If the hand, or even two fingers can be introduced within the womb, the adhering substance can be gently separated by pressing the edge of the fingers along the inner wall and by rotating the cord and membranes, remove it without any tearing or wounding of the parts.

BARRENNESS IN COWS.

Barrenness is doubtless most frequent as a sequel of abortion.

Inflammatory action within the uterus or the small tubes which receive and convey the ovules from the ovaries to the uterine cavity may result in producing adhesions of the surfaces of the lining membrane, thereby making obstructions or strictures which prevent either the semen or ovules from entering the organ, so that a union of the male and female germs is impossible.

The only cure for such a condition is effectual dilatation of all the strictures.

Dr. A. D. Newell, of New Brunswick, N. J., has devoted much attention to the study and surgical treatment of barrenness, and has treated several cases successfully.

I quote from an article published by him in the *Jersey Bulletin* of February 15th, 1885 :

" In almost every herd there are one or more cows that their owners fail to get with calf, even after the cow has calved once, and often using various bulls, large and small, usually throwing the blame on the bull. I am of the opinion that it is seldom the fault of the bull, but almost always the relative location of the male germ and ovum in the cow. The male germ must meet the ovum beyond the *os internum* or conception will not take place. I will mention only two of the main causes and opposite conditions of the *cervix uteri, os tincæ* and *os internum* that I find prevent conception. There are other minor causes. Conception cannot take place if either of these two conditions exist. One is where the *cervix uteri* is patulous or relaxed and lets out the male germ and ovum before it makes vital connection with the internal mucous membrane of the womb. The other is where the *os tincæ* or the *os internum* is closed, or so small as not to admit the male germ to the womb easily, and thus cannot reach the ovum to impregnate it in the womb. The usual length of the cervix of a cow is about one and a half inches. In a post-mortem examination of a cow killed for beef I found the *cervix uteri* full five inches long from *os tincæ* to *os internum*, a very unusual length. I have found quite a number that measure three and four inches, and with the *os internum* completely closed, some with *os internum* open and *os tincæ* closed. This great distance of *cervix uteri* to *os internum*, and its firm closure, with open *os tincæ*, has deceived me, and no doubt others, the *os tincæ* often being easily opened with the finger, and the extra depth of the *cervix* causing the operator to think he was through both *sphincters* and into the womb.

TREATMENT OF CLOSURE OF THE OS TINCÆ AND OS INTERNUM.

"Extract of belladonna will relax the *cervix uteri* when the tube is pervious, but no medicine will open the internal *os* when closed by a cicatrix caused by abortion or the rupture and tear of the mucous membrane near the *os internum* at natural calving.

"The whole mucous membrane that lines the womb is thrown off every time a cow aborts or calves, except just at the internal neck. I believe this torn condition of the membrane and its healing causes this cicatrix and closure.

"The canal to the womb must be opened by mechanical means. The parts are of a very delicate structure, and this must be done by very gradual easy dilators and a day or two before the cow comes into heat.

"I have not been able to find any dilators or sponge tents that will answer the purpose fully. The sponge tents were too soft, and gave before they could be got inside.

"The instrument had to be used with one hand, and that in the vagina, and so could not handle the instrument and at the same time keep the finger at the *os tincæ*, and thus prevent the instrument from catching into the folds and fossæ, and could not use gradual continuous pressure, and was uncertain when the canal was tortuous.

"To overcome these defects I made a metallic bougie two feet long, the end of flexible metal that could be bent to any sweep by the end of the right forefinger acting as a live guide to the *os tincæ*. With an arrangement at the end out of the vagina I can make the flexible point sweep to any course, and at the same time keep up a steady, continuous pressure at the obstructions.

"Some points are made of soft material, strengthened by internal broken joints that adjust themselves to any course by a simple rotation, so that there is no danger of wounding the canal. As soon as the canal is pervious I introduce sponge tents to make the canal larger and remain open.

"These tents should be made of tough sponge well saturated with gum-arabic and bound tight over a steel knitting-needle, to be removed when dry.

"Many of the worst cases of barren cows can be made to breed."

The above-mentioned instruments, consisting of a soft metal stylet, and a hollow flexible bougie, the doctor kindly sent to me for inspection. In the hands of a skilful operator and with the use of properly prepared sponge tents, doubtless many barren cows could be made to breed. But not until we have colleges which will develop the genius for surgical skill that lies dormant in possible veterinary surgeons, and a new order of surgical instrument-makers has been trained, can we look for many cures of these internal deformities. Under the best conditions, with trained surgeons, ingenious devices, rare instruments, and consummate skill in applying them, it may be possible that bovine uterine surgery will yet become a popular art.

In all cases of barrenness arising from functional derangement the cure must be sought by either sanitary or medicinal treatment, or both.

If all the conditions of healthful air, feed, exercise, warmth, dryness and sunlight are secured, and it is found that there is no stricture in the uterus or Fallopian ducts, the breeder must resort to medicines.

We have much to learn in this department of medication, but will offer to suggest a few remedies for investigation.

ALETRIS FARINOSA; DAMIANA; LACHESIS; PULSATILLA; RUTA GRAVEOLENS; SABINA; SECALE CORNUTUM; SEPIA; USTILAGO MAYDIS; VIBURNUM OPULUS; VIBURNUM PRUNIFOLIUM, and XANTHOXYLUM.

Use only one remedy at a time. Give Aletris, Damiana and the Viburnums in the mother tincture, from one to three drops daily, as uterine stimulants.

Give Xanthoxylum where the animal does not come into heat, using five drops of the *first* or third dilution.

Give Sabina, 30, where you suspect abortion in the first to the third month.

Give Secale, 30, and Ustilago, 30, in lean, scrawny, sickly animals.

Give Ruta, 30, in all cases where you suspect a tendency to a persistent habit of abortion.

Give Lachesis, 30, and Sepia, 30, for fetid vaginal discharges or suspected diseases of the mucous membranes. Always mix the medicine in a little water, and insert it in the mouth of the cow by the injector, if she will not drink water.

A cow that is barren from an enormous accumulation of fat may perchance become fruitful by reducing her to the condition of flesh requisite in a milking animal. This may be accomplished by abundant exercise and suitable feeding with hay and straw.

PROLAPSUS UTERI—EXTRUSION OF THE WOMB.

This is a displacement which is sometimes very troublesome, and unless properly treated may cause the death of the animal or become a chronic ailment.

It most frequently occurs in an aggravated form in those cows having a badly formed rump. The ligaments of the uterus, from various causes, become relaxed or stretched, the vagina loses its elasticity, and the uterus during gestation almost protrudes from the vulva. Parturition may be passed safely and the uterus completely extruded within a few hours afterward.

The uterus should always be immediately returned by a hand and arm well anointed with vaseline and pressed forward to its place.

The cow should stand with the hips elevated. Hot-water injections at 130° should be given once a day, cleaning out the vagina, while the discharge lasts. Several quarts may be used each time.

The cow should receive internal medical treatment for several months, and ought not to be admitted to service again within nine months.

REMEDIES.

Calc. carb. For general flabbiness and relaxed condition. One dose daily of the thirtieth dilution.

Conium maculatum. Chronic enlargement and hardening of the womb.

Helonias dioica. A very important remedy for chronic *prolapsus uteri*, with enlarged uterus. Give the thirtieth dilution, a dose of ten drops once a day.

Nux vomica. Constipation, or alternate diarrhœa and constipation accompanying the conditions. Use, as above, the thirtieth dilution, to give tone to the uterus.

Tabacum. Excessive relaxation of the whole system; it seems impossible to keep the organs in place. Use one dose daily of the thirtieth dilution, and apply a bandage or truss if necessary.

Viburnum prun. Chronic hemorrhage from womb.

HÆMATURIA—REDWATER—BLACKWATER.

Bloody urine is common among cattle in certain localities where the land is wet and the pasture poor, and is especially prevalent in rainy seasons with animals that are badly nourished. It is characterized by an impoverished condition of the system.

The disease may also be caused by eating many plants that have an inflammatory action upon the kidneys and bladder.

Oftentimes inflammatory action with this condition of urine may result from a mechanical injury, by sprain or by blows, or various other causes.

Acute Redwater is always an inflammatory condition of the kidneys resulting from one of the above-named causes.

Chronic Redwater, a still more common disease, is characterized by inflammation of the kidneys, and is more difficult to remedy.

Acute Redwater may result from injury or neglect during calving, or bad results following delivery.

It occurs rarely on well-drained lands with well-fed cattle.

SYMPTOMS OF ACUTE REDWATER.

Fever; rapid breathing; cold ears, feet and legs; dry, hot muzzle; tenderness over loins; loss of appetite and cud; bent back; straining to discharge urine, which is very scanty or bloody.

There is at first a bloody diarrhœa or dysentery, afterward obstinate constipation of the bowels and pure bloody discharge from the bladder, which gradually becomes darker or quite black, and may become fetid because of gangrene of the parts.

SYMPTOMS OF CHRONIC REDWATER.

Jaundice; languor; collapsed belly; animals want to be alone; the ears are cold and drooping; eye turgid and yellow; quick pulse; diarrhœa, followed by constipation; emaciation; urine at first yellow-brown, then red, dark brown, and finally black; the discharge is by a fine stream, but copious, with or without straining; milk brown-yellow and lessened, with bad flavor. Sudden remissions and recurrences may continue for months.

TREATMENT.

Remove the animal to dry, comfortable quarters, and give good rations, accompanied with linseed gruel, three times daily.

Arsenicum. In advanced stages with fetid diarrhœa.

Camphor. For chilliness and prostration give drop doses of the third dilution every hour.

Cannabis sativa. For bright, bloody discharges of urine use drop doses of the third dilution every hour.

Cantharis. Terrible straining, with bloody urine.

Terebinth. Bloody urine. Third dilution, one drop every two hours.

TREATMENT OF DIFFICULT PARTURITION.

The too-long-continued pains, the convulsive violence of the efforts, the straining after delivery, excessive hemorrhage, and any other irregularities call for medical treatment.

Pulsatilla. When the pains are slow in developing, or there is fear of a mal-presentation in the first stage of parturition, give ten-drop doses of the third, sixth or thirtieth dilution every hour to facilitate delivery. It is always safe and often greatly aids delivery.

Chamomilla. If the animal is irritable because of the pains and the labor is very slow, give, after Pulsatilla, a dose of the thirtieth dilution.

Opium. If the pains are very sluggish or cease for very long intervals give the sixth dilution.

Viburnum prunifolium. For lack of tone in the uterus, or for excessive hemorrhage resulting therefrom, give five-drop doses of the tincture every three hours.

Viburnum opulus. This may be preferred by many to the above.

Secale cornut. The pains are accompanied with excessive straining. Or the pains are intense with straining after the placenta has been several hours delivered (after-pains). Give doses of the thirtieth dilution after each pain.

Pulsatilla and **Secale** in the thirtieth dilution promote delivery of the afterbirth. Any serious delay in the delivery of the calf or of the placenta requires manual interference, which should be given with the utmost care and gentleness. If the womb is inverted or extruded from the vulva it should be very gently returned with a well-oiled hand. The cow should be kept standing for many hours or a supporting bandage applied. Doses of arnica should be always administered after parturition, ten drops of the third dilution every three hours, and if the vulva has been bruised or lacerated lotions of arnica should be applied to the parts. Sixteen parts of hot water to one part arnica tincture.

Helonias. This remedy should follow arnica for extrusion or falling of the womb. Use the third or sixth dilution.

COLOSTRUM APOPLEXY—"MILK FEVER."

If the cow has been properly fed and not too fat, and the digestive organs are in full health and free from constipation and flatulence, and care is given to protect from taking cold, she is not liable to colostrum fever or apoplexy.

Watch the pulse by placing the finger on the temporal artery near the outer angle of the eye, or by applying the hand to the left side of the chest beneath and behind the elbow. The normal pulse of the cow may vary from thirty-five to forty-two beats a minute. If the pulse rises rapidly to fifty or sixty beats there is much constitutional disturbance; if it rises to ninety or one hundred beats the case indicates peril. Apply the fever thermometer to the rectum; if the temperature is $101\frac{1}{4}°$ or 102° there is no danger, but if it suddenly rises, and the rise is progressive, the danger increases with each degree and fraction of a degree. The udder should be relieved of its colostrum after the calf has sucked, by a thorough stripping.

HOT-WATER TREATMENT.

If the udder is hard and swollen it should be immersed in a bag containing hot water at a temperature of 125°. Hot water should be applied to the crown and nape of the neck and the spine by saturated sponges or cloths at a temperature of 140° or as hot as can be used.

DRY CALORIC TREATMENT.

The following is communicated for this work by Mr. F. Loeser, New York, who translated it from the German:

"The Hanover *Agricultural Gazette* contains the following article regarding the treatment of milk fever from Mr. von Rhedn:

"'A few days ago, in one of my stables one of my best cows was taken with milk fever, a violent and apparently hopeless case. I called veterinary Meinberg Grönau. He ordered treatment used by a veterinary with success in Baden, and published in the *Veterinary Journal*.

"'The cow was covered with a woollen blanket, and a common smoothing-iron heated very hot was passed along the spine, and repeated continuously at the highest degree of heat, from 10 A.M. until 8 P.M., when the cow arose and commenced to eat.

"'During the night, in spite of the constant use of the iron, a relapse occurred, but the persistent use of the treatment was successful in the recovery of the cow.

"'The success is explained by the supposition that there is a collection of fluid along the spine, and that this excess of fluid is dissipated by the heat. The woollen blanket is essential to protect the cow from injury. Veterinary Grönau treated four cases successfully. The Baden veterinary claims a recovery of three fourths of all cases thus treated.'"

MEDICAL TREATMENT.

Aconite. If the cow is an extraordinary milker watch her closely; if she refuses food, the horns become hot, the muzzle dry and hot, and the pulse increases, continue to apply the hot water to the head and the udder, and give *Aconite* every fifteen minutes, five drops of the sixth dilution.

Ammonium causticum. If the rumen becomes distended and the breathing difficult, the pulse weakens, and the cow seems in pain, give this remedy, ten drops every fifteen minutes until the swelling subsides. Use the first watery dilution.

Belladonna. If the cow seems wild or furious and the pupils dilate give ten drops of the sixth dilution every fifteen minutes.

Gelsemium. If the pupils are widely dilated, so that the animal cannot see, and the eyes are bloodshot, give every ten minutes a dose of the third dilution. The medicine is to be administered in a spoonful of water without elevating the cow's head too high, for fear of strangling.

Arsenicum. In the drowsy stage; insensible to pain; glassy eyes; open mouth; inability to hold up the head. Give a dose of the thirtieth dilution every fifteen minutes. In all cases if decided improvement follows a dose lengthen the time of administering medicines.

Rhus toxicodendron. For a total suppression of the lochia, or discharge from vulva, or for signs of paralysis of back and legs, give frequent doses of the sixth or thirtieth of this valuable remedy. It should restore the discharge in a little while. If the discharge afterward lasts too long, and becomes offensive and ichorous, or bloody, give the Rhus again.

Nux vomica. If the fever has subsided, and the cow lies comparatively at ease, but with loss of muscular power, give a dose of the sixth dilution in water every four hours.

SPECIAL DIRECTIONS.

The cow should be treated in a well-bedded box-stall, with abundance of fresh air of the right temperature. If in the first stages the fever and temperature threaten to reach a high figure the cow should be enveloped in a rubber blanket and wet with water at 102°, or the normal temperature, while water at 140° is applied to the head and also to the udder.

All excretions should be removed at once.

The temperature of the stable should be kept at about 65°.

The udder should be milked out every four hours.

If the cow cannot pass urine the catheter must be used every six hours.

Her head should be supported with bundles of clean straw.

In no case should she be allowed to get cast or to lie extended with the legs stretched out; but she should be placed in such position as to favor easy respiration.

The water if used as directed at the right temperature will greatly assist in allaying congestion, restoring a normal circulation, and abating the fever. Especially excellent is the application of hot water to the head and neck in conjunction with such remedies as *Aconite, Gelsemium* and *Rhus.* Hot injections will sometimes prove useful.

MILK DISEASES—RED MILK.

Galactohæmia is an imperfect secretion wherein milk or colostrum and a red secretion are commingled. From some defect in the udder, or from a diseased condition resulting from over-stimulation by improper food, the secretion becomes imperfect and the organs are unable to secrete milk by the proper transformation of the blood.

It is the theory of some physiologists that food which is radically deficient in potash may be a cause of the disorder. The cow should always be well fed on the most wholesome food in order to avoid the development of this disease. When the disease occurs put her on a diet of good clover or cow-pea hay, cut and moistened, and give with it twice a day one quart of rye bran and one pint of linseed-cake meal, or pea or bean meal.

Argentum nitricum. The calf does not thrive or refuses the milk.

Asafœtida. Deficiency of milk with over-sensitiveness.

Borax veneta. Milk curdles soon after being drawn, tastes badly or has an offensive odor. Give third to sixth dilution three times a day.

Calcarea carbonica. Deficient or very scanty milk, with a distended udder.

Calcarea phosphorica. Milk tastes saltish; milk acid, thin, watery, neutral; udder sore to the touch; teats sore on pressure.

Causticum. Milk almost disappears from fatigue after long driving. Sixth dilution.

Chelidonium. Milk diminished.

Cinchona. Debility from excessive flow of milk. Third dilution.

Dulcamara. Suppressed milk from taking cold.

Ferri phosphoricum. Debility; want of appetite; cough.

Ignatia. Milk suppressed; homesickness; lowing for loss of calf. Third dilution.

Kali hydriodicum. Bloody milk, with wasting or diminishing of the udder. Third dilution.

Kali carbonicum. Give ten drops three times daily of the first dilution in the water drank, as long as the milk is bloody in appearance.

Lachesis. Milk thin and blue. Thirtieth dilution.

Millefolium. Total suppression of milk. Drop doses of the mother tincture.

Phosphoric acid. Scanty milk, with apathy and great dulness; debility from excessive milking. Give drop doses thrice daily in the water drank.

Phytolacca. Stringy milk; offensive odor in milk. Always give in garget or threatened garget. Give first or third dilution three times a day, ten drops.

Pulsatilla. Sudden suppression of milk. Sixth dilution.

Urtica urens. Entire want of secretion of milk after parturition.

CONSTIPATION.

Constipation is a term applied to a loss of power in the intestines by which the stools are difficult, or altogether obstructed.

The dung may become dry, hardened or impacted, or it may be soft and adhesive.

Among the sources of constipation in cattle is a diseased condition of the intestine caused by poisoning or frequent irritation by the use of "cathartic" drugs. One dose is sometimes a sufficient cause to induce the habit.

Constipation may be habitual in an animal of feeble constitution. It may occur as a result of many acute diseases. It is frequently caused by improper feeding, as giving dry rations of woody hay, or in the reaction after an excess of laxative food. It may be induced from impure air in a close, dark stable, from insufficient exercise, or from any cause that impairs the nervous force of the animal. Cows are especially prone to constipation in the last month of gestation. This should be avoided, as it is one of the conditions tending to produce apoplexy in the *colostrum* period, or the three days after calving. Give a sufficient amount of laxative food, such as sweet grass or green forage plants or cabbages, carrots and sugar beets, and linseed meal.

MEDICINES FOR CONSTIPATION.

Aconite. Dryness of the nose; much thirst; constant restlessness, especially if the animal has had a fright. Dose, ten drops of thirtieth dilution.

Alumina. Great straining, with soft adhesive stool; torpor of the rectum.

Belladonna. Congestion to the head; injected eyes; intolerance of noise and light. In acute diseases. Dose, ten drops of thirtieth dilution.

Bryonia. Stools dark, dry and hard; much thirst. Especially if the animal is lame or dreads to move because of soreness of any part, or in rheumatism or simple fever. Dose, ten drops of the sixth or thirtieth dilution.

Lycopodium. Great gurgling of wind in the bowels. The thirtieth dilution.

Nux vomica. After use of cathartics; alternate constipation and diarrhœa.

Opium. Stools very small, hard and black; general torpor of the system. Use the thirtieth or two hundredth dilution.

Plumbum acet. Stools compacted like sheep's dung, accompanied with violent colic pains. Use the two hundredth dilution.

Pulsatilla. Obstinate constipation after a severe attack of diarrhœa, especially in calves.

Ratanhia. Most obstinate and long-continued constipation. Use the sixth dilution.

Selenium. Stool so large and hard that it has to be removed by mechanical aid; shreds of mucus that look like hair in stool.

Sepia. Terrible straining; stool covered with mucus. Especially in calves or for cows in last month of gestation. Thirtieth dilution.

Silicea. The stool is lumpy and requires a number of severe efforts before it can be expelled. Thirtieth dilution once a day.

Sulphur. In all cases where other remedies fail to act give once a day a dose of the thirtieth dilution.

Zinc. Stools remarkably dry, hard and insufficient.

It may sometimes be found necessary to remove fecal matter by a small scoop or by the introduction of the hand, well oiled. Or an injection of warm water and molasses may be thrown well into the rectum by a long flexible tube attached to the syringe. But it is best to so care for the health of all animals that such severe cases of constipation shall never occur. With a sufficient knowledge of feeding and by avoiding the old-fashioned cruel dosing it will be rare to meet with a severe case of constipation. Nothing more laxative than a little boiled flaxseed should ever be allowed in the treatment of cattle, and that with caution. Use carrots and other roots when they are in season.

The drugs for constipation are given in minute doses and at long intervals. They induce reaction by gently stimulating those portions of the nervous and muscular systems that are impaired or lacking in tone.

Extreme care, guided by knowledge and experience, with good judgment and prompt decision, are necessary factors in a good stock feeder. If, added to this, he can gain a fair knowledge of drugs and their use as *cures* he becomes well fitted to be the friend of good animals.

RHEUMATISM.

Rheumatism is a disease which attacks cattle more frequently than other domestic animals. The malady arises from malarial blood-poisoning in conjunction with a cold, moist atmosphere, or cold basement stables. The conditions are identical with those of the same disease in the human subject. A person who perspires easily and lives in a damp dwelling will scarcely escape rheumatism in some form. To prevent the disease is better than to try its cure. To guard against rheumatism in choice Jerseys, have only dry stables above ground. A barn with a basement will do more annual damage to a good herd, by causing rheumatism and other maladies, than it would cost to build an expensive sanitary stable.

CHARACTER AND SYMPTOMS.

Rheumatism irritates and inflames the joints, muscles, tendons, sheaths of nerves, and particularly the heart and heart-sac and pleura. It is painful in the highest degree, and may attack the healthiest animals, and become chronic. It is commonly of the chronic form in cattle, owing chiefly to lack of proper care and right treatment. In the acute form it may prove speedily fatal, especially if it attacks the heart. In acute rheumatism the animal is very restless, loses appetite, has dry skin, constipation, and apparent stiffness of joints and muscles.

The force of the disease may be expended chiefly upon one joint, with more or less painful swelling. The disease may move to other joints and muscles or suddenly change from part to part. This sudden transition from one part to another is characteristic of acute rheumatism.

Chronic rheumatism causes extensive structural changes in one or more joints, and sometimes causes abscesses, especially in the knee, when the joint may become enormously enlarged.

TREATMENT.

Aconite. Very useful in the acute form when there is much fever and soreness, especially if the disease is ushered in by shivering with great disturbance of the pulse. Give five drops of the third dilution every two hours until improvement begins.

Ammon. phos. Pain in the joints and spine, with great nervous irritability.

Give drop doses in water every four hours.

Arsenicum. Where there is great debility, with anxiety and restlessness, especially at night, and when hot applications relieve; *profuse perspiration*, and change to the heart. Use the third, thirtieth or two hundredth dilution.

Belladonna. High fever, hot, dry skin, and extreme soreness to the touch. Use the third to thirtieth dilutions.

Bryonia. Pain that is continually worse from the slightest motion; pains affect the legs, shoulders and ribs; thirst; stools very dry; breathing short; urine very red; *great dread of moving.* Use the third or thirtieth dilution in frequent doses. A very valuable remedy.

Comocladia. Great languor; painful swellings. Sixth dilution.

Cimicifuga. Very important for pains in the side and chest, as well as in all the joints. Use the first, third or thirtieth dilution.

Calcarea carb. Useful, and is needed in chronic rheumatism of the joints.

Calcarea phosph. Needful to complete a cure after other remedies have failed.

Chamomilla. Muscular pains, with great irritability of disposition, and especially to restore the milk which is usually suppressed in cows. Give the thirtieth dilution.

Gelsemium. Often a great relief for the *severe pains at night*, for partial paralysis, or great loss of muscular power; rheumatism of the legs, with great weakness, and little fever.

Give the first, third or thirtieth dilution, ten drops three times daily.

Phytolacca. Chronic rheumatism in damp weather; swollen glands, bone pains worse at night. Give the first or third dilution, three times daily, in ten-drop doses.

Rhus tox. Pains caused by wet weather or damp stables or from straining or injuries; the pains are worse during rest or when first beginning to move; better from exercise, warmth, external applications. *There is always great languor and excessive restlessness.*

Give the sixth or thirtieth dilution, in ten-drop doses.

Rhododendron. Bone pains in stormy weather.

Ruta. Pains in spine and legs.

Silicea. Chronic swelling of the joints and for abscesses or diseases of the bones in old cases. Give the thirtieth dilution once a day.

Spigelia. Inflammation of the heart and heart-sac.

In all cases the animal should be put into a dry, warm stable and have extraordinary care. The food should be very light until recovery is complete.

HOOVE—TYMPANITIS.

Ammonium causticum. Violent spasms of the stomach; difficult swallowing; panting breathing; violent trembling; rapid pulse; sudden starting. Give from five to ten drops of the watery solution in half a pint of water every fifteen minutes until relief is produced.

In all cases use hot injections at 140° Fahr.

From two to eight quarts of hot water may be gently injected, using an elastic syringe with a long flexible tube, to be passed into the rectum. This will speedily allay the pain and inflammation.

Colchicum. Swelling and puffiness of the belly; alternate heat and coldness; scanty, red urine; stool very hard and dry or loose, with mucus and blood, preceded by severe colic pains. Uneasy, constantly changing position; pawing the ground; stamping the hind feet; lies down and gets up, turns from side to side; the hair stands on end; great distress when the animal attempts to urinate; tender to the touch; full of wind; great distention. Give ten drops of the first or third dilution every fifteen minutes.

Colocynth. Paroxysms of violent colic every half hour or at shorter intervals. Loose, thin stools.

Give the sixth dilution, ten drops at each paroxysm.

Hyposulphite of Soda. Give first trituration for fermentation of food in stomach.

Lycopodium. Weak stomach; animal has frequent attacks of indigestion; great rumbling and rolling of wind in the bowels. Give a dose of ten drops of the thirtieth dilution every half hour in acute cases, and once a day for chronic indigestion. This remedy should cause flatulence to be discharged freely.

Nux vomica. Give during last stages or for chronic indigestion, especially if the animal has had cruel treatment by excessive dosing with violent medicines. One dose a day at night of the third, sixth or thirtieth dilution.

When remedies fail to relieve the hoove, and the rumen remains inflated or the tympanitis increases, a trocar and canula must be inserted, after making a small cut through the skin, and penetrate to the interior of the rumen and allow the gas to escape. The opening should be made midway between the last rib and the hip, and about nine inches below the transverse lumbar bones.

ACCLIMATION FEVER.

Aconite. Simple fever; restlessness; thirst; timidity. A dose of the thirtieth dilution daily.

Nux vomica. Depraved, fastidious or capricious appetite, with constipation.

SUN-STROKE.

Glonoine. If a cow seems dizzy or falls off in her milk, or stops her cud after being a few hours exposed to the sun in very sultry weather, there is danger of great injury. She should be treated precisely as for sun-stroke in the human subject, by applying, every five or ten minutes, to the crown and neck sponges saturated with water at from 130° to 140° temperature, and a dose of Glonoine, thirtieth dilution, every fifteen minutes until all signs of danger are passed.

GARGET.

Phytolacca. Udder distended, hard and hot. Give every two or three hours ten drops of the first, third, or sixth dilution in a little water. Wash the bag with a lotion of hot water and Phytolacca, using one teaspoonful of the tincture to a pint of water. After bathing the udder for ten or fifteen minutes with gentle friction and manipulation of the milk-glands, immerse the whole udder in hot water (125°) by means of a rubber bag. The process of bathing and immersing should be repeated several times a day until the udder recovers a normal condition. Milk-tubes may be used to advantage in some cases.

SORE TEATS.

Arnica. The teats have been scratched or bruised. Give the first dilution internally and a weak lotion externally. Milking tubes may be necessary.

Chamomilla. The teats are inflamed and very tender; hard, knotty tumors in the udder. Give a dose of the sixth dilution three or more times a day.

Hydrastis. Ulcers on the teats which will not heal. Use the sixth dilution internally, and for a lotion one part of Hydrastis tincture to a hundred parts warm (102°) water. Apply three or more times daily, and follow with Vaseline.

WARTS ON TEATS.

Thuya. Give a dose of the sixth or thirtieth dilution daily. Apply a lotion of one part Thuya tincture to sixteen parts warm (102°)water daily. Always give Thuya internally for warts, and continue the remedy for six weeks, one dose daily. If the warts do not disappear they may be carefully touched with chromic acid or with nitric acid, using the point of a small wooden toothpick, always taking care that the teat is covered by slipping a piece of kid leather over the wart. The warts may also be destroyed by applying a concentrated solution of wood-ash lye in the same way.   Always use the Thuya internally, however.

Ol. Ricini.   Castor Oil applied externally twice a day is often effective.

INDIGESTION IN CALVES.

Owing to the artificial methods which obtain in the rearing of calves it will readily be seen that there are many causes for indigestion.   From the use of artificially prepared food and the too rapid swallowing, and the numerous incidental changes that accrue, disordered assimilation and indigestion may depend upon, 1. *Insufficient or altered saliva ;*   2. *Deficient action of the gastric juice ;*   3. *Deficient action of the pancreatic juice ;*   4. *Disordered liver ;*   5. *Deficient action of the intestinal juice ;*   6. *Nervous irritation ;*   7. *Altered blood supply.*

PREVENTION OF INDIGESTION IN CALVES.

It is of the highest importance that the digestive organs of the young Jersey should always remain in the normal state of perfect health.

The breeder, to be successful in the management of his young stock, cannot neglect this point without suffering disastrous consequences, in the loss of his most valuable animals and the deterioration in quality and vigor of his whole herd.

The powers of digestion and assimilation must have their full development in the young calf through good management.   Serious diseases, accruing from carelessness and ignorance, will surely follow even slight neglect.

Milk, according to the eminent chemist, E. Duclaux, is only assimilated by animals after it has received treatment by two ferments—*rennet* and *casease*.

This noted chemist has not only demonstrated this proposition by numerous experiments, but practical breeders and others have shown that calves may be kept from indigestion, and the violent diseases resulting therefrom, by the punctual addition of a small quantity of prepared rennet after the milk has been warmed for feeding.   It is well to have some arrangements by which the calf will be compelled to drink the milk slowly, so as to mix it with the secretions of the mouth.

For this purpose an artificial teat, made of rubber, and attached to a wooden float, is placed upon the surface of the milk, and the calf sucks the fluid at leisure.

It is necessary, however, to pay strict attention to the cleansing of this instrument, by washing and scalding, after each feeding.

The calf must also be placed in a very dry, well-ventilated stall, provided with plenty of soft bedding and an abundant admission of sunlight.   If all these requirements are met and steadfastly followed by all Jersey breeders the results will be of immense benefit to every one of them, and the Jersey interest in America will be greatly promoted and rendered highly remunerative.

With the above treatment of the milk, calves will require full rations, according to size and rapidity of growth. But if from climatic or other unknown causes gastric disorders appear, diminish the ration from one half to two thirds, and give, in addition to the rennet, the medicines prescribed for indigestion, constipation or diarrhœa, according to the indications given in those sections of this work relating thereto. In addition to the foregoing, let it be remembered that calves require the free use of soft, pure water to quench thirst, and this must be amply provided for them, and always at a temperature of about from 65° to 70° Fahr.

In addition to the former simple preparation of rennet I would suggest that a preparation be made which can be used in the form of a powder, making a permanent, long-keeping article, which may be styled LACTO-RENNETINE.

FORMULA.

Pure milk sugar	40 ounces.
Pure rennet	10 ounces.
Pancreatine	5 ounces.
Ptyalin	4 drachms.
Lactic acid	5 fluid drachms.
Hydrochloric acid	5 fluid drachms.

I believe that such a preparation, if made of pure articles, would excel, in beneficial effects, the rennet alone, and produce always uniform results.

DIARRHŒA IN CALVES.

Aconite. Stool bloody, slimy, mucus, small, frequent; worse after exposure to cold, dry winds; after fright; after being overheated; after getting wet in rain; in summer after cool nights; restlessness; great thirst; dry heat; quick pulse. Dose of sixth dilution after each stool.

Aloe. Involuntary stools after feeding; pain in stomach; loud gurgling in the belly, like water running out of a bottle. Sixth dilution.

Arsenicum. Stool thin, watery, frequent, scanty; *worse after feeding with cold milk; great restlessness; great thirst;* weakness; emaciation; rapid exhaustion. Thirtieth dilution.

Baptisia. Stool of pure blood; low states; fevers. Tincture, five drops.

Benzoic acid. Stool watery and white; copious; very offensive and pungent odor; urine dark colored, and very strong smelling.

Bryonia. Undigested stools; aggravated by warm water and by moving about. Desire to be quiet.

Camphor. Stools involuntary; attack very sudden; worse from hot sun; sudden collapse, with coldness of the whole body. Use first or third dilution.

Capsicum. Stool of tenacious mucus, frequent, small; worse after feeding, with cutting colic; difficult urination; shivering after drinking.

Carbo vegetabilis. Stools thin, frequent, putrid; worse from cold milk or cold water; flatulent distention of the belly; coldness; collapse.

Chamomilla. Stools hot, small, frequent, with smell like rotten eggs; worse from taking cold; colic.

Cinchona. Undigested stools; frequent, involuntary, painless; worse after feeding; colic from gas in belly; distention of the belly; great weakness; sweating.

Cina. Diarrhœa, with pin-worms.

Cistus. Thin, hot, squirting stools.

Colchicum. Stools watery or mixed with white mucus; profuse; worse in hot, moist weather; straining; colic; distention of the belly with gas; weakness; prostration.

Cotoin. Where Arsenicum temporarily relieves; persistent chronic diarrhœa.

Colocynth. Frothy, liquid stools; sour, putrid, musty; worse from cold milk. *Cutting, violent colic, which makes the animal bellow with pain.* Give frequent doses of the third or sixth dilution until colic is relieved. Use injections of hot water at 130°.

Croton tiglium. Watery stool, *coming out like a shot; worse after drinking cold milk.* Better from a moderate quantity of hot milk at from 120° to 130°. Colic, with writhing pains; soreness of the intestines. Add rennet to milk at time of feeding.

Dioscorea. Watery stools; profuse; violent, twisting colic, occurring in regular paroxysms with remissions; colic relieved by pressure on the belly, by walking, and by rubbing. Use injections per rectum at 130°.

Hamamelis. Stool of pure blood. Tincture, five drops.

Hepar sulph. Whitish, sour-smelling stools; painless; *indigestion; chronic diarrhœa.* Use the sixth or thirtieth dilution. A grand remedy. Add a little rennet before feeding the milk.

Lycopodium. Stools thin, fetid, painless; worse after a feed of milk; the belly fills with gas from very little food; rumbling of wind in the belly; weakness, emaciation, prostration.

Nux vomica. Stools thin or bloody, alternating with constipation; worse from too much drugs; urging and straining constant; colic and griping; much gas in belly; emaciation.

Opium. Offensive, involuntary stools; worse from fright or any excitement. Use third or sixth dilution.

Phosphoric acid. White, watery diarrhœa, painless. Dilute acid, one tenth in water. Add a little rennet to the warm milk.

Podophyllum. Profuse, frequent, gushing stools.

Pulsatilla. Stools *watery, greenish, yellow or very changeable; very frequent; loss of appetite; emaciation; chilliness; worse at night; worse from poor milk;* painful, rumbling flatulence.

Sepia. Worse from boiled milk; rapid exhaustion; chronic diarrhœa.

Silicea. Liquid, slimy, frothy stools; worse from exposure to cold air; better from warmth; milk not digested; *hard, hot, distended belly; emaciation.*

Sodæ hyposulph. For fermented stools give the first trituration.

Sulphur. Watery, undigested stools; changeable, frothy, sour, fetid. Expulsion sudden or involuntary; worse in the early morning; worse after a feed of milk; colic; straining; should be given when other remedies fail to produce their usual effects. Give doses of the thirtieth dilution in a little water.

Thuya. Stool forcibly expelled; copious; gurgling, like water from a bunghole; worse after feeding; rapid exhaustion; rapid emaciation.

Veratrum album. Stools frequent, profuse, thin; worse at night in hot weather; *preceded by pinching colic;* great sinking and weakness; skin cold; prostration; collapse.

Calves should be put upon a smaller allowance of milk as soon as indigestion or diarrhœa is indicated. *The milk should always be sweet, the feeding vessels scoured, rinsed, scalded, rinsed again, and dried in the sun.* If colic occurs the animal should be fed but very lightly twice a day. Every effort should be made to bring the calf back to good health as soon as possible. As soon as the first signs of diarrhœa occur a dose of sulphur (sixth dilution) may be given, unless the symptoms indicate some other remedy, and a raw egg may be beaten and mixed with the milk. If the case of any calf seems desperate do not give it up till every effort fails. If there is violent colic and collapse apply hot-water bottles to the belly, legs and back. Rub the belly with a roller and give the remedies as described above as long as the calf will swallow. You will probably save the calf by such persistent effort and careful nursing, with the medicines given as directed. Keep the calf blanketed in cold weather.

SELF-ABUSE IN THE BULL—SPERMATORRHŒA.

This bad habit of self-abuse in bulls is very common, and the results are disastrous to the breeder. The conditions are owing to an excessive irritability of the seminal vesicles, or to a general disorder of the sexual system.

Among the causes of this form of spermatorrhœa are too early service; excessive service; solitude in dark stable; insufficient daily exercise in the open air; worms in the intestines or rectum. The animal becomes so excitable that the sight of another animal or the presence of a human being causes a sudden effort and an ejaculation. Such a frequent drain from the system of this important vital secretion soon produces a change of character in the animal. If he does not become speedily impotent he is an uncertain server and slow, his disposition becomes either very sluggish or very desperately ugly and fierce, and his progeny are necessarily inferior. He becomes susceptible to acute and fatal diseases, and sooner or later becomes permanently impotent.

TREATMENT.

Avoid all causes of self-abuse, and stop the cause that induces it as soon as it can be ascertained. As soon as the first indications of such a habit present themselves prepare and adjust a broad leather girth around the loins, with a piece of fine steel chain about a foot long set in the girth so as to cross the spine in front of the hips. This must have a buckle and be so arranged as to fit just tight enough to bring a strong pressure of the chain upon the back at the first effort.

MEDICINES.

I would recommend trial of the following remedies to assist in curing the bad habit at its very beginning:

Camphora. Great depression and lack of power. Give first, third and thirtieth dilutions in rotation daily.

Cantharides. Great irritability of the sexual organs. Give ten drops third dilution three times a day.

Cina. Irritability from worms in rectum. Give first, third or thirtieth dilution daily. Also inject into the rectum a pint of warm milk in which has been mixed a drachm of tincture of aloe; once a week it may be repeated.

Damiana. Great seminal excitability. Use the sixth dilution.

Gelsemium. Want of irritability; relaxation; lack of tone. Give ten drops of first or third dilution three times a day.

Nux vomica. Indigestion; constipation; lack of vital force. Give first third and thirtieth dilutions in rotation daily.

Phosphoric Acid. Very frequent emissions on the slightest excitement. Use the pure acid, one part to one hundred parts of pure water, ten drops of mixture twice a day.

Picrate of Zinc. When the lack of energy and true vigor threaten impotency. Give the sixth or thirtieth dilution in ten-drop doses twice a day.

Sulphur. When irritability increases in spite of the use of any remedy, or when the system does not respond at once to the remedy selected. Give the first and third or thirtieth *triturations* in rotation once a day.

The Butcher is the last remedy and a sure cure.

COUGHS.

BRONCHITIS.

Lung diseases are too common among dairy cattle. A proper attention to the ventilation and temperature of stables, and the protection of cattle from rain and cold winds, would prevent the majority of cases of bronchitis, pneumonia and tuberculosis, and also render cattle less susceptible to contagious pleuro-pneumonia.

Ample provision must be made for a perpetual supply of pure, unadulterated, ozonized air in the stable for every hour of every day of the year, and also for rendering such air of the proper temperature, so that the animals may never be chilled by cold currents or oppressed by too high a temperature.

Bronchitis is acquired from chilling air impinging upon animals while standing in their stalls, or from exposure to cold rains or to dry cold winds in autumn and winter. The cough is at first short, hard and dry, but soon becomes moist and more prolonged, with a varying degree of mucous secretion. This may become, if not properly treated, a chronic cough and last for months or years. Oftentimes a chronic dry cough is the result of nervous irritation, sympathetic or otherwise, and may be the result of worms or other parasites that excite a reflex action in various nerves.

TREATMENT.

Aconite. Fever; dry nose; restlessness; use in the first stages, especially if caused by exposure to cold, dry winds. Use the third, sixth or thirtieth dilution.

Apis mellifica. The cough is suffocative, painful, with much difficulty of breathing.

Arsenicum album. Cough remaining after influenza or catarrh; dry cough, with watery discharge from the nostrils. Give the sixth or thirtieth dilution.

Belladonna. Cough in the larynx or in the windpipe; painful cough from inflammation of the bronchial membranes, with fever and depression; dry cough.

Bryonia. Cough in larynx, windpipe and the large bronchial tubes; dry cough from irritability of the upper air passages, especially in morning; cough from pressure on windpipe; from exposure to cold wind or from the least exercise; cough that causes pain; breathing quick; phlegm frothy. Thirtieth dilution.

Causticum. Persistent hacking cough. Thirtieth dilution.

Hydrastis. Debility; mucus thick and ropy. Use the third or sixth dilution.

Ipecac. Rattling, convulsive cough, with difficult breathing.

Drosera. Deep, hollow, groaning cough.

Kali carb. Chronic cough.

Cina. Cough arising from worms. Animal presses its nose against the wall.

Iodide of Arsenic. Windpipe cough, with thin discharge from nostrils. Sixth or thirtieth dilution.

Mercurius sol. Catarrh of all the respiratory mucous membranes, but especially the larynx and nasal region. Third, sixth or thirtieth dilution.

Nux vomica. Nervous, spasmodic cough; a chronic cough, arising from irritability of the digestive organs; discharge of flatulence from the rectum while coughing. Third or sixth dilution.

Opium. Convulsive, dry cough in paroxysms at night.

Pinus. Chronic, short, feeble, hacking or grunting cough. Use drop doses of the mother tincture, or a decoction of white pine needles and the tree bark.

Phosphorus. Cough arising from inflammation of the small bronchial tubes or lung substance. Chronic cough; dry, short, frequent, racking cough; distress from difficult breathing, with discharge of reddish or yellowish mucus. Third, sixth or thirtieth dilution.

Populus balsamifera. Chronic catarrhal cough. Use drop doses of the saturated tincture.

Rumex crispus. Cough in throat and windpipe; frequent cough from the slightest exertion.

Sulphur. When other remedies do not produce the expected result. Third. sixth or thirtieth dilution.

Spongia. Sharp, shrill, ringing cough, or dry, hollow, barking and hooping cough.

Tartar emetic. Chronic cough, where the whole respiratory organs seem loaded with a loose, rattling mucus.

PNEUMONIA.

Bulls seem to be especially prone to attacks of pneumonia. Alternate heat and coldness of the ears and horns; costiveness or diarrhœa; short, oppressed breathing; dry muzzle; dry, harsh, frequent cough; loss of cud; intense thirst; lassitude; discharge of water and mucus from the nostrils, and later on bloody or rusty discharges; brilliancy of the eyes; sensitive tenderness of the spine; continual change of heat, with or without shivering; partial or general sweating. In severe cases the panting becomes laborious; the flanks heave; the nostrils expand, emitting discolored, fetid mucus; the strength fails; the legs are drawn under the belly, which is contracted and puckered; the evacuations become putrid; the eyes have an offensive discharge; the pupils are dilated; the breath becomes cold, and the animal sinks.

TREATMENT.

Aconite. In the first stage, especially if the animal has been exposed to cold, dry wind, and is very restless.

Arsenicum. Great thirst ; prostration.

Belladonna. Congestion of brain ; dilated pupils ; drowsiness, with frequent starting, as from fear.

Bromine. When the lungs become solid like liver. Give the second, third or sixth watery dilution, ten drops every hour.

Bryonia. Pain and dread upon the slightest motion ; great thirst for large draughts.

Carbo vegetabilis. Rattling in lungs; great prostration ; fetid discharges, especially in last stage. Use third or sixth trituration.

Cuprum nitrate. Suffocative spells ; diarrhœa.

Ferri phos. In first stage. A very important remedy. Use third or sixth trituration.

Lycopodium. Sweat without relief ; fan-like movement of the nostrils ; rumbling of wind in the bowels. Use sixth or thirtieth dilution.

Phosphorus. In catarrhal pneumonia. Give the sixth or thirtieth dilution, ten drops every two or four hours.

Sanguinaria. Extreme difficulty of breathing ; tough, rust-colored mucus; pulse weak; extremities cold. Use the third, sixth or thirtieth dilution, with frequent doses.

Tartar emetic. Great rattling of mucus; much coughing; great suffocation. Sixth or thirtieth dilution, ten drops every hour.

Veratrum viride. When the pulse is hard and *very slow*. Sixth dilution.

Sulphur. When there is heard, by applying the ear to the chest, a *fine crackling* or *crepitant rattle*.

Give the sixth trituration, a teaspoonful of the powder every two or three hours.

TUBERCULOSIS—CONSUMPTION.

Lung tubercle, abortion and apoplexy are the three scourges of the dairy cattle-breeder.

When the diagnosis of consumption is clear the animal should be slaughtered and buried deep in dry soil. Such animals must not be used for breeding, as the defect would thereby be propagated, while the milk and flesh will be liable to communicate the disease to the human subject. This disease is most frequently generated by close, dark, non-ventilated stables, especially in malarial regions.

To prevent lung tubercle supply the stables with proper ventilation ; every animal requires twelve hundred cubic feet of fresh air each hour. The disease is contagious to a certain degree.   Quarantine doubtful cases, but do not fail to use the knife in every instance of tubercular development.

Among the remedies to be used upon doubtful cases are the following :

Calcarea carbonica.   Loose, rattling cough ; dulness of lung upon percussion.   Thirtieth dilution.

Ferri phos.   Congestion of the lungs, with dulness and frequent cough.

Hepar sulphuris.   Cough, excited by cold air.   Thirtieth dilution.

Iodine.   Emaciation ; cough.   Thirtieth dilution.

Iodide of Potash.   Dulness of lung ; cough, with thin discharge from nostrils.   Third dilution.

Jaborandi.   Very profuse sweating.   Third dilution.

Lycopodium.   Fan-like motion of nostrils ; rattling of flatulence in bowels; dulness of lung.   Thirtieth dilution.

Pinus.   Chronic hacking, or racking, dry cough.

Phosphorus.   Dry, tight, tormenting cough ; loose stools ; sweat.

Sambucus.   Profuse sweat.

CONTAGIOUS PLEURO-PNEUMONIA.

This is not a common disease, and under our efficient quarantine regulations it is not probable that the country will ever suffer very seriously from this much-dreaded malady.

Gamgee gives the following description of the symptoms of this disease : "From the time that an animal is exposed to the contagion to the first manifestation of the symptoms a certain period elapses : this is the period of incubation.   It varies from a fortnight to forty days, or longer.   The first signs proving that the animal has been seized can scarcely be detected by any but a professional man ; though, if a proprietor were extremely careful, and had painstaking individuals about his stock, he would invariably notice a slight shiver usher in the disorder, which for several days, even after the shivering fit, would limit itself to slight interference with the breathing, detected readily on auscultation (by the ear).

"Perhaps a cough might be noticed, and the appetite and milk secretion diminish. The animal becomes costive and the shivering fits recur.   The cough becomes more constant and oppressive, the pulse full and frequent, usually numbering about 80 per minute at first, and rising to upward of 100.   The temperature of the body rises, and all the symptoms of acute fever set in.   A moan or a grunt, in the early part of the disease, indicates a dangerous attack, and the alæ nasi, or nasal cartilages, rise spasmodically at each inspiration ; the air rushes through the inflamed windpipe and bronchial tubes, so as to produce a loud, coarse, respiratory murmur ; and the spasmodic action of the abdominal muscles indicates the difficulty the animal experiences in the act of expiration.   Pressure over the intercostal spaces and pressing on the spine induce the pain so characteristic of pleurisy, and a deep moan not infrequently follows such an experiment.

"The eyes are bloodshot, the mouth clammy, skin dry and tightly bound to the sub-cutaneous textures, and the urine is scanty and high-colored.

"On auscultation the characteristic, dry, sonorous râle of ordinary bronchitis may be detected along the windpipe and in the bronchial tubes. A loud sound of this description is not unfrequently detected at the anterior part of either side of the chest, while the respiratory murmur is entirely lost posteriorly, from consolidation of the lung. A decided leathery friction-sound is detected over a considerable portion of the thoracic surface. As the disease advances, and gangrene, with the production of cavities in the lungs, ensues, loud, cavernous râles are heard, which are more or less circumscribed, occasionally attended by a decided metallic noise. When one lung alone is affected the morbid sounds are confined to one side, and on the healthy side the respiratory murmur is uniformly louder all over.

"By carefully auscultating diseased cows from day to day interesting changes can be discovered during the animal's life-time. Frequently the abnormal sounds indicate progressive destruction; but at other times portions of lung that have been totally impervious to air become the seat of sibilant râles, and gradually a healthy respiratory murmur proves that, by absorption of the materials that have been plugging the lung-tissue, resolution is fast advancing.

"Unfortunately we often find a rapid destruction of lung tissue and speedy dissolution. In other instances the general symptoms of hectic or consumption attend lingering cases, in which the temperature of the body becomes low; the animal has a dainty appetite, or refuses all nourishment. It has a discharge from the eyes and a fetid, sanious discharge from the nose, but unfrequently it coughs up disorganized lung tissue and putrid pus. Great prostration, and, indeed, typhus symptoms set in. There is a fetid diarrhœa, and the animal sinks in the most emaciated state, often dying from suffocation, in consequence of the complete destruction of the respiratory structures."

Dr. James Moore thus describes the disease:

"**First stage.** It begins in one of three ways: Firstly, it may attack the cow suddenly, and run a rapid course in spite of all treatment; secondly, it may come on slowly and insidiously, the cow appearing not to be very ill, while the lungs are becoming diseased beyond the hope of restoration; and, thirdly, it sometimes begins with violent purging, followed by great weakness and loss of flesh. The majority of cases, however, present the following symptoms: a short, dry, husky cough, which is heard only occasionally; it is highly characteristic of this disease, and when once heard cannot be mistaken again. The owner says, perhaps, that he has heard this 'hoose' for two or three days, but thought no more about it. On inquiry it will be found that the beast does not give as much milk as usual, and that has a slightly yellowish tinge ; the appetite is not much worse, yet still she is careless about her food, and does not lick her dish clean ; when at rest the breathing may not show any departure from its healthy play, but when the animal is moved and walked some distance it becomes more frequent, labored and difficult ; the pulse is often healthy in character, although sometimes it is weak and slightly increased in frequency ; the bowels may either be confined or purged, or quite regular ; the body is sometimes hot, sometimes cold.

" The cow appears dull and listless ; when at grass she separates herself from the others, and lies on the ground while they are browsing.

" Second stage. The cough is now more frequent, and thick, frothy phlegm dribbles from the mouth ; the breathing is short when the air is taken into the lungs, and long when it is pressed out of them ; the inward breathing is attended with much pain, which causes the animal to grunt and to grate her teeth ; the grunt is heard when the animal is pressing the air out from the lungs ; the pain is much increased by coughing and change of position, and to lessen it the cough is now suppressed, or held back and short, and the cow stands fixed in one place. The pain is owing to the pleura being inflamed, and the position of the diseased place may be ascertained by pressing the side, between the ribs, with the point of the thumb ; when pressed on the animal will flinch and grunt.

" The pulse is quickened and oppressed ; the skin is hard, tight, and bound to the ribs ; the horns are hotter and the muzzle dryer than usual ; the head is lowered and thrust forward, with the nose poked out ; the back is raised up ; little or no food is eaten ; the cud is seldom or never chewed ; the milk is stopped ; the bowels are bound, and, when moved, the dung is in hard, dry lumps.

"Third stage. The breathing is much quickened, very difficult, labored and even gasping ; the breathing is carried on partly through the mouth, partly through the nostrils ; the breath has a bad smell ; a stringy, frothy fluid constantly dribbles from the mouth ; the cow groans loudly and frequently, while the grunt is either gone or subdued ; the pulse is quick, weak, and in some cases imperceptible or intermittent ; the horns, ears and legs are cold, the skin covered with cold sweat, the head and neck stretched out, with the nose poked into the corner of the manger ; the fore legs are separated from one another and fixed in one place, unless the cow is restless and uneasy ; sometimes the hind legs are crossed over each other, or the hind fetlock joints are knuckled forward ; the stoppage of milk is complete ; the animal is reduced to a skeleton ; the strength is also of course greatly impaired, and the beast can scarcely cough ; insensibility sometimes steals over her ; the urine is very highly colored ; toward the last violent purging comes on, the discharged matter being quite watery, blackish, highly offensive, and sometimes mixed with blood ; eventually the cavity of the chest becomes so full of fluid, or so much of the lung is condensed, that the breathing, from being more difficult and frequent, at last ceases, and the animal is dead."

TREATMENT.

Aconite. Pulse hard and quickened; shivering or trembling, attended with coldness of the legs or the horns, and dry heat of the skin; breathing short, painful, anxious, attended with open mouth and groans. Give the third or sixth dilution, ten drops every hour.

Ammonium causticum. Quick, difficult breathing, with rattle; inhalation of air very short, from pain; frequent cough, with discharge of mucus; great languor and listlessness; pulse feeble and quick; frequent shivering; skin at first hot and dry, afterward moist.

Give five drops of the watery solution in a little water every two or three hours till improvement begins.

Arsenicum. Wheezing; hurried breathing; small, quick pulse; great weakness; cold, clammy sweats; frequent short cough; purging in every stage.

Give ten drops of the sixth or thirtieth dilution every two hours.

Baptisia. Stupor; listlessness; restless, but too lifeless to move; eyes congested, look red and inflamed; thick mucus from nose; fetid odor from mouth; can only swallow water; distended abdomen; mushy stools or dark, very offensive stools; great prostration; urine scanty and dark; oppressed breathing, with cough; pulse at first accelerated, afterward slow and faint; pain along the back; restless, uneasy, or drowsy and stupid; chilly; great prostration, with tendency of the fluids to decompose. Discharges and exhalations fetid; ulceration of mucous membranes, especially of mouth, with tendency to putrescence; intolerance of pressure; constant change of position.

Give drop doses of the saturated tincture, or of the first and third dilution, every two hours, in a little water.

Bryonia. Frequent, short, suppressed cough, which seems to cause sharp pain in the chest; breathing short, with characteristic grunt; when the ribs are pressed by a hand or a finger the cow flinches and utters the short grunt, as if the pain were very acute; the animal dreads to move from pain.

Give ten drops of the third, sixth or thirtieth dilution every two to four hours.

Lycopodium. Fan-like motion of the nostrils; loose stools; rumbling of flatulence in the bowels. Give ten drops of the thirtieth dilution every two to four hours.

Phosphorus. Difficult, obstructed breathing; pains in chest; pain between ribs; frequent short cough, with slimy phlegm, sometimes mixed with blood; violent purging, like gushes of water; wasting, weakness and prostration. Give ten drops of the sixth or thirtieth dilution every two, three, or four hours.

Sulphur. During convalescence, or when the other remedies do not act promptly.

Give ten drops of the sixth or thirtieth dilution in a little water three times a day.

The diet should be oat-meal gruel and boiled carrots. All animals should be quarantined most securely for ninety days. No healthy animals that have been in the same stable should be allowed to mingle with healthy cattle within a period of ninety days. A cow, whether in sickness or health, needs one thousand two hundred cubic feet of air per hour, and the system of ventilation should admit of a constant change of air to that amount without subjecting the animal to chilly currents.

LOCKJAW—TETANUS.

This disease is usually the result of a slight wound, and in some localities is very common to man and beast, especially if the wound is caused by some blunt instrument and does not bleed much. Exposure to wet and cold increases the liability of attack. There is also some uninvestigated source of aggravation, as evidenced by the varying prevalence of the disease or its aptitude for certain localities.

Among cattle the pulse may be at first apparently normal, presenting but little disturbance until the disease has become perilous; the muzzle, horns and ears are also normal; the animal stands rigidly fixed, or appears afraid to move; the head is extended horizontally forward; the nostrils dilated; the eyes bulging outward, or sunken and retracted, the membrane at the corners partly covering the eyeball; the tail elevated and tremulous; the legs splayed out; the quarters depressed; all the muscles rigidly fixed, so that the animal cannot bend; the muscles of the belly and neck tense, stiff, and in hard ridges; the teeth convulsively clinched or slightly parted. *The Country Gentleman* of July 31st, 1884, gives this case, which conveys a moral: " Beware of Pitchforks.—Died, of lockjaw, July 11th, 1884, Signalda 2d 6748. Signal 1170 is close on both sides of his pedigree, and he was no mean representative of that long line of fine breeding. About a month before his death he was pricked with a pitchfork for breaking up his water-tub. He was kind and gentle."

TREATMENT.

Arnica. After all wounds, however slight, give this remedy, either in drop doses of the tincture or the third, sixth or thirtieth dilution.

Arsenicum. Great thirst; restlessness; rigidity remaining after the use of other remedies.

Give drop doses of the thirtieth dilution on the tongue three or more times a day.

Belladonna. Where wounds are greatly inflamed, with great heat and much fever.

Bryonia. The animal dreads to move or be touched.

Camphor. Great prostration; languor. Give the tincture or first dilution, one drop every hour upon the tongue or by olfaction.

Gelsemium. Great debility; convulsive spasm; congestion to head. Give the tincture or first dilution, one drop every half hour upon the tongue.

Nitrate of Amyl. This remedy may be used as a palliative where the spasms are desperately rigid. It is given by inhalation, ten to twenty drops upon a small sponge enclosed in a napkin folded in conical form.

Nux vomica. This is the most important and distinctively homœopathic remedy. Give the sixth or thirtieth dilution, ten drops upon the tongue every three or four hours. Some cases may require the saturated tincture in doses of from one to five drops, while others may do better when given the two hundredth or one thousandth dilution.

Passiflora incarnata. A very important remedy in the first stage. Give one drop of tincture every hour, or as Nux vomica.

In all cases where practicable immerse an inflamed limb in hot water (130° to 140°), or apply saturated cloths or sponges of the same temperature. There is nothing like hot water to relieve inflammation, congestion, and the agonizing pain resulting from wounds.

BROKEN HORN.

The Jersey horn is fine and fragile, rendering it liable to fracture and casting of the shell upon slight provocation. When fighting, or even rubbing against a fence or tree, a shell may be broken or knocked off, and followed by a severe hemorrhage from the vessels at or near the base of the horn, and sometimes a very slow recovery.

TREATMENT.

To check hemorrhage apply the solution of subsulphate of iron (*Liq. Ferri subsulph.*) to the bleeding parts until a clot is formed. When the clot falls off and there is no more bleeding apply bandages saturated with a dilution of tincture *Calendula officinalis* in water, one part tincture to sixteen parts of water. The *Calendula* is the best lotion for all forms of lacerated wounds in any part of the body. Continue the application until the healing of the parts, which will be rapid if the cow is otherwise in good health.

"Styptic cotton," or cotton saturated with the subsulphate of iron, is convenient for application to most conditions of hemorrhage.

LOTIONS AND UNGUENTS.

Calendula. For lacerated wounds this is a rapidly healing wash. Dilute with twenty parts water and keep the parts wet.

Hamamelis, or Witch Hazel. For inflamed surfaces and inflamed veins the best application. Also valuable in hemorrhages. Apply diluted, one to ten, with hot water.

Arnica. For all bruises and sprains without laceration. Dilute with hot water, one part tincture to ten parts water.

Phytolacca. For garget. Dilute with hot water, one to sixteen.

Hydrastis. For old ulcers, for eruptions by poisoning, like poison sumach, or rhus-poisoning. Dilute with hot water, one part tincture to one hundred parts water.

Thuja. Use for warts at any time the pure tincture.

Castor Oil. For warts apply pure when the cows are dry.

Listerine. Use for vaginal injections and for deodorizer.

Calendula Vaseline. Apply to all wounds or burns.

Mutton Tallow. Apply melted to sore teats and ulcers.

Vaseline. Very useful for many eruptions and sores.

Crude Petroleum. Useful in alternation with Phytolacca for garget.

DISINFECTANTS.

The best disinfectant is absolute cleanliness.

Remove all discharges and wash the stalls each day.

If there is any evidence of contagious disease the cattle that are infected should be quarantined at a distance from all others.

DISINFECTION WITH PURE CHLORINE GAS.

For the destruction of the germs of disease in buildings where contagious pleuro-pneumonia and other dreaded destructive maladies have existed, the most effective method of disinfection is probably with pure chlorine gas. This method is recommended by Professor Doremus for old hospitals where the walls are permeated with filth, and for the destruction of the cholera germ, and other disease elements.

Dr. Doremus says: " The gas must be used in large quantities. We spread out large sheets of lead and turned up the edges so that they would hold the chemicals for generating the gas. I would have three or four assistants, and when ready the word was given to ' pour.' Then all would run out and the door would be fastened, and the gas penetrated everything. To have entered the room during the time would have been certain death." Chlorine gas may be made in large quantities, by pouring slightly diluted sulphuric acid upon a mixture composed of common salt and oxide of manganese in large leaden vessels. Its development requires care. It is safer to have the pouring done automatically than to risk human life in the experiment.

GERMICIDES.

The method of **Dr. Doremus** is expensive and only advisable in special cases.

Mr. John C. Pennington, chemist, of Paterson, N. J., who has made thorough and persistent experiments in the propagation of various forms of bacteria, in his studies of the germ-theory of disease, has found that the salts of mercury are the most effective germicides or disinfectants. He uses the *bichloride of mercury* in a solution of one part to a thousand of water, sprinkling it in the air and upon walls with a brush or whisk-broom. He also uses the *hyposulphite of mercury* in the same manner, with similar results, in annihilating the bacteria which float in the air or contaminate almost every substance. It must ever be borne in mind that these powerful mineral salts are very violent agents, and must be used in very dilute form, and with great caution. Never venture to apply any of the mineral salts where they will fall upon the bedding or feed of animals. Mr. Pennington finds that the burning of sulphur will not destroy bacteria, and therefore condemns the use of any such means, as well as all the so-called germicides which by experiment have proved to be less effective than the hyposulphite of mercury and the corrosive sublimate.

These mineral salts must not be used for injections in the treatment of any disease unless further diluted.

If your walls are tinted with "alabastine" once in two or three years, a mild blue color, they will be very pleasant for the eyes of animals, a cheap and tasty finish for a brown wall. If not, apply lime whitewash semi-annually.

The agents that destroy every form of disease germs in the air, in all discharges, and in the walls of buildings, are the only true disinfectants. The number of these is very limited. For ordinary use the HYPOSULPHITE OF MERCURY is probably the best. Prepare a solution in the proportion of $\frac{1}{1000}$th in water, or sixty (60) grains of the hyposulphite to one gallon of the latter, and sprinkle it in the air and upon the walls of the building, and apply to all the fetid excretions of diseased organs.

The chlorine gas method of Professor Doremus may be adopted wherever there has been infection of *pleuro-pneumonia* and the *Texas fever*. The authorities should see that this method of disinfection is employed, and a competent chemist given charge of the work.

Caution: These germicides are fatal poisons to every living organism, and should be used intelligently in every instance, knowing that chlorine gas is deadly if inhaled, and the mercurial salts unsafe to be taken into the stomach of any animal except in a smaller quantity than used above. USE NO VESSEL CONTAINING THEM FOR ANY OTHER PURPOSE WHATSOEVER.

HYPOCHLORITE OF SODA.*

Hypochlorite of soda may be used as a germicide, and can be safely applied to ulcers and putrid eruptions, in dilution of 1 to 60 of water, or used as an injection

---

* Reed & Carnrick.

for diseased mucous membranes, 1 to 100 of water. As a germicide apply to all infectious matter 1 to 16.

DEODORIZERS.

For every offensive odor in the stable seek out and remove the cause. Keep the air as sweet as a pasture-field. There is no better deodorizer than a hot roasting pan of *coffee-beans* carried through the building so as to freely give the fumes of parched coffee, while the grains also act as an absorbent. Among the commercial compounds *Listerine* will prove useful and pleasant as a deodorizer for the hands after operations.

GENERAL SUMMARY FOR DISINFECTION.

A radical discrimination must be made between deodorizers, disinfectants, and germicides.

Use each of these for special purposes in the stable, just as in the human dwelling and hospital.

Listerine, a fragrant antiseptic mixture, may be found useful in deodorizing the foul discharges that follow abortion, parturition, and those excretions accompanying various diseases. It is a mixture of oils and extracts from Thyme, Eucalyptus, Baptisia, Gaultheria and Mentha arvensis. Each drachm contains two grains of refined benzo-boracic acid.

The Listerine when well diluted is useful as an injection where the vaginal discharges are very fetid. It is useful for deodorizing when a thorough washing with hot water and soap fails to remove offensive odors.

SUMMARY OF PRACTICAL USE OF DISINFECTANTS.

FOR EXCRETIONS.

1. Chloride of Mercury in solution, 1 to 500.
2. Hyposulphite of Mercury in solution, 1 to 500.

FOR INFECTED CLOTHS OR SPONGES.

1. Destruction by fire if of little value. The combustion must be total and complete.
2. Boiling one hour.
3. Immersion in a solution of Chloride of Mercury of the strength of 1 to 1000 four hours.

FOR CLOTHING OF ATTENDANTS.

1. Exposure to dry heat at a temperature of 230° F. for two hours.
2. Destruction by fire if of little value and badly infected with contagion spores.

3. Immersion in boiling water for one hour.

4. Immersion in solution of Chloride of Mercury of 1 to 2000 for four hours.

FOR THE PERSON.

1. Wash the hands and surface of the body in a ten per cent. solution of Chlorinated Soda.

2. Wash the hands in a solution of Chloride of Mercury or Hyposulphite of Mercury 1 to 1000.

FOR THE WALLS OF THE STABLE.

1. Wash all surfaces, while occupied, with a solution of Chloride of Mercury of 1 to 1000.

2. When vacated use Prof. Doremus's method with Chlorine Gas.

3. For instruments and all metallic surfaces a solution of the Hyposulphite of Mercury of 1 to 1000.

EUDORA 1863.

AT 18 YEARS OLD.

BILLINGS HERD.

FREDERICK BILLINGS, WOODSTOCK, VERMONT.

LA FINANCIERE 11970.

AT 8 YEARS OLD.

Grey King Type.

FAIRVIEW HERD.

G. AND H. B. CROMWELL, NEW DORP, P. O. STATEN ISLAND, N. Y.

SULTANE 2d 11,373.

Sultan Type.

TESTED IN 3 DAYS, 10 LBS. 1 OZ.

M. H. MESSCHERT, DOUGLASSVILLE, PA.

SULTAN OF ST. SAVIOUR'S 5328.
Sultan Type.
M. H. MESSCHERT, DOUGLASSVILLE, PA.

HAZEN'S NORA 4791.

AT 8 YEARS OLD.

Rajah—Bismarck—Splendid Type.

GREEN MOUNTAIN HERD.

MOULTON BROTHERS, WEST RANDOLPH, VERMONT.

PROCTOR'S REGINA 35,665.

AT 2 YEARS OLD.

Rex—Cetewayo Type.

BAGGS HOTEL HERD.

T. R. PROCTOR, UTICA, NEW YORK.

MISS SHARPLESS 24,352.

Khedive Type.

HIGHLAND HERD.

James N. Smith, Litchfield, Connecticut.

OXFORD KATE 13,646.

AT 5 YEARS OLD.

Khedive Type.

R. S. ANDREWS, BALTIMORE, MARYLAND.

MISS COOPER 5869.

Alphea Type.

HOLLY GROVE HERD.

JOHN I. HOLLY, PLAINFIELD, NEW JERSEY.

DOMINO OF DARLINGTON 2459.

Alphea Type.

BRIARCLIFF HERD.

JAMES STILLMAN, SING SING, NEW YORK.

PEDRO ALPHEA 13,889.

AT 4 YEARS OLD.

Test, One Day, 3 lbs. 11 oz.

Eurotas Type.

FAIRVIEW HERD.

G. AND H. B. CROMWELL, NEW DORP, P. O. STATEN ISLAND, N. Y.

DUKE OF DARLINGTON 2460.

Alphea-Rioter Type.

DARLINGTON HERD.

A. B. DARLING, RAMSEY'S, NEW JERSEY.

MERCURY 432.

AT 12 YEARS OLD.

Alpha Type.

SIMPSON HERD.

WILLIAM SIMPSON, 51 CHATHAM STREET, NEW YORK.

ALPHEA 171.
A Fountain Head.

MATIN'S GLORY 9135.

AT 2 YEARS OLD.

Matin — Lille Bonne — Favorite Type.

Average Tests of Dam and Grandams, 17 lbs. 13¾ oz.

BILLINGS HERD.

FREDERICK BILLINGS, WOODSTOCK, VERMONT.

MATIN 7768.

AT 8 YEARS OLD.

Brown Prince Type.

BILLINGS HERD.

FREDERICK BILLINGS, WOODSTOCK, VERMONT.

POGIS CHIEF 3998.

Stoke Pogis—Marjoram Type.

BRYN MAWR HERD.

F. C. SAYLES, PAWTUCKET, RHODE ISLAND.

MARJORAM 3239.

AT 12 YEARS OLD.

BRYN MAWR HERD.

F. C. SAYLES, PAWTUCKET, RHODE ISLAND.

HILDA D. 6683.

AT 9 YEARS OLD.

VERNA HERD.

FREDERIC BRONSON, SOUTHPORT, CONNECTICUT.

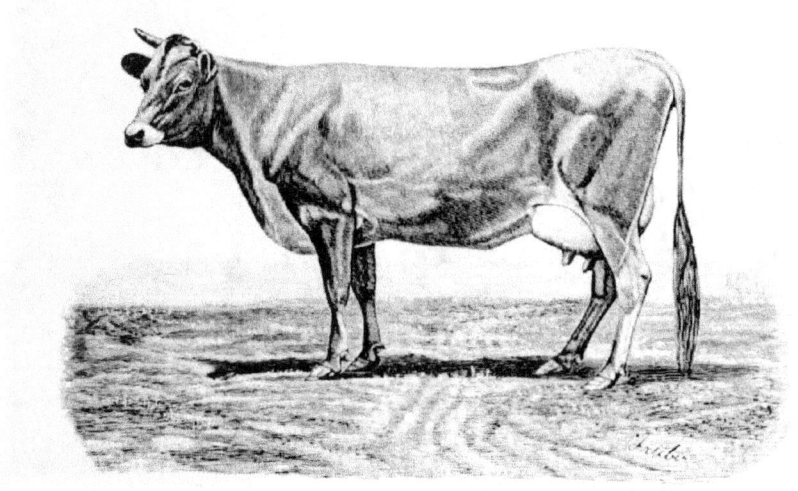

EVELINA OF VERNA 10,971.

AT 6 YEARS OLD.

Signal Type.

VERNA HERD.

FREDERIC BRONSON, SOUTHPORT, CONNECTICUT.

FOOTSTEP 5163.

Signal Type.

VERNA HERD.

FREDERIC BRONSON, SOUTHPORT, CONNECTICUT.

TAOMA 7200.

St. Helier Type.

BRIARCLIFF HERD.

JAMES STILLMAN, SING SING, N. Y.

THALMA 4288.

St. Helier Type.

OAKLANDS HERD.

VALANCEY E. FULLER, HAMILTON, ONTARIO, CANADA.

"BRIARCLIFF FARM."

PROPERTY OF JAMES STILLMAN,

Sing Sing, N. Y.

PRINCE POGIS 10,682.

AT 1 YEAR OLD.

Mary Anne of St. Lambert—Rob Roy—Splendid Type.

OAKLANDS HERD.

Valancey E. Fuller, Hamilton, Ontario, Canada.

MARY ANNE OF ST. LAMBERT 9770.

AT 5 YEARS OLD.

Stoke Pogis—Marjoram—Victor Hugo Type.

OAKLANDS HERD.

VALANCEY E. FULLER, HAMILTON, ONTARIO, CANADA.

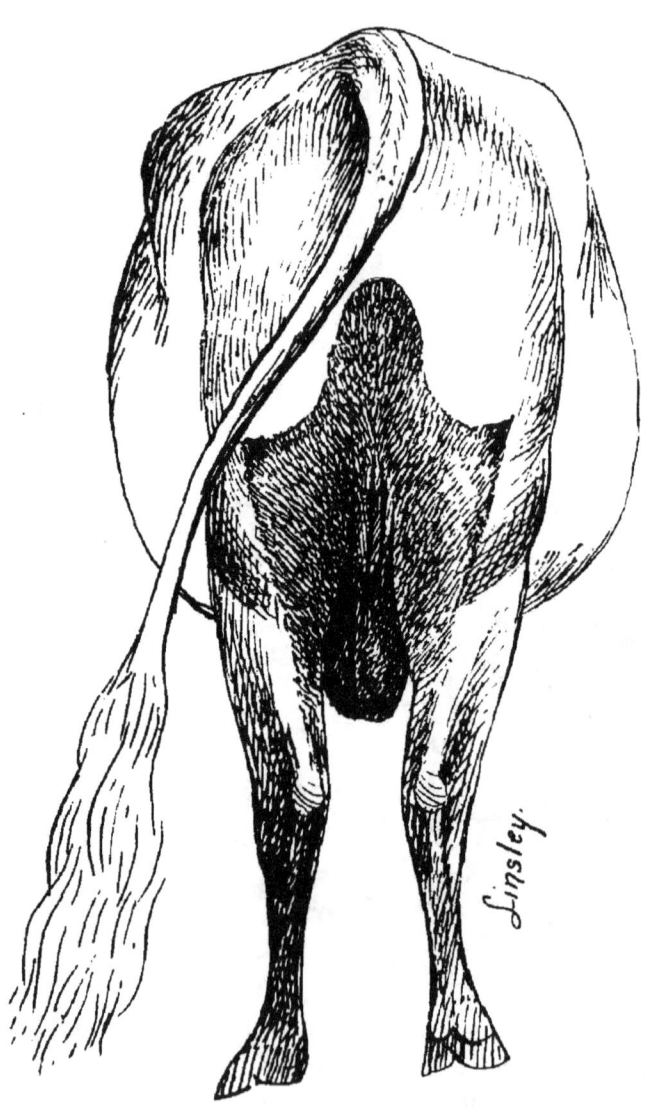

CURVELINE ESCUTCHEON OF THE BULL YOUNG MERCURY 7485.

BRED AND OWNED BY WILLIAM SIMPSON, NEW YORK.

The fore escutcheon of this bull covers more than half the belly.

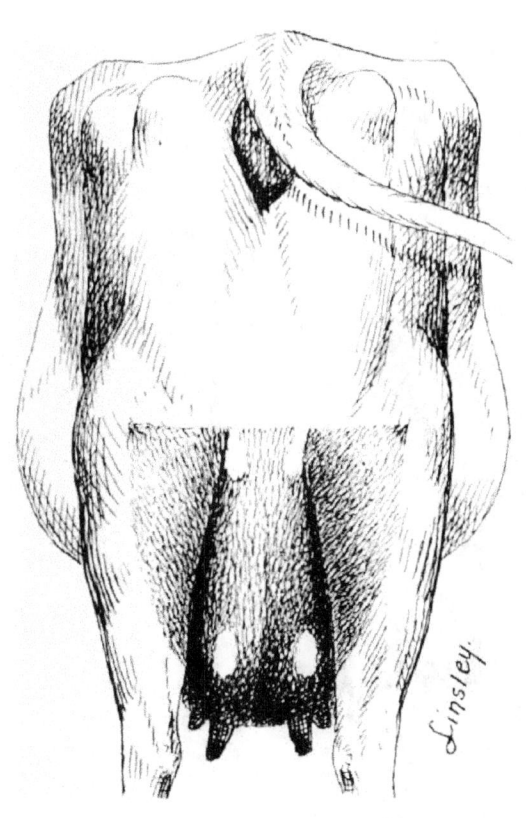

CARRESINE OR LEVEL ESCUTCHEON.

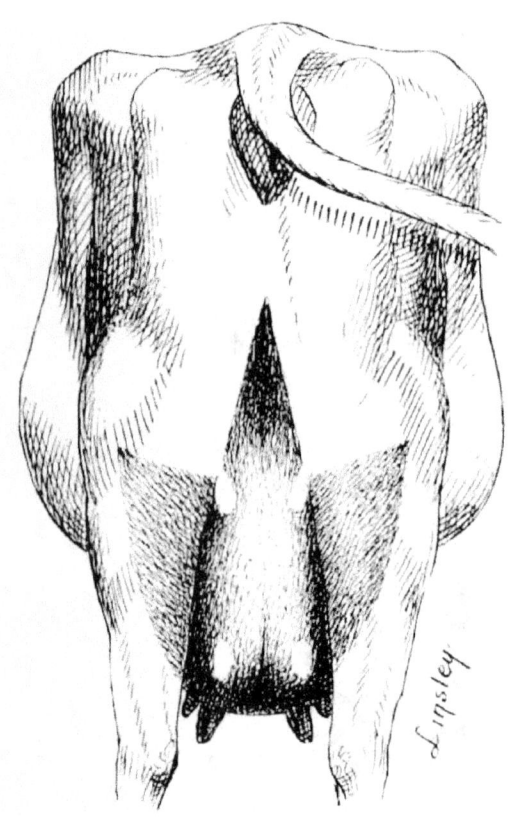

LIMOUSINE ESCUTCHEON.

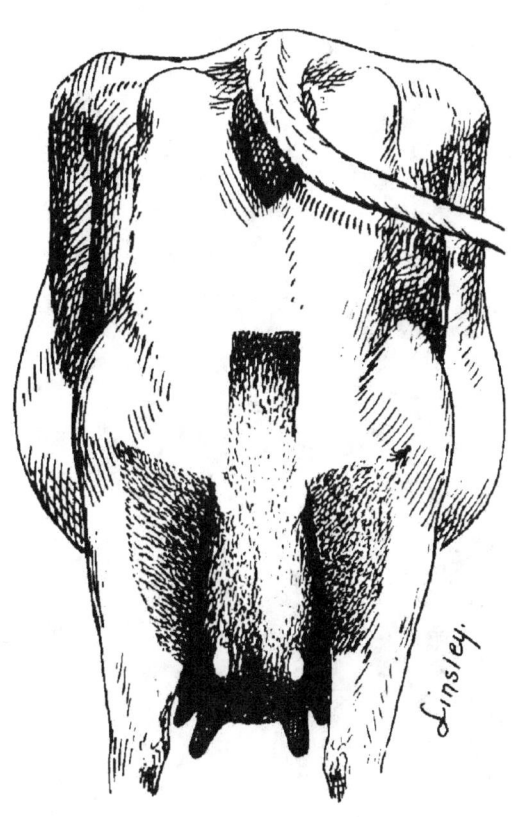

DEMIJOHN ESCUTCHEON.

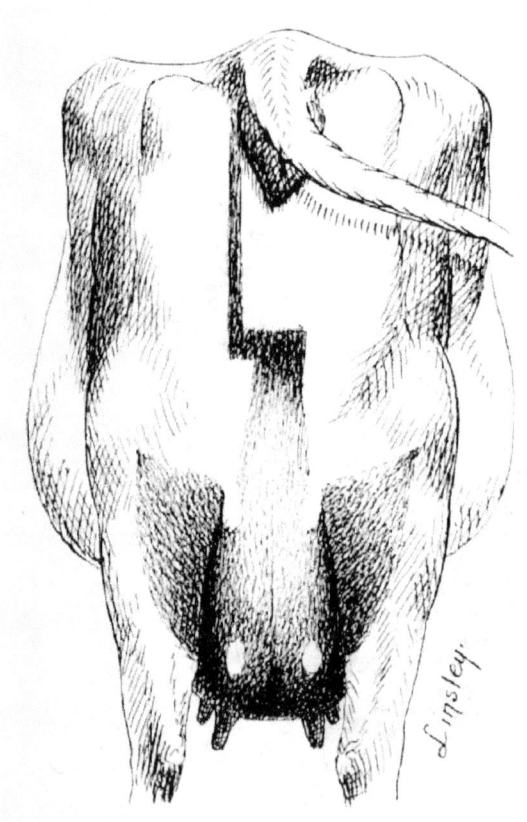

SQUARE ESCUTCHEON.

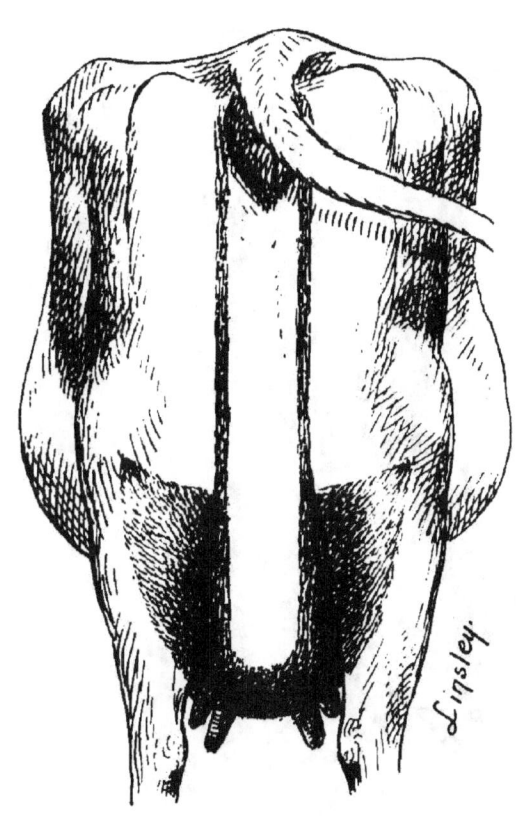

DOUBLE SELVEDGE ESCUTCHEON.

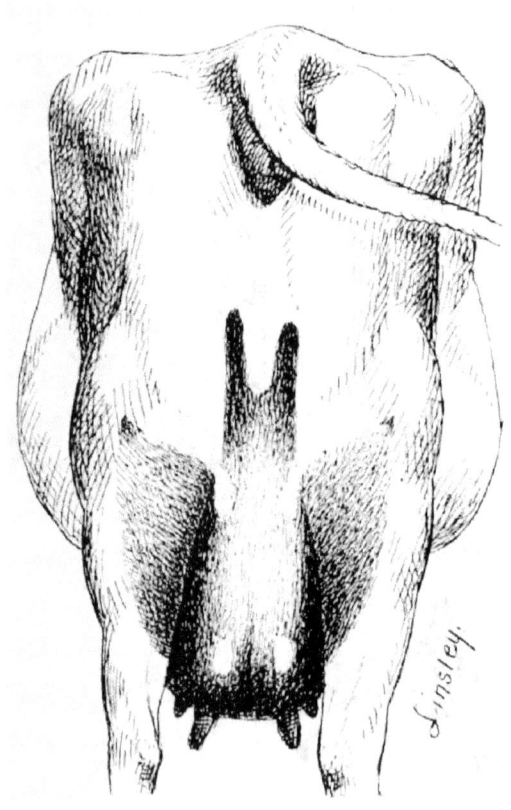

BICORN ESCUTCHEON.

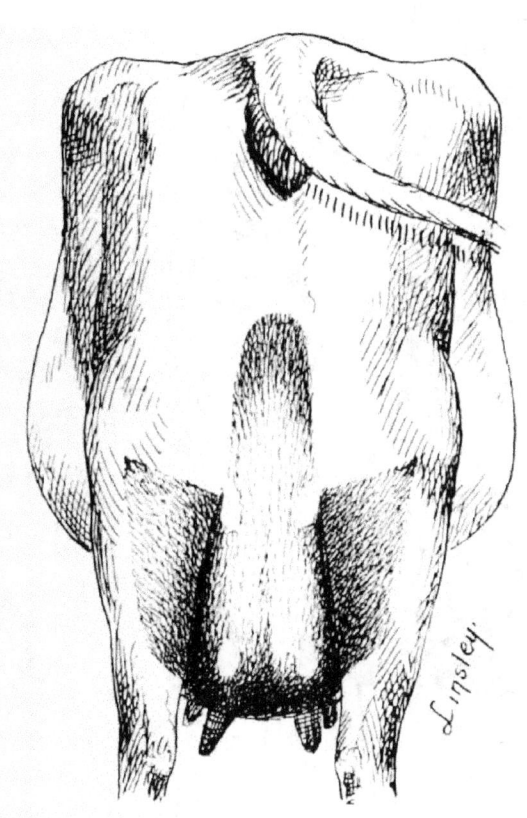

CURVELINE ESCUTCHEON.

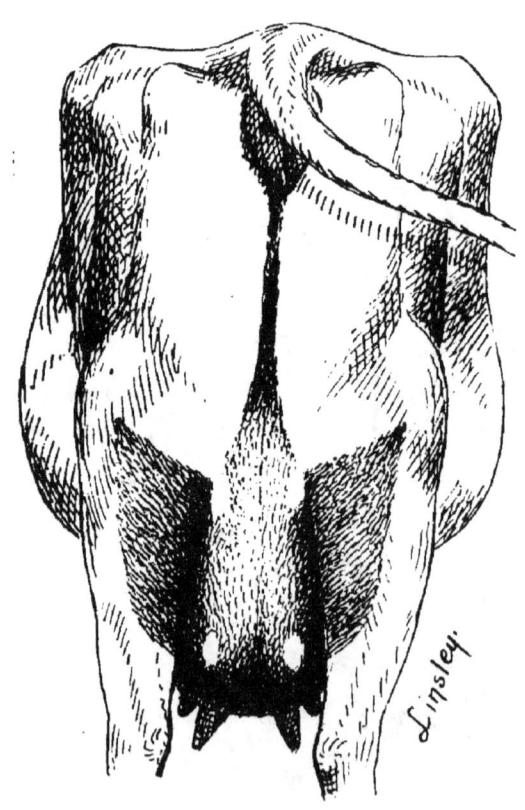

SELVEDGE ESCUTCHEON.

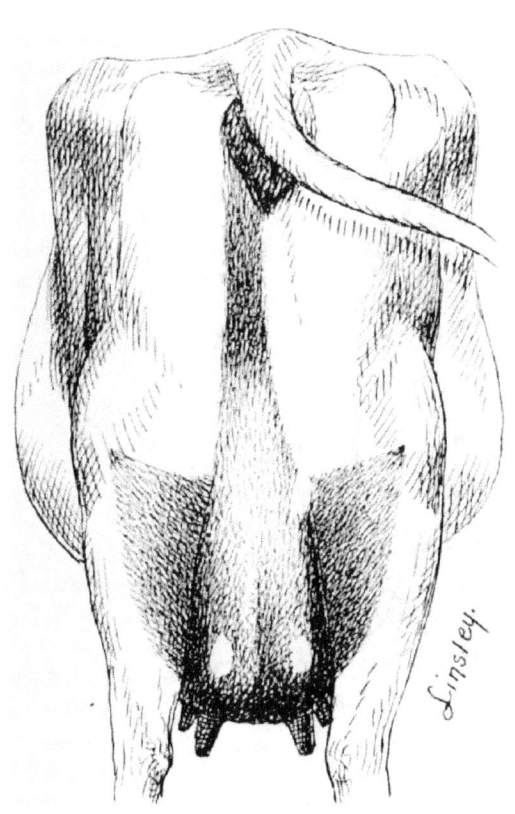

LEFT FLANDRINE ESCUTCHEON.

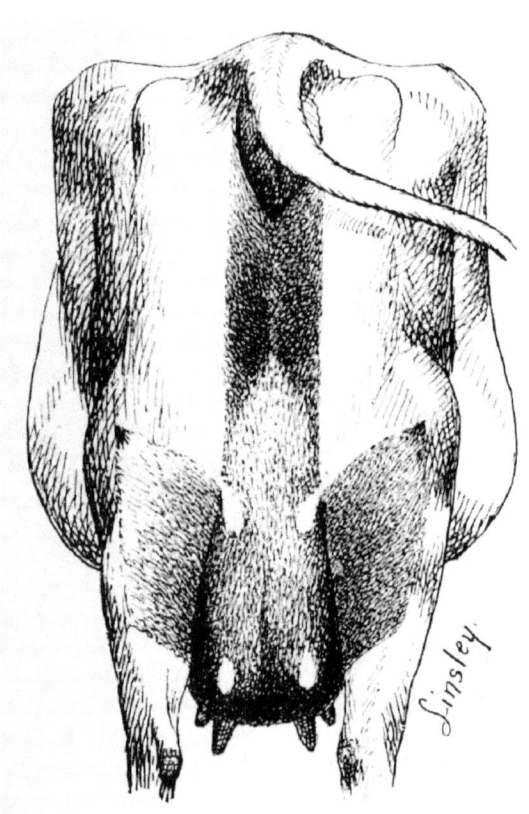

FLANDRINE ESCUTCHEON.

FILLPAIL 2d 24,388.

AT 4 YEARS OLD.

Imported by T. S. Cooper.

Khedive—King—Fillpail Type.

A. N. MARTIN, SUMMIT, NEW JERSEY.

HIPPARCHUS 11,672.
AT 2 YEARS OLD.
Orange Peel-King Type.
BRIGHTSIDE HERD.
R. M. HOE, 504 GRAND STREET, NEW YORK.

LADY BUCKINGHAM 11,670.

Pierrot Type.

HIGHLAND HERD.

JAMES N. SMITH, LITCHFIELD, CONNECTICUT.

BELMEDA 6229.

AT 9 YEARS OLD.

Superb Type.

FAIRVIEW HERD.

G. AND H. B. CROMWELL, NEW DORP, P. O. STATEN ISLAND, N. Y.

PRIDE OF MOUNTAINSIDE 7118.

AT 3 YEARS OLD.

Eurotas—Belle Dame Type.

FAIRVIEW HERD.

G. AND H. B. CROMWELL, NEW DORP, P. O. STATEN ISLAND, N. Y.

EUROTAS' BLACK PRINCE 14,384.

AT 17 MONTHS OLD.

Eurotas Type.

FAIRVIEW HERD.

G. AND H. B. CROMWELL, NEW DORP, P. O. STATEN ISLAND, N. Y.

EUROTAS 2454.

Rioter-Alphea Type.

DARLINGTON HERD.

A. B. DARLING, RAMSEY'S, NEW JERSEY.

ULTISSIMA 24,633.

AT 2 YEARS OLD.

Jersey Belle—Eurotas Type.

GREEN MOUNTAIN HERD.

MOULTON BROTHERS, WEST RANDOLPH, VERMONT.

ROMANO 11,806.

AT 20 MONTHS OLD.

Couch's Lily—Jersey Belle—Eurotas Type.

GREEN MOUNTAIN HERD.

MOULTON BROTHERS, WEST RANDOLPH, VERMONT.

JERSEY BELLE OF SCITUATE 7828.
AT 10 YEARS OLD.
Victor Type.
THE THOROUGHBRED MODEL.